Communications in Computer and Information Science 3093

Series Editors

Rationale

The CCIS series is devoted to the publication of proceedings of computer science conferences. Its aim is to efficiently disseminate original research results in informatics in printed and electronic form. While the focus is on publication of peer-reviewed full papers presenting mature work, inclusion of reviewed short papers reporting on work in progress is welcome, too. Besides globally relevant meetings with internationally representative program committees guaranteeing a strict peer-reviewing and paper selection process, conferences run by societies or of high regional or national relevance are also considered for publication.

Topics

The topical scope of CCIS spans the entire spectrum of informatics ranging from foundational topics in the theory of computing to information and communications science and technology and a broad variety of interdisciplinary application fields.

Information for Volume Editors and Authors

Publication in CCIS is free of charge. No royalties are paid, however, we offer registered conference participants temporary free access to the online version of the conference proceedings on SpringerLink (http://link.springer.com) by means of an http referrer from the conference website and/or a number of complimentary printed copies, as specified in the official acceptance email of the event.

CCIS proceedings can be published in time for distribution at conferences or as post-proceedings, and delivered in the form of printed books and/or electronically as USBs and/or e-content licenses for accessing proceedings at SpringerLink. Furthermore, CCIS proceedings are included in the CCIS electronic book series hosted in the SpringerLink digital library at http://link.springer.com/bookseries/7899. Conferences publishing in CCIS are allowed to use our online conference service (Meteor) for managing the whole proceedings lifecycle (from submission and reviewing to preparing for publication) free of charge.

Publication process

The language of publication is exclusively English. Authors publishing in CCIS have to sign the Springer CCIS copyright transfer form, however, they are free to use their material published in CCIS for substantially changed, more elaborate subsequent publications elsewhere. For the preparation of the camera-ready papers/files, authors have to strictly adhere to the Springer CCIS Authors' Instructions and are strongly encouraged to use the CCIS LaTeX style files or templates.

Abstracting/Indexing

CCIS is abstracted/indexed in DBLP, Google Scholar, EI-Compendex, Mathematical Reviews, SCImago, Scopus. CCIS volumes are also submitted for the inclusion in ISI Proceedings.

How to start

To start the evaluation of your proposal for inclusion in the CCIS series, please send an e-mail to ccis@springer.com

Nurfadhlina Mohd Sharef · Azreen Azman ·
Tutut Herawan
Editors

Information Retrieval and Knowledge Management

7th International Conference, CAMP 2026
Putrajaya, Malaysia, August 12–13, 2026
Proceedings, Part I

Springer

Editors
Nurfadhlina Mohd Sharef
Universiti Putra Malaysia
Serdang, Malaysia

Azreen Azman
Universiti Putra Malaysia
Serdang, Malaysia

Tutut Herawan
Universiti Malaya
Kuala Lumpur, Malaysia

ISSN 1865-0929 ISSN 1865-0937 (electronic)
Communications in Computer and Information Science
ISBN 978-981-92-4301-3 ISBN 978-981-92-4302-0 (eBook)
https://doi.org/10.1007/978-981-92-4302-0

This Springer imprint is published by the registered company Springer Nature Singapore Pte Ltd.
The registered company address is: 152 Beach Road, #21-01/04 Gateway East, Singapore 189721, Singapore

Preface

The Seventh International Conference on Information Retrieval and Knowledge Management (CAMP 2026) was held on 12–13 August 2026 at the Bangi Resort Hotel, Malaysia. Organized by the Society of Information Retrieval and Knowledge Management Malaysia (PECAMP), CAMP serves as a platform for researchers, practitioners, and industry experts to share advances in information retrieval, knowledge management, computational linguistics, and data science.

The theme of CAMP 2026, "Beyond Retrieval: Information, Knowledge, and Intelligence," reflects the transformation of intelligent information systems from traditional retrieval mechanisms toward knowledge discovery, reasoning, prediction, and decision support. Recent advances in artificial intelligence, large language models, retrieval-augmented generation, multimodal learning, and knowledge-driven systems continue to reshape how information is transformed into actionable intelligence across diverse domains.

CAMP 2026 featured four technical tracks: Information Retrieval, Computational Linguistics, Knowledge Management, and Data Science. The conference received 113 submissions from researchers worldwide. Following a rigorous single-blind peer-review process conducted by an international program committee, 27 submissions were accepted for publication as regular papers, resulting in an acceptance rate of 24%.

This volume presents selected contributions that illustrate the convergence of information retrieval, knowledge management, computational linguistics, and data science in creating intelligent systems that support knowledge discovery and decision-making. These include retrieval-augmented and knowledge-enhanced artificial intelligence, multimodal learning, conversational systems, multilingual language technologies, machine learning, digital twins, intelligent decision support, and human-centered AI. The papers also demonstrate the application of information and knowledge technologies to address real-world challenges in healthcare, agriculture, education, cybersecurity, transportation, environmental sustainability, and public services. Collectively, the papers demonstrate how advances in information, knowledge, and intelligence can generate meaningful impact across research, industry, and society.

We sincerely thank all authors for their submissions and contributions. We also extend our appreciation to the Program Committee members, reviewers, track chairs, keynote speakers, special session organizers, sponsors, and organizing committee members, whose dedication and expertise ensured the success of CAMP 2026 and the quality of these proceedings. We also extend our sincere appreciation to Springer Nature as the publisher of the proceedings, for their invaluable support and guidance throughout this publication project.

We hope that the contributions presented in this volume will stimulate further research, foster interdisciplinary collaboration, and inspire future innovations in intelligent information systems.

June 2026

Nurfadhlina Mohd Sharef
Azreen Azman
Tutut Herawan

Organization

Honorary Chair

Tengku Mohd Tengku Sembok	PECAMP, Malaysia

General Co-chairs

Nurfadhlina Mohd Sharef	Universiti Putra Malaysia, Malaysia
Roslina Othman	IIUM, Malaysia

Conference Co-chairs

Shahrul Azman Noah	Universiti Kebangsaan Malaysia, Malaysia
Tutut Herawan	Universiti Malaya, Malaysia

Technical Program Chair

Azreen Azman	Universiti Putra Malaysia, Malaysia

Track Chairs

Mas Rina Mustaffa	Universiti Putra Malaysia, Malaysia
Shaiful Bakhtiar Rodzman	Universiti Teknologi MARA, Malaysia
Nor Liyana Mohd Shuib	Universiti Malaya, Malaysia
Sri Devi Ravana	Universiti Malaya, Malaysia

Publication Co-chairs

Mohd. Helmy Abdul Wahab	Universiti Tun Hussein Onn Malaysia, Malaysia
Haslizatul Fairuz Mohamed Hanum	Universiti Teknologi MARA, Malaysia
Aslina Baharum	Taylor's University, Malaysia
Suhaila Shahul Hameed	UniKL, Malaysia
Nurul Aiman Abdul Rahim	Taylor's University, Malaysia

Publicity Chair

Fakhrul Hazman Yusoff	UNITAR International University, Malaysia

Tutorials and Workshops Chairs

Azliza Yacob	University College TATI, Malaysia
Zirawani Baharum	UniKL, Malaysia

Finance Co-chairs

Masnizah Mohd	Universiti Kebangsaan Malaysia, Malaysia
Siti Syakirah Sazali	INTI International College, Malaysia

Sponsorship Chair

Dayang Nurfatimah Awang Iskandar	Universiti Malaysia Sarawak, Malaysia

Logistics Co-chairs

Azrul Hazri Jantan	Universiti Putra Malaysia, Malaysia
Alfian Abdul Halin	Universiti Putra Malaysia, Malaysia

Secretariat

Nurul Amelina Nasharuddin	Universiti Putra Malaysia, Malaysia
Nurul Sharefa Haneem Shahri	Universiti Putra Malaysia, Malaysia

Special Sessions Chairs

Annisa Ristya Rahmanti	Universitas Gadjah Mada, Indonesia
Nurfadhlina Mohd Sharef	Universiti Putra Malaysia, Malaysia
Brian Halubanza	Mulungushi University, Zambia
Musab Al Rawi	University of Technology and Applied Sciences, Oman
Vivi Indra Amelia Nasution	Universitas Terbuka, Indonesia
Shahrinaz Ismail	Universiti Sains Malaysia, Malaysia
Pradorn Sureephong	Chiang Mai University, Thailand
Suepphong Chernbumroong	Chiang Mai University, Thailand
Mohsin Iftikhar	Higher Colleges of Technology, UAE
Nidal Nasser	Alfaisal University, Saudi Arabia

Program Committee

Adhi Susilo	Universitas Terbuka Indonesia, Indonesia
Aggy Pramana Gusman	Universitas Putra Indonesia YPTK, Indonesia
Aldilas Achmad Nursetyo	ASEAN Biological Threats Surveillance Centre, Indonesia
Alfian Abdul Halin	Universiti Putra Malaysia, Malaysia
Amja Manullang	Middlesex University London, UK
Annisa Rahmanti	Universitas Gadjah Mada, Indonesia
Ardiansyah Dores	Universitas Mercu Buana, Indonesia
Ari Prayogo Pribadi	Middlesex University London, Indonesia
Asad Shah	University of Jeddah, Saudi Arabia
Aslina Baharum	Taylor's University, Malaysia
Azreen Azman	Universiti Putra Malaysia, Malaysia
Bagas Suryo Bintoro	Center of Health Behavior and Promotion, Indonesia
Brian Halubanza	Mulungushi University, Zambia
Dayang Nur Fatimah Awang Iskandar	Universiti Malaysia Sarawak, Malaysia
Devia Kartika	Universitas Putra Indonesia YPTK, Indonesia
Diva Kurnianingtyas	Universitas Brawijaya, Indonesia
Eissa Alshari	IBB University, Yemen
Fakhrul Hazman Yusoff	Sohar University, Oman
Farikin	Diponegoro University, Indonesia
Fatimah Khalid	Universiti Putra Malaysia, Malaysia
Firdaus Hafidz As Shidieq	Liverpool School of Tropical Medicine, UK

Hamdan Daniyal	Universiti Malaysia Pahang Al-Sultan Abdullah, Malaysia
Hanifah Wulandari	Universitas Gadjah Mada, Indonesia
Haslizatul Mohamed Hanum	Universiti Teknologi MARA, Malaysia
Javid Iqbal Thirupattur	Sunway University, Malaysia
Jing Fu	Chengdu University, China
Krittika Kantawong	University of Phayao, Thailand
Lailatul Qadri Zakaria	Universiti Kebangsaan Malaysia, Malaysia
Liu Ziyan	Universiti Putra Malaysia, Malaysia
Lucky LhauraVan FC	Universitas Lancang Kuning, Indonesia
Mas Rina Mustaffa	Universiti Putra Malaysia, Malaysia
Masnizah Mohd	Universiti Kebangsaan Malaysia, Malaysia
Masrah Azrifah	Universiti Putra Malaysia, Malaysia
Moch. Nafkhan Alzamzami	Institut Teknologi Sepuluh Nopember, Indonesia
Mohd Helmy Abd Wahab	Universiti Tun Hussein Onn Malaysia, Malaysia
Mohd Nordin Abd Rahman	Universiti Sultan Zainal Abidin, Malaysia
Mohsin Iftikhar	Higher Colleges of Technology, UAE
Muhammad Sam'an	Universitas Muhammadiyah Semarang, Indonesia
Muhd Eizan Shafiq Abd Aziz	Universiti Teknologi MARA, Malaysia
Musab Al Rawi	University of Technology and Applied Sciences, Oman
Myat Noe Win	University Malaya, Malaysia
Nabila Wardah Zamani	Universiti Putra Malaysia, Malaysia
Nasa Zata Dina	Universitas Airlangga, Indonesia
Nor Liyana Mohd Shuib	University of Malaya, Malaysia
Noraini Che Pa	Universiti Putra Malaysia, Malaysia
Nurfadhlina Mohd Sharef	Universiti Putra Malaysia, Malaysia
Nurul Amelina Nasharuddin	Universiti Putra Malaysia, Malaysia
Peer Azmat Shah	Higher Colleges of Technology, UAE
Pradorn Sureephong	Chulalongkorn University, Thailand
Pruet Putjorn	Mae Fah Luang University, Thailand
Puteri Suhaiza Sulaiman	Universiti Putra Malaysia, Malaysia
Qais Qassim	University of Technology and Applied Sciences, Oman
Rabiah Abdul Kadir	Universiti Kebangsaan Malaysia, Malaysia
Raihani Mohamed	Universiti Putra Malaysia, Malaysia
Razali Yaakob	Universiti Putra Malaysia, Malaysia
Rima Gema	Universitas Putra Indonesia YPTK, Indonesia
Roslina Othman	International Islamic Universiti Malaysia, Malaysia
Rosmayati Mohemad	University of Malaysia Terengganu, Malaysia
Rozi Nor Haizan Nor	Universiti Putra Malaysia, Malaysia

Ruqaya Subhi Tawfeeq	Al-Iraqia University, Iraq
Saffa Raihan Zainal Abidin	Universiti Teknologi MARA, Malaysia
Santichai Wicha	Mae Fah Luang University, Thailand
Shahrinaz Ismail	Universiti Sains Malaysia, Malaysia
Shahrul Azman Noah	Universiti Kebangsaan Malaysia, Malaysia
Shaiful Bakhtiar Rodzman	Universiti Teknologi MARA, Malaysia
Sharifah Sakinah Syed Ahmad	Universiti Teknikal Malaysia Melaka, Malaysia
Siti 'Aisyah Sa'dan	Universiti Teknologi MARA, Malaysia
Siti Syakirah Sazali	HELP University, Malaysia
Sonia Soans	University of Technology and Applied Sciences, Oman
Sri Devi Ravana	Universiti Malaya, Malaysia
Tutut Herawan	Universiti Malaya, Malaysia
Venkataraman Meenakshi	University of Texas at Austin, USA
Vivi Indra Amelia Nasution	Universitas Terbuka Indonesia, Indonesia
Xin Liu	Knowledge and Innovation Research Laboratory, Thailand
Yootthapong Tongpaeng	Mae Fah Luang University, Thailand
Yusmadi Yah Jusoh	Universiti Putra Malaysia, Malaysia
Zamri Abu Bakar	Universiti Teknologi MARA, Malaysia
Zulkifli Halim	Universiti Teknologi MARA, Malaysia

Contents

Computational Linguistics

Clustering Noisy Malay Text Type Using K-MEANS with Word Representation Approaches

Helmi Ashraf Bin Ahmad and Azilawati Azizan(✉)

Faculty of Computer and Mathematical Sciences, Universiti Teknologi MARA, Perak Branch, Tapah Campus, Perak, Malaysia
azila899@uitm.edu.my

Abstract. Social media text commonly contains informal and non-standard word forms, including misspellings, abbreviations, repeated characters, digit-based expressions, and mixed-language usage. These variations create challenges for processing Malay user-generated content because the observed words may differ structurally from standard vocabulary. This study presents a K-Means-based clustering approach for grouping candidate noisy Malay words according to their structural patterns. A total of 6,988 Facebook and Instagram comments were collected and preprocessed into 13,014 valid word tokens. Following noisy word filtering, 5,097 candidate noisy Malay words were retained for analysis. Four word representation approaches were compared: character-level TF-IDF, FastText subword embedding, binary noise-type features, and RLE with TF-IDF. For a consistent comparison, all approaches were clustered using K-Means with a common setting of k = 6 and evaluated using Silhouette Score, Davies-Bouldin Index, and Calinski-Harabasz Score. The results showed that binary noise-type features produced the strongest cluster structure, achieving a Silhouette Score of 0.7451, a Davies-Bouldin Index of 0.9383, and a Calinski-Harabasz Score of 3776.5049. The resulting clusters also demonstrated identifiable structural patterns, including repeated-letter forms, shortened words, and abbreviation-like forms. These findings indicate that explicit structural features are effective for grouping candidate noisy Malay words and may support further development of Malay noisy text normalisation procedures.

Keywords: Malay noisy text · unsupervised clustering · K-MEANS · FastText · TF-IDF · social media text

1 Introduction

User-generated content from social media platforms has become an important source of text data for natural language processing. However, this type of text is often informal, inconsistent, and noisy. Sohn et al. (2025) explain that social media text commonly contains spelling variation, abbreviations, repeated characters, code-mixing, and out-of-vocabulary forms (Wahid et al., 2025).

N. Mohd Sharef et al. (Eds.): CAMP 2026, CCIS 3093, pp. 3–13, 2027.
https://doi.org/10.1007/978-981-92-4302-0_1

In the Malay language context, noisy social media text is particularly challenging because users may combine standard Malay with English words, slang, shortened spellings, dialectal expressions, and repeated characters. Yong et al. (2024) discuss this issue in Rojak Malay, where Malay-English mixed expressions require specific language processing treatment. Malay noisy text studies such as (Azizan et al., 2025) have mainly focused on normalisation, while the structural grouping of candidate noisy Malay words remains less explored.

This study investigates the clustering of candidate noisy Malay words according to their observable structural patterns. Different noisy forms, such as abbreviations, misspellings, repeated-character words, and digit-based reduplication, may require different representations because they do not share the same surface structure. Mansur et al. (2024) show that repeated-letter forms require specific handling, while (Dankolo et al., 2023) note that different non-standard word forms may require different processing strategies.

The objective of this study is to compare four word representation approaches under K-Means clustering: character-level TF-IDF, FastText subword embedding, binary noise-type features, and RLE with TF-IDF. Each representation is applied to the same set of 5,097 candidate noisy Malay words derived from Facebook and Instagram comments. Hoti et al. (2024) and Khan et al. (2025b) show that clustering quality in low-resource text settings is influenced by preprocessing and vector representation choices.

The main contribution of this study is a comparative internal evaluation of four K-Means representation approaches for grouping candidate noisy Malay words. Rather than proposing a new clustering algorithm, the study examines how different representations affect cluster structure under the same K-Means configuration. The clusters are evaluated using the Silhouette Score, Davies-Bouldin Index, and Calinski-Harabasz Score, which (Ikotun et al., 2025a) describe as useful for assessing cohesion and separation. Chicco et al. (2025) further emphasise the need to consider multiple internal metrics, as each captures a different aspect of cluster structure.

2 Related Work

Social media text presents challenges for natural language processing because users frequently write in informal and non-standard forms. Sohn et al. (2025) describe spelling changes, abbreviations, repeated characters, code-mixing, and out-of-vocabulary forms as common issues in social media processing (Wahid et al., 2025; Mustofa, 2025) also highlight the need for preprocessing when analysing informal online text.

These issues are particularly relevant for Malay user-generated content, where informal writing may include shortened words, Malay-English mixing, repeated characters, slang, and non-standard spelling. Yong et al. (2024) examined Rojak Malay and showed that mixed Malay-English expressions require specific language processing treatment. In noisy Malay text normalisation, Noor Allia Noor Ariffin et al. (2020) applied rule-based techniques with Levenshtein distance, while (Azizan et al., 2024) examined similarity measures for handling non-standard Malay forms. However, these studies mainly focus on normalisation rather than grouping noisy words according to their structural similarities.

Noisy word patterns may require different forms of representation because noisy forms are structurally different. Mansur et al. (2024) showed that repeated-letter patterns in social media text require specific handling, while (Dankolo et al., 2023) reported that non-standard words may require different normalisation strategies depending on their form. Therefore, this study compares four representation approaches: character-level TF-IDF, FastText subword embedding, binary noise-type features, and RLE with TF-IDF.

Representation selection is important because clustering depends on how textual patterns are transformed into numerical features. Keinan (2024) and Setiawan et al. (2024) demonstrated the usefulness of TF-IDF and n-gram features for irregular online text, while (Ravi & Kulkarni, 2023) showed that embedding representations can influence clustering quality.

K-Means is relevant to this study because it provides a consistent clustering method for comparing different word representations on the same noisy Malay word dataset. Previous studies show that clustering quality is affected by preprocessing, vector representation, and feature selection (Guleria et al., 2025; Hoti et al., 2024; Khan et al., 2025b). Internal validation metrics are used to assess cluster cohesion and separation. Ikotun et al. (2025a) explain the usefulness of validity indices for comparing clustering outcomes, while (Chicco et al., 2025) support the use of validity indices for interpreting clustering outcomes.

3 Methodology

This study compares four word representation approaches for clustering noisy Malay words using K-Means. The methodology consists of six main stages: data collection, data preprocessing, noisy word filtering, word representation, K-Means clustering, and internal evaluation.

3.1 Data Collection

Data for this study were collected from selected Facebook and Instagram public posts using the ESUIT Comments Exporter browser extension. The exported records were saved in Excel format, where each row represented one comment.

The exported dataset contained comment text and several metadata fields, including user identifiers, author names, reaction counts, and timestamps. However, only the comment text column was retained because this study focuses on word-level noisy text analysis. Metadata fields were excluded before preprocessing, and no personal identifiers were used in the clustering analysis. After consolidation, the raw dataset contained 6,988 comments from Facebook and Instagram. Social media comments are suitable for noisy text analysis because user-generated content often contains informal spelling, abbreviations, repeated characters, slang, and mixed-language expressions (Sohn et al., 2025).

3.2 Data Preprocessing

The raw comment data were preprocessed in two stages: comment-level cleaning and word-level tokenisation. Comment-level cleaning was performed using Google Apps

Script on the consolidated Excel file. The cleaning procedures included removing URLs, mention tags, hashtags, emoji symbols, non-Latin characters such as Jawi and Arabic script, repeated punctuation, isolated punctuation, HTML entities, long numeric strings, and extra whitespace. These steps were applied to reduce irrelevant textual noise before word-level analysis, since preprocessing and tokenisation are commonly used to convert raw text into units suitable for numerical representation in NLP tasks (Siino et al., 2024).

After cleaning, the dataset retained only Latin-script text. The cleaned dataset consisted of 4,189 Facebook comments and 2,799 Instagram comments. These cleaned comments were then tokenised at the word level, where each word was placed into a separate row. This word-level format was required because the embedding and clustering stages in this study operate on individual noisy word tokens rather than full sentences. Tokenisation is important because it splits text into smaller processing units, and the resulting token list can then be mapped into numerical representations for machine learning or clustering (Siino et al., 2024). In this study, word-level tokenisation also allowed surface-level noisy patterns such as misspellings, abbreviations, character repetitions, and alphanumeric substitutions to be examined independently. This is suitable for social media text because informal writing often produces out-of-vocabulary forms and textual variations that make processing more challenging (Mansur et al., 2024).

The tokenisation process produced 13,015 rows. After removing one invalid missing-value entry, 13,014 valid word tokens were retained and passed to the noisy word filtering stage.

3.3 Noisy Word Filtering

Following word-level tokenisation, an exclusion-based filtering procedure was applied to isolate candidate noisy Malay words from the collected social media text. Text preprocessing is important in social media analysis because informal content frequently contains irregular word forms and non-relevant tokens that may affect later numerical representation and clustering (Siino et al., 2024).

The filtering process removed entries that did not represent the noisy Malay word patterns targeted in this study, including invalid entries, numeric-only items, duplicate tokens, single-character items, standard Malay words, English words, and detected person names. Standard Malay and English lexical items were excluded so that the subsequent analysis focused on non-standard Malay word candidates, consistent with Malay noisy text studies that distinguish informal or noisy terms from standard and mixed-language expressions (Azizan et al., 2025; Dankolo et al., 2023; Yong et al., 2024). Detected person names were also excluded using Named Entity Recognition because named entities represent personal references rather than noisy word structures (Keraghel et al., 2024).

From the 13,014 valid word tokens derived from Facebook and Instagram comments, 5,097 candidate noisy Malay words were retained after the filtering process. These retained noisy words formed the final dataset used to compare the four word representation approaches under K-Means clustering in this study.

3.4 Word Representation Methods

Four word representation methods were compared to capture different structural properties of noisy Malay words: character-level TF-IDF, FastText subword embedding, binary noise-type features, and RLE with TF-IDF. These methods were selected because non-standard words may appear as misspellings, abbreviations, repeated-character forms, digit-based reduplication, or phonetic-like spellings. Previous studies show that noisy and non-standard words may require different processing strategies depending on their form (Azizan et al., 2020, 2025).

Character-level TF-IDF using 2–3 character n-grams was used to capture partial spelling and character-level similarity in irregular online text (Keinan, 2024; Setiawan et al., 2024). FastText subword embedding was included to examine whether subword-level similarity could group noisy Malay words with similar internal patterns, as representation choice is important for Malaysian language data (Zolkepli et al., 2024). Binary noise-type features were used to directly encode visible noisy-word characteristics, including elongation, digit presence, abbreviation-like structure, consonant clusters, and phonetic suffixes. This is relevant because repeated-letter patterns require specific handling in social media text (Mansur et al., 2024). RLE with TF-IDF was included to capture repeated and character-shape patterns rather than exact spelling. The four representation methods are summarised in Table 1.

Table 1 Word Representation Methods

Representation	Description	Purpose
Character-level TF-IDF	TF-IDF using 2–3 character n-grams	Captures lexical and character-level similarity
FastText subword	Subword embedding trained on noisy words	Captures subword-level similarity
Binary noise-type features	Eleven handcrafted binary features	Captures explicit noisy word patterns
RLE + TF-IDF	Character-class RLE representation with TF-IDF	Captures repeated and structural word patterns

3.5 K-Means Clustering Design

K-Means was applied to all four word representation approaches to compare how different feature spaces affect the grouping of noisy Malay words. Using the same clustering algorithm and number of clusters across all representations allows the comparison to focus on representation differences rather than changes in clustering configuration. Previous studies show that clustering quality is influenced by preprocessing and vector representation (Hoti et al., 2024; Khan et al., 2025b).

The number of clusters, k, was considered by comparing candidate values from $k = 2$ to $k = 20$ using Silhouette Score, Davies-Bouldin Index, and Ca-linski-Harabasz Score. Schubert (2023) cautions that the elbow criterion alone is not sufficient for selecting k in K-Means, while (Chicco et al., 2025; Ikotun et al., 2025a) show that internal metrics provide complementary evidence of cluster structure. The initial comparison showed that different representations favoured different k values. Therefore, k = 6 was adopted as a common comparative setting to maintain equivalent cluster granularity across all four approaches. This value is not claimed as the optimal k for every representation, but it provides a consistent setting for comparing the feature spaces.

For reproducibility, K-Means was configured with a fixed random state of 42 and 10 centroid initialisations. The four K-Means approaches are summarised in Table 2.

Table 2 K-Means Word Representation Approaches

Approach	Word Representation	Clustering Method	Number of Clusters
A1	Character-level TF-IDF using 2–3 character n-grams	K-Means	6
A2	FastText subword embedding	K-Means	6
A3	Binary noise-type features	K-Means	6
A4	RLE + TF-IDF character-class representation	K-Means	6

3.6 Internal Evaluation

The quality of the clusters generated by the four K-Means approaches was evaluated using Silhouette Score, Davies-Bouldin Index, and Calinski-Harabasz Score. Ikotun et al. (2025b) describe internal cluster validity indices as useful measures for evaluating clustering structure, while (Chicco et al., 2025) discuss the suitability of Silhouette Score and Davies-Bouldin Index for interpreting cluster quality.

The Silhouette Score measures how closely each word is associated with its assigned cluster relative to other clusters, where a higher value indicates stronger cohesion and separation. The Davies-Bouldin Index evaluates similarity between clusters, where a lower value indicates better separation. The Calinski-Harabasz Score measures separation between clusters relative to dispersion within clusters, where a higher value indicates stronger cluster structure. Since different validity indices may reflect different aspects of clustering quality, all three metrics were used together to compare the four word representation approaches (Chicco et al., 2025; Ikotun et al., 2025a).

Character-level TF-IDF, binary noise-type features, and RLE + TF-IDF produced deterministic representations in this study. FastText subword embeddings were generated through unsupervised training, and repeated training produced slight variation in the resulting metric values. Therefore, the FastText approach was evaluated over five training runs and reported using the mean and standard deviation, while the other three approaches were reported using their deterministic evaluation values (Table 3).

Table 3 Internal Evaluation Metrics

Metric	Desired Direction	Purpose
Silhouette Score	Higher is better	Measures cohesion and separation among clusters
Davies-Bouldin Index	Lower is better	Measures similarity between clusters
Calinski-Harabasz Score	Higher is better	Measures between-cluster separation relative to within-cluster dispersion

4 Results and Discussion

This section presents the results of the four word representation approaches clustered using K-Means with the same setting of k = 6. The clustering outputs were evaluated using Silhouette Score, Davies-Bouldin Index, and Calinski-Harabasz Score.

4.1 Internal Evaluation of the Four K-Means Representation Approaches

Table 4 presents the internal evaluation results for the four word representation approaches. Since all approaches used the same K-Means configuration, the comparison focuses on the effect of representation choice on noisy Malay word grouping.

Table 4 Internal Evaluation Results of the Four K-Means Representation Approaches at k = 6

Approach	Representation	Number of Clusters	Silhouette Score ↑	Davies-Bouldin Index ↓	Calinski-Harabasz Score ↑
A1	Character-level TF-IDF	6	0.0094	6.4604	26.7879
A2	FastText subword embedding	6	0.0737 ± 0.0105	2.4666 ± 0.1681	1088.6374 ± 190.8787
A3	Binary noise-type features	6	0.7451	0.9383	3776.5049
A4	RLE + TF-IDF	6	0.2686	1.8332	728.3922

The results show that binary noise-type features produced the strongest cluster structure across all three internal metrics, with the highest Silhouette Score, the lowest Davies-Bouldin Index, and the highest Calinski-Harabasz Score. RLE + TF-IDF produced stronger results than character-level TF-IDF and FastText for Silhouette Score and Davies-Bouldin Index, while character-level TF-IDF produced the weakest clustering structure. FastText results are reported as mean ± standard deviation because repeated unsupervised training produced slight variation between runs.

4.2 Comparison by Representation Method

The character-level TF-IDF approach produced the weakest clustering structure, with a Silhouette Score of 0.0094 and a Davies-Bouldin Index of 6.4604. Although character n-grams can represent partial spelling patterns in informal text (Keinan, 2024; Setiawan et al., 2024), the result suggests that character overlap alone was insufficient to distinguish the main noisy-word structures in this dataset.

The FastText subword embedding approach improved over character-level TF-IDF, with a mean Silhouette Score of 0.0737 ± 0.0105 and a mean Calinski-Harabasz Score of 1088.6374 ± 190.8787. This indicates that subword information captured some internal similarity among noisy word forms, although the clusters remained weakly separated compared with the structural feature-based approaches.

The binary noise-type feature approach produced the strongest cluster structure because it directly represented visible noisy-word characteristics, including repeated characters, abbreviation-like shapes, digit-based patterns, consonant clusters, and phonetic suffixes. Mansur et al. (2024) showed that repeated-letter forms require specific handling in social media text, while (Khan et al., 2025a) describe spelling variation, abbreviation, and character repetition as common textual variations in online communication. However, this result should be interpreted carefully because the binary features were manually designed to capture visible noisy-word structures.

The RLE + TF-IDF approach produced meaningful clustering results, with a Silhouette Score of 0.2686 and a Davies-Bouldin Index of 1.8332. This suggests that character-class and repetition-based representation can capture useful surface structures in noisy Malay words, although it remains less discriminative than binary noise-type features.

4.3 Cluster Interpretability of the Binary Noise-Type Feature Approach

The binary noise-type feature approach formed six clusters from the 5,097 candidate noisy Malay words. Its largest clusters provide examples of the structural patterns captured through K-Means, as shown in Table 5.

Table 5 Examples of Major Clusters Produced by Binary Noise-Type Features with K-Means

Cluster	Size	Example Words	Observable Surface Characteristic
2	1,577	abangg, abng, abt, acc, acct., act, adults, aff	Shortened or consonant-dominant forms with mixed abbreviation-like patterns
1	1,201	aaamiiinnn, aaaminnn, aahhh, aamiinnn, aaminnn, aannnggg, aarrgghhh, abaikn	Character elongation and repeated-letter forms
4	952	aa, aaa, aamiiin, ab, abik, abil, abis, aboh	Short forms and vowel-extension patterns

The examples show that the strongest K-Means representation was able to organize several visible noisy word characteristics into identifiable structural groups. In particular, Cluster 1 contains clear repeated-letter and elongated forms, which supports the relevance of explicit surface features for informal text analysis. This observation is consistent with studies that identify repeated characters and other non-standard forms as important patterns in social media text (Dankolo et al., 2023; Mansur et al., 2024; Sohn et al., 2025). However, some clusters still contain mixed forms, particularly in larger groups. Therefore, the clusters are interpreted as structural groupings of candidate noisy Malay words rather than fixed linguistic categories.

4.4 Overall Discussion

The K-Means comparison shows that word representation choice strongly affects noisy Malay word clustering. Under the same $k = 6$ setting, binary noise-type features produced the strongest cluster structure across all three internal validation metrics because they directly represented visible surface patterns, including repeated characters, abbreviation-like forms, and digit-based patterns.

RLE + TF-IDF also showed potential by capturing character-shape and repetition-based similarities. FastText improved over character-level TF-IDF, but its clusters remained weakly separated, while character-level TF-IDF produced the least separated structure. Overall, binary noise-type features produced the most internally evaluated and structurally interpretable K-Means groupings, which may support further linguistic inspection and targeted Malay noisy text normalisation for future normalisation rules and more detailed manual analysis of noisy word patterns.

5 Conclusion

This study presented a K-Means-based approach for grouping candidate noisy Malay words according to their structural patterns. Four word representation approaches were compared: character-level TF-IDF, FastText subword embedding, binary noise-type features, and RLE with TF-IDF. Each representation was applied to the same set of 5,097 candidate noisy Malay words and evaluated under a common K-Means setting of $k = 6$ using Silhouette Score, Davies-Bouldin Index, and Calinski-Harabasz Score.

The results showed that binary noise-type features produced the strongest cluster structure among the four representation approaches. This representation achieved a Silhouette Score of 0.7451, a Davies-Bouldin Index of 0.9383, and a Calinski-Harabasz Score of 3776.5049. Its clusters also demonstrated observable structural patterns, including repeated-letter forms, shortened words, abbreviation-like forms, and vowel-extension patterns. These findings indicate that explicitly representing visible noisy word characteristics is effective for organising candidate noisy Malay words using K-Means.

RLE + TF-IDF also showed potential for capturing repetition-based and character-shape similarities, while FastText and character-level TF-IDF produced weaker cluster separation under the common setting applied in this study. Overall, binary noise-type features produced the strongest internally evaluated and structurally interpretable K-Means groupings among the four representations considered. Future work may refine the structural feature set, include expert annotation or cluster purity analysis, compare K-Means with other clustering algorithms, and test the approach on additional Malay social media datasets to further validate the linguistic usefulness of the generated clusters.

Acknowledgments. This research is based upon work supported by Universiti Teknologi MARA and Faculty of Computer and Mathematical Sciences, Universiti Teknologi MARA, Shah Alam, Selangor, Malaysia.

Competing Interests. The author declares no competing interests.

References

Azizan, A., Saidin, N.A., Khairudin, N., Ismail, R.: An application of Malay short-form word conversion using Levenshtein distance. Math. Sci. Inf. J. **1**(2), 34–42 (2020)

Azizan, A., et al.: Normalization of Malay Noisy text in social media using Levenshtein distance and rule-based techniques (2024). https://doi.org/10.47772/IJRISS

Azizan, A., et al.: Evaluating similarity measures for Malay Noisy text normalization: performance and threshold analysis (2025). Malays. J. Comput. Sci. 38, 1-13 (2025). https://doi.org/10.22452/mjcs.vol38spc.1

Chicco, D., Campagner, A., Spagnolo, A., Ciucci, D., Jurman, G.: The Silhouette coefficient and the Davies-Bouldin index are more informative than Dunn index, Calinski-Harabasz index, Shannon entropy, and gap statistic for unsupervised clustering internal evaluation of two convex clusters. PeerJ Comput. Sci. **11**, e3309 (2025). https://doi.org/10.7717/peerj-cs.3309

Dankolo, N., et al.: Systematic review on text normalization techniques and its approach to non-standard words. Int. J. Comput. Appl. **185**(33) (2023) https://www.researchgate.net/publication/374160515

Guleria, P., Frnda, J., Srinivasu, P.N.: NLP based text classification using TF-IDF enabled fine-tuned long short-term memory: an empirical analysis. Array. **27**, 100467 (2025). https://doi.org/10.1016/j.array.2025.100467

Hoti, M., Rexhepi, A., Hoti, A., Rexha, B.: Challenges and solutions in clustering low-resource language social media text: an evaluation using unsupervised algorithms. (2024). https://doi.org/10.21203/rs.3.Rs-4913729/v1

Ikotun, A.M., Habyarimana, F., Ezugwu, A.E.: Benchmarking validity indices for evolutionary K-means clustering performance. Sci. Rep. **15**(1), 21842 (2025a). https://doi.org/10.1038/s41598-025-08473-6

Ikotun, A. M., Habyarimana, F., & Ezugwu, A. E. (2025b). Cluster validity indices for automatic clustering: a comprehensive review. Heliyon (Vol. 11, 2). Elsevier Ltd. doi:https://doi.org/10.1016/j.heliyon.2025.e41953

Keinan, R.: Sexism Identification in Social Networks using TF-IDF Embeddings, PreProccessing, Feature Selection, Word/Char N-Grams and Various Machine Learning Models In Spanish and English (2024)

Keraghel, I., Morbieu, S., Nadif, M.: Recent advances in named entity recognition: a comprehensive survey and comparative study. http://arxiv.org/abs/2401.10825. (2024)

Khan, J., Ahmad, K., Jagatheesaperumal, S.K., Sohn, K.A.: Textual variations in social media text processing applications: challenges, solutions, and trends. Artif. Intell. Rev. **58**(3), 89 (2025a). https://doi.org/10.1007/s10462-024-11071-z

Khan, T.F., Hussain, M., Arslan, M., Saeed, M., Khan, L., Chang, H.T.: LLM-based enhanced clustering for low-resource language: an empirical study. CMES – Comput. Model. Eng. Sci. **145**(3), 3883–3911 (2025b). https://doi.org/10.32604/cmes.2025.073021

Mansur, Z., Omar, N., Tiun, S., Alshari, E.M.: A normalization model for repeated letters in social media hate speech text based on rules and spelling correction. PLoS One. **19**, e0299652 (2024). https://doi.org/10.1371/journal.pone.0299652

Mustofa, B.A.: View of Use of Natural Language Processing in Social Media Text Analysis (2025)

Noor Allia Noor Ariffin, S., Tiun, S., Kebangsaan Malaysia, U., Selangor, B.: Rule-based text normalization for malay social media texts. Int. J. Adv. Comput. Sci. Appl. **11**(10) (2020) www.ijacsa.thesai.org

Ravi, J., Kulkarni, S.: Text embedding techniques for efficient clustering of twitter data. Evol. Intel. **16**(5), 1667–1677 (2023). https://doi.org/10.1007/s12065-023-00825-3

Schubert, E.: Stop using the elbow criterion for k-means and how to choose the number of clusters instead (2023)

Setiawan, Y., Maulidevi, N.U., Surendro, K.: The optimization of n-gram feature extraction based on term occurrence for cyberbullying classification. Data Sci. J. **23**(1) (2024). https://doi.org/10.5334/dsj-2024-031

Siino, M., Tinnirello, I., La Cascia, M.: Is text preprocessing still worth the time? A comparative survey on the influence of popular preprocessing methods on transformers and traditional classifiers. Inf. Syst. **121**, 102342 (2024). https://doi.org/10.1016/j.is.2023.102342

Sohn, K.-A. et al.: Textual variations in social media text processing applications: challenges, solutions, and trends. Artif. Intell. Rev. **58**(3), 1–48 (2025). https://doi.org/10.1007/S10462-024-11071-Z

Wahid, J.A. et al.: AI-driven social media text analysis during crisis: a review for natural disasters and pandemics. In: Applied Soft Computing, vol. 171. Elsevier Ltd. (2025). https://doi.org/10.1016/j.asoc.2025.112774

Yong, J.Z.H., Koh, J.Y., Liew, J.X., Tan, C.W.: Linguistic harmony in diversity: lemmatizing Rojak Malay for global communication. In: ICDXA 2024 - Conference Proceedings: 2024 3rd International Conference on Digital Transformation and Applications, pp. 6–10 (2024). https://doi.org/10.1109/ICDXA61007.2024.10470819

Zolkepli, H., Razak, A., Adha, K., Nazhan, A.: Multi-lingual malaysian embedding: leveraging large language models for semantic representations. http://arxiv.org/abs/2402.03053. (2024)

Feature-Group Influence on Student Performance Prediction across Heterogeneous Learning Platforms

Abdelatti Blg(✉), Muhammad Haziq Lim Abdullah, and Mohamad Lutfi Dolhalit

Faculty of Information and Communication Technology, Universiti Teknikal Malaysia Melaka (UTeM), Durian Tunggal, Melaka, Malaysia
p032420010@student.utem.edu.my

Abstract. Student performance prediction has received sustained attention in learning analytics. Yet, much of the literature remains model-centred and offers limited insight into which categories of learner data actually drive predictive performance. This study adopted a data-centric perspective and examined how academic, behavioural, and cognitive-proxy feature groups influenced student performance prediction across two heterogeneous learning environments, OULAD and EdNet. Using a different feature group's experimental design under full-information conditions, logistic regression models were estimated progressively to evaluate the marginal contribution of different feature groups within each dataset and to compare dominant predictive patterns across platforms. The findings demonstrated that integrating multiple feature groups, rather than using them separately, enhanced prediction performance; each dataset had a different most informative combination. With an AUC of 0.796, a PR-AUC of 0.813, and a Recall of 0.644, behavioural and cognitive-proxy characteristics performed the best in EdNet. With an AUC of 0.962, an F1-score of 0.894, and an Accuracy of 0.898, the combination of academic and behavioural data with cognitive-proxy traits yielded the best results in OULAD. Across both datasets, interaction-enhanced models produced no additional gain beyond the main integrated representations. These findings indicate that feature-group influence is context-dependent rather than universal and highlight the importance of dataset-aware feature engineering for interpretable, lightweight, and portable student performance prediction in e-learning

Keywords: student performance · learning analytics · feature-group · behavioural engagement · cognitive proxies · cross-dataset

1 Introduction

With the rising use of learning management systems (LMSs) and online learning platforms, student performance prediction has grown in importance as a field of study in educational data mining and learning analytics. Large amounts of assessment data, interaction logs, and behavioural traces are constantly produced by these environments, which

N. Mohd Sharef et al. (Eds.): CAMP 2026, CCIS 3093, pp. 14–24, 2027.
https://doi.org/10.1007/978-981-92-4302-0_2

present beneficial chances to use machine learning for early risk identification of students and for more prompt educational intervention (Alhazmi and Sheneamer 2023; Ahmed 2024; Christou et al. 2023). Even with this expansion, a major portion of the current research is still model-centric. Many studies compare algorithms like deep learning architectures, support vector machines, and ensemble methods; these studies frequently report better predictive performance within a single dataset or institutional context (Alkan et al. 2025; Huang and Chen 2024; Junejo et al. 2025). Although these comparisons are helpful, they frequently overlook a more crucial question: what kinds of learner data actually have the biggest impact on predictive performance? In many instances, it is still unclear whether the predictive value of particular feature categories, such as academic, behavioural, or derived indicators, or the model selection or dataset structure are the primary driver of high outcomes (Roy and Farid 2024; Wang and Luo 2024).

Recent evidence suggests that the composition of the data may be just as important as the choice of algorithm. While behavioural engagement features derived from LMS activity and clickstream logs have also demonstrated strong predictive value across a wide range of online learning settings, academic indicators frequently remain highly informative when formal assessment records are available (Latif et al. 2023; Liu et al. 2023; Santos and Henriques 2023). It is challenging to compare results between platforms or to make more general inferences about the relative value of feature groups, though, because these data are frequently provided individually. This study addresses that gap by adopting a comparative, data-centric perspective on student performance prediction. Rather than proposing a new predictive model, it examines how academic, behavioural, and cognitive-proxy feature groups contribute to prediction across two heterogeneous datasets, OULAD and EdNet (Kuzilek et al. 2017; Choi et al. 2020). By comparing the experiments across these contrasting environments, the study seeks to identify which feature groups contribute most strongly to predictive performance and whether their influence remains stable across different e-learning contexts.

2 Related Work

Research on student performance prediction has expanded rapidly alongside the growth of LMSs and other online learning platforms. As these environments generate increasingly rich records of learner activity, they have created favourable conditions for the use of machine learning in educational settings. Much of this work has focused on developing predictive models that can identify students at risk, support timely intervention, and improve learner monitoring (Ahmed 2024; Alhazmi and Sheneamer 2023; Wang and Yu 2025).

A large part of the literature has taken a model-centred approach. Because they work well on structured educational data and provide a useful mix between efficiency and interpretability, traditional approaches like Logistic Regression, Random Forest, and boosting techniques continue to be widely employed (Ahmed 2024; Alkan et al. 2025; Thakur and Rawat 2025). Simultaneously, more sophisticated methods have been devised to capture relational and sequential patterns in learner behaviour, such as graph-based and temporal deep learning models (Huang and Chen 2024; Junejo et al. 2025). Although these studies have improved predictive performance, they often treat the input data as fixed and place greater emphasis on algorithmic comparison than on the informational value of the data itself.

This is significant since new research indicates that learner data composition influences prediction accuracy in addition to the model. In organised higher education environments, academic characteristics like test results and official accomplishment records are frequently quite predictive. Behavioural traits derived from LMS logs and clickstream data, such as activity frequency, persistence, and regularity, have also demonstrated strong predictive value in a variety of online learning environments (Latif et al. 2023; Liu et al. 2023; Santos and Henriques 2023). Additionally, several studies have included developed indicators intended to capture learner stability, self-regulation, and larger engagement dynamics in order to offer a more comprehensible explanation of learning behaviour (Hoq et al. 2024; Sarker et al. 2024). This line of inquiry is further supported by recent advancements in deep sequential architectures; for instance, Ghazvini et al. (2024) investigated early course grade prediction using optimised Long Short-Term Memory (LSTM) frameworks built upon combinations of demographic indicators, explicit academic assessments, and weekly learning management system (LMS) activity logs. Their empirical evaluation across dozens of distinct feature configurations underscored how granular, early-semester behavioural traces significantly alter model capacity, emphasising a direct intersection between architectural efficiency and feature-set composition.

Nevertheless, these feature categories are frequently studied separately, and comparatively few studies assess their contribution across disparate datasets using a uniform analytical methodology. In light of this, the current study takes a more overtly data-centric stance. Rather than adding another algorithmic comparison, it investigates how academic, behavioural, and cognitive-proxy feature groups contribute to prediction across two heterogeneous datasets, OULAD and EdNet. Through different feature groups, the study aims to provide a clearer understanding of feature-group influence and a stronger basis for cross-context student performance prediction research.

3 Methodology and Experimental Setup

This study adopted a comparative, data-centric experimental design to examine how different categories of learner data influenced student performance prediction across heterogeneous online learning environments. Rather than focusing on algorithmic novelty, the study used predictive modelling as an analytical instrument for assessing the relative contribution of feature groups under a consistent modelling framework. All experiments were conducted under full-information conditions, meaning that all relevant predictors available within each dataset were considered during modelling.

3.1 Datasets, Outcome Definition, and Feature Design

The Open University Learning Analytics Dataset (OULAD) and EdNet, two publicly accessible educational datasets, were chosen to illustrate disparate online learning environments. These datasets were chosen because they differ substantially in scope, platform structure, and learner evidence, thereby providing a suitable basis for examining whether feature-group influence remains stable across heterogeneous data environments.

OULAD represents a structured university distance-learning environment from the Open University in the United Kingdom. It contains records for 32,593 students across 22 module presentations and covers learner activity from 2013 to 2014. The dataset includes demographic information, course registrations, assessment outcomes, and Virtual Learning Environment (VLE) interaction data, including 10,655,280 click records summarised by day (Kuzilek et al. 2017). EdNet, in contrast, represents a large-scale platform-based learning environment collected from Santa, a self-study platform supported by an artificial-intelligence tutoring service. It comprises 784,309 students and 131,441,538 interactions collected over a period of more than two years, documenting specific learner behaviour across activities, including watching lectures, responding to questions, and other types of interaction (Choi et al. 2020) as shown in Table 1. As a result, the two datasets were examined separately and compared only once, after within-dataset modelling was finished. Student performance was modelled as a binary outcome in both datasets while preserving the logic of each environment. In OULAD, learners were classified as successful if their final result was Pass or Distinction, and unsuccessful otherwise. In EdNet, where no directly comparable final-grade variable was available, performance was defined at the user level using mean correctness, with learners classified according to whether their score was at or above the dataset median. Predictors were grouped into three categories: academic, behavioural, and cognitive-proxy features. Academic features captured direct evidence of formal performance, including assessment scores, weighted achievement, submission behaviour, and lateness-related measures where available. Behavioural features described platform activity, such as interaction volume, participation frequency, active periods, and engagement with learning resources. Cognitive-proxy features formed a derived group constructed from longitudinal patterns in learner activity and performance to reflect tendencies such as regularity, stability, persistence, and progression over time. These variables were treated as proxies rather than direct cognitive measures.

Table 1. Summary of dataset scale and engineered feature groups

Dataset	Learners/users	Raw data scale	Academic	Behavioural	Cognitive-proxy	Total
OULAD	32,593	10,655,280 VLE records; 173,912 assessment records	5	5	4	14
EdNet	30,000 sampled users from 784,309 available user files	User-level interaction histories	2	5	4	11

3.2 Preprocessing, Modelling, and Evaluation

Data preparation was done separately for OULAD and EdNet because the two datasets had different structures. For OULAD, learner information, assessment records, and VLE activity logs were combined at the student-presentation level. From the assessment data, five academic features were created: MeanScore, MedianScore, SubmitCount, LateRatio, and WeightedScore. From the VLE logs, five behavioural features were created: TotalClicks, ActiveDays, ResourceDiversity, WeeklyEngagementMean, and EngagementEntropy. Four additional cognitive-proxy features were also derived to describe learner stability, improvement, consistency, and persistence over time.

For EdNet, the data were prepared at the user level using interaction histories from 30,000 sampled learners. Two academic-like features were retained: MeanScore and AttemptCount. Five behavioural features were created, namely TotalAttempts, ActiveWeeks, WeeklyMean, EngagementEntropy, and ResourceDiversity. Four cognitive-proxy features were also generated: Stability, Improvement Slope, Consistency, and Persistence. Records without usable correctness information were removed, and learner activities were arranged in chronological order before feature construction.

For both datasets, non-informative identifiers were removed before modelling. Only variables that would be available at the prediction stage were kept to reduce the risk of target leakage. All variables were converted into numeric tables, and non-finite values were replaced or removed where needed. Numerical predictors were standardised after the train-test split, so the scaling values were learned from the training set only and then applied to the test set. To assess the relative contribution of feature groups, a modelling framework was implemented. Rather than entering all predictors simultaneously from the outset, the design progressively expanded the feature groups so that the marginal contribution of each category could be examined more clearly. Modelling configurations were adapted to each dataset while preserving the same analytical logic, moving from constrained baselines to more integrated combinations of academic, behavioural, and cognitive-proxy features, with a final step including interaction-enhanced representations where appropriate.

Binary Logistic Regression (BLR) was selected as the modelling framework. BLR was used as the common predictive framework because it offers a parsimonious, interpretable, and stable basis for comparing feature sets on tabular educational data. It is widely recognised in learning analytics for its rigorous balance between predictive power and domain interpretability, allowing for the extraction of odds ratios that can directly inform institutional early-warning systems (Chung & Lee, 2019). Furthermore, its probabilistic nature makes it uniquely suited for an incremental, block-wise evaluation framework, enabling a clear assessment of the marginal contributions of distinct behavioural, temporal, and cognitive feature dimensions without the confounding opacity of black-box ensembles (Morsu, 2025). The aim was not to benchmark multiple classifiers, but to maintain a fixed modelling family so that performance differences could be attributed primarily to feature-group composition rather than algorithmic variation. Stratified hold-out splits were used to preserve class proportions, with an 80:20 train-test split for OULAD and a 70:30 split for EdNet. The test sets were reserved strictly for final evaluation, and where more complex specifications introduced numerical instability, regularised estimation was used without changing the modelling framework.

Evaluation metrics were selected according to dataset characteristics. For EdNet, AUC, PR-AUC, and Recall were used to reflect class-sensitive performance in an interaction-driven environment. For OULAD, AUC, F1-score, and Accuracy were used to capture discrimination, balance, and overall classification quality in a more structured academic setting. A fixed decision threshold of 0.5 was applied throughout to ensure consistent comparison across different feature groups.

4 Results

This section reports the empirical findings from the modelling experiments conducted on EdNet and OULAD. The analysis examined how progressively adding feature groups influenced student performance prediction under the framework (M1–M5). To keep the presentation concise, the results are organised into three parts: EdNet results, OULAD results, and a final cross-dataset comparison. Each part is summarised through one table and one figure where appropriate.

4.1 EdNet Results

Figure 1 shows the general trend across modelling phases, whereas Table 2 presents the EdNet results. With an AUC of 0.779, PR-AUC of 0.798, and a recall of 0.424, the behavioural-only model (M1) offered a helpful starting point. While PR-AUC and Recall slightly increased, AUC decreased to 0.761 when temporal engagement factors were introduced to M2. In M3, stability-related features raised AUC to 0.780, PR-AUC to 0.813, and Recall to 0.593, suggesting a more notable improvement. With behavioural and cognitive-proxy features producing an AUC of 0.796, a PR-AUC of 0.813, and a Recall of 0.644, M4 produced the best results. No further gain was observed in M5, indicating that interaction terms did not improve performance beyond the main integrated representation.

Taken together, the EdNet results show that fundamental activity volume is not as significant for prediction in an interaction-rich environment as when the feature set exhibits ordered behavioural regularity, stability, and broader learning dynamics.

Table 2: Predictive performance comparison across different feature groups using the EdNet dataset

Model	Feature Groups	AUC	PR-AUC	Recall
M1	Behavioural	0.779	0.798	0.424
M2	Behavioural and Temporal Engagement	0.761	0.805	0.441
M3	Behavioural and Stability	0.780	0.813	0.593
M4	Behavioural and Cognitive	0.796	0.813	0.644
M5	Integrated Model and Interaction Terms	0.796	0.813	0.644

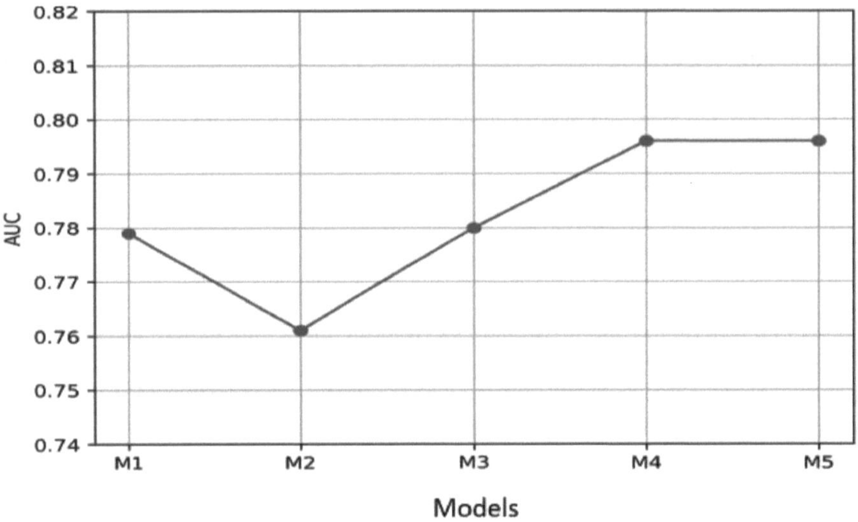

Fig. 1: Model AUC performance across different feature groups of EdNet dataset.

4.2 OULAD Results

Table 3 displays the OULAD results, while Fig. 2 summarises the performance trend throughout selected feature groups. In comparison to EdNet, the academic-only model (M1) has already shown exceptional predictive ability with an AUC of 0.9405, an F1-score of 0.8653, and an accuracy of 0.8716. This implies that there is a strong predictive value for explicit assessment-related information in an organised institutional framework. With an accuracy of 0.8518, an F1-score of 0.8490, and an AUC of 0.9268, the behavioural-only model (M2) fared a little worse. Combining academic and behavioural data significantly improved M3 performance, yielding an accuracy of 0.8968, an F1-score of 0.8940, and an AUC of 0.9602. The AUC rose to 0.9620, the F1-score to 0.8944, and the accuracy to 0.8980 when cognitive-proxy characteristics were added to M4. M5 produced no additional improvement, just like EdNet.

These results show that academic variables remain the strongest single predictive foundation in OULAD, while behavioural and cognitive-proxy features provide complementary value and improve the final integrated model.

Table 3: Predictive performance comparison across different feature groups using the OULAD dataset.

Model	Feature Groups	AUC	F1	Accuracy
M1	Academic	0.9405	0.8653	0.8716
M2	Behavioural	0.9268	0.8490	0.8518
M3	Academic and Behavioural	0.9602	0.8940	0.8968

(continued)

Table 3: (*continued*)

Model	Feature Groups	AUC	F1	Accuracy
M4	Academic and Behavioural and Cognitive	0.9620	0.8944	0.8980
M5	All + Interaction	0.9620	0.8944	0.8980

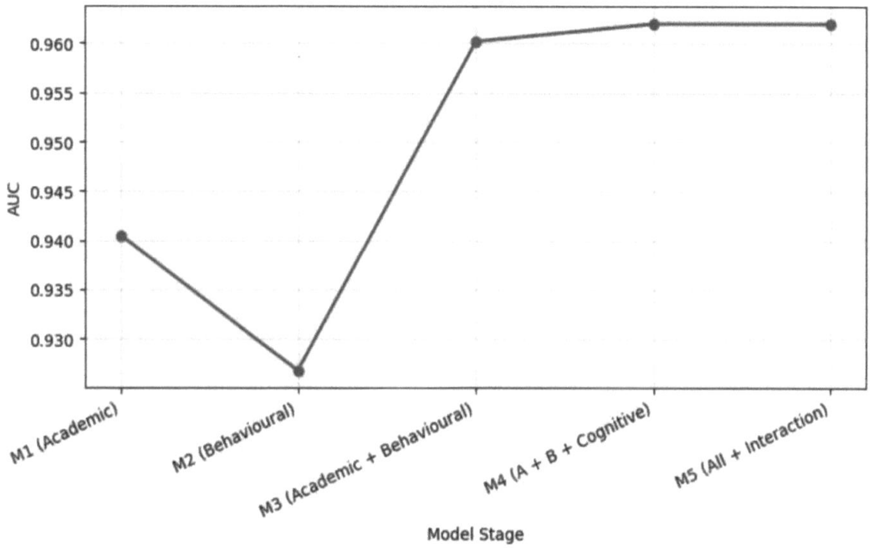

Fig. 2: Model AUC performance across different feature groups of OULAD dataset.

4.3 Synthesis of Findings

The top configurations from the two datasets are compared in Table 4 and Fig. 3. Clear differences can be observed between the two datasets. The combination of behavioural and cognitive-proxy traits produced the greatest results in EdNet, with an AUC of 0.796. The integration of academic, behavioural, and cognitive-proxy traits produced the greatest results in OULAD, with an AUC of 0.962. This distinction suggests that the feature-group effect is not universal but rather context-dependent. In interaction-driven environments such as EdNet, predictive performance is shaped more strongly by behaviour-derived regularity and stability patterns. In structured academic settings like OULAD, direct assessment-related variables remain essential, and the final model is reinforced by behavioural and proxy-based traits.

Overall, the comparative findings support two main conclusions. First, predictive performance improves when multiple feature groups are integrated rather than used in isolation. Second, the most informative combination depends on the dataset structure. These findings therefore support a dataset-aware view of student performance prediction, in which the value of a feature group is shaped by the learning environment in which the model is developed and applied.

Table 4: Comparison of dominant feature groups and best predictive performance

Dataset	Dominant Feature Group	Best AUC
EdNet	Behavioural and Cognitive-Proxy	0.796
OULAD	Academic and Behavioural and Cognitive-Proxy	0.962

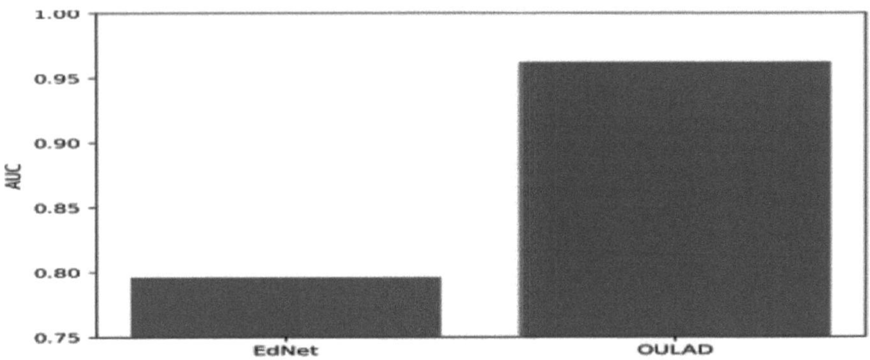

Fig. 3: Comparison of the best predictive performance obtained on EdNet and OULAD.

5 Discussion

This study examined how different feature groups shape student performance prediction across two contrasting online learning datasets, EdNet and OULAD. The findings demonstrate that integrating several feature groups enhanced predictive accuracy; the most informative combination varied depending on the structure of the dataset. Behavioural and cognitive-proxy features produced the best results in EdNet, with an AUC of 0.796, a PR-AUC of 0.813, and a Recall of 0.644. Combining academic, behavioural, and cognitive-proxy traits produced the greatest results in OULAD, with an accuracy of 0.8980, an F1-score of 0.8944, and an AUC of 0.962.

These results imply that the link between feature representation and the learning environment, rather than just feature type, shapes predictive value. Prediction in interaction-rich environments like EdNet is more dependent on whether the feature set captures significant patterns of regularity, stability, and learning dynamics than it is on basic activity volume. In more structured institutional settings such as OULAD, academic variables remain the strongest single predictive foundation, while behavioural and cognitive-proxy features provide complementary improvement. The results also highlight the limits of increasing feature complexity. In both datasets, predictive performance improved as more informative feature groups were introduced, but the final interaction-enhanced configuration produced no further gain beyond the preceding integrated model. This encourages a more frugal strategy where careful feature-group design could be more beneficial than more complicated models. However, there are limitations to the study. The cognitive-proxy variables were indirect proxies rather than direct measurements of cognition, and it was based on just two datasets. Therefore, rather than being conclusions that can be applied to all situations, the results should be seen as comparative evidence within the current design.

6 Conclusion

This study compared the influence of academic, behavioural, and cognitive-proxy feature groups on student performance prediction across OULAD and EdNet. The findings demonstrated that while combining different feature groups enhanced prediction accuracy in both datasets, the most informative combination differed depending on the context. With an AUC of 0.796, behavioural and cognitive-proxy features yielded the best results in EdNet. Combining academic, behavioural, and cognitive-proxy traits produced the greatest results in OULAD, with an accuracy of 0.8980, an F1-score of 0.8944, and an AUC of 0.962.

Overall, the study shows that effective student performance prediction depends not only on model choice but also on the composition of learner data. This underlines the importance of dataset-aware feature engineering in e-learning. Future work should extend the same framework to additional datasets and earlier prediction windows to test the robustness of these patterns across different institutional and temporal settings.

References

Ahmed, E.: Student performance prediction using machine learning algorithms. Appl. Comput. Intell. Soft Comput. **2024**, 4067721 (2024). https://doi.org/10.1155/2024/4067721

Alhazmi, E., Sheneamer, A.: Early prediction of students' performance in higher education. IEEE Access. **11**, 27579–27592 (2023). https://doi.org/10.1109/ACCESS.2023.3250702

Alkan, B.B., Kuzucuk, S., Alkan, N., Sinan, A.: Using machine learning to predict student outcomes for early intervention and formative assessment. Sci. Rep. **15**(1), 39797 (2025). https://doi.org/10.1038/s41598-025-23409-w

Choi, Y., Lee, Y., Shin, D., Cho, J., Park, S., Lee, S., et al.: EdNet: a large-scale hierarchical dataset in education. In: Artificial Intelligence in Education, pp. 69–73. Springer, Cham (2020). https://doi.org/10.1007/978-3-030-52240-7_13

Christou, V., Tsoulos, I., Loupas, V., Tzallas, A.T., Gogos, C., Karvelis, P.S., et al.: Performance and early drop prediction for higher education students using machine learning. Expert Syst. Appl. **225**, 120079 (2023). https://doi.org/10.1016/j.eswa.2023.120079

Hoq, M., Brusilovsky, P., Akram, B.: Explaining explainability: early performance prediction with student programming pattern profiling. J. Educ. Data Min. **16**(2), 115–150 (2024). https://doi.org/10.5281/zenodo.14246435

Huang, Q., Chen, J.: Enhancing academic performance prediction with temporal graph networks for MOOCs. J. Big Data. **11**, 52 (2024). https://doi.org/10.1186/s40537-024-00918-5

Junejo, N.U.R., Nawaz, M.W., Huang, Q., Dong, X., Wang, C., Zheng, G.: Accurate multi-category student performance forecasting at early stages of online education using neural networks. Sci. Rep. **15**, 16251 (2025). https://doi.org/10.1038/s41598-025-00256-3

Kuzilek, J., Hlosta, M., Zdrahal, Z.: Open University learning analytics dataset. Sci. Data. **4**, 170171 (2017). https://doi.org/10.1038/sdata.2017.171

Latif, G., Abdelhamid, S.E., Fawagreh, K.S., Ben Brahim, G., Alghazo, R.: Machine learning in higher education: students' performance assessment considering online activity logs. IEEE Access. **11**, 69586–69602 (2023). https://doi.org/10.1109/ACCESS.2023.3287972

Liu, Y., Fan, S., Xu, S., Sajjanhar, A., Yeom, S., Wei, Y.: Predicting student performance using clickstream data and machine learning. Educ. Sci. **13**(1), 17 (2023). https://doi.org/10.3390/educsci13010017

Roy, K., Farid, D.M.: An adaptive feature selection algorithm for student performance prediction. IEEE Access. **12**, 75577–75599 (2024). https://doi.org/10.1109/ACCESS.2024.3406252

Santos, R.M., Henriques, R.: Accurate, timely, and portable: course-agnostic early prediction of student performance from LMS logs. Comput. Educ. Artif. Intell. **5**, 100175 (2023). https://doi.org/10.1016/j.caeai.2023.100175

Sarker, S., Paul, M.K., Thasin, S.T.H., Hasan, M.A.M.: Analyzing students' academic performance using educational data mining. Comput. Educ. Artif. Intell. **7**, 100263 (2024). https://doi.org/10.1016/j.caeai.2024.100263

Thakur, S.N., Rawat, B.: Prediction of student performance in e-learning environment using machine learning techniques. Procedia Comput. Sci. **259**, 1416–1425 (2025). https://doi.org/10.1016/j.procs.2025.04.096

Wang, J., Yu, Y.: Machine learning approach to student performance prediction of online learning. PLoS One. **20**(1), e0299018 (2025). https://doi.org/10.1371/journal.pone.0299018

Wang, S., Luo, B.: Academic achievement prediction in higher education through interpretable modelling. PLoS One. **19**(9), e0309838 (2024). https://doi.org/10.1371/journal.pone.0309838

Chung, J.Y., Lee, S.: Dropout early warning systems for high school students using machine learning. Child Youth Serv. Rev. **96**, 346–353 (2019). https://doi.org/10.1016/j.childyouth.2018.11.030

Morsu, H.R.: Machine learning models for predicting student dropout, enrollment, and graduation (2025). https://doi.org/10.20944/preprints202512.0166.v1

Ghazvini, A., Mohd Sharef, N., Sidi, F.B.: Prediction of course grades in computer science higher education program via a combination of loss functions in LSTM model. IEEE Access. **12**, 30220–30241 (2024). https://doi.org/10.1109/ACCESS.2024.3351186

Sarcasm Markers for Malay Social Media Text: Establishing a Classification Baseline

Siti Salwa Hasbullah[1,2(✉)], Lailatul Qadri Zakaria[1], Maslida Yusof[3], Nazlia Omar[1], and Rita Zaharah[4]

[1] Faculty of Information Science and Technology, Centre for Artificial Intelligence Technology, Universiti Kebangsaan Malaysia, 43600 Bangi, Selangor, Malaysia
p135128@siswa.ukm.edu.my, sitisalwah@unikl.edu.my, {lailatul.qadri,nazlia}@ukm.edu.my

[2] Foundation and Fundamental Computing Section, Malaysian Institute of Information Technology, Universiti Kuala Lumpur, 1016 Jalan Sultan Ismail, Kuala Lumpur, Malaysia

[3] Centre for Research in Malay Language, Literature and Culture, Universiti Kebangsaan Malaysia, 43600 Bangi, Selangor, Malaysia
maslida@ukm.edu.my

[4] Technical Foundation Section, Malaysian Institute of Aviation Technology, Universiti Kuala Lumpur, Jalan Lapangan Terbang Subang, U3, 47200 Shah Alam, Selangor, Malaysia
ritazaharah@unikl.edu.my

Abstract. Sarcasm detection in low-resource Malay social media text remains a challenging problem in Natural Language Processing (NLP), largely due to the lack of linguistically grounded sarcasm markers and the absence of established computational baselines for the language. This study aims to develop and validate a linguistically grounded baseline classification for Malay sarcasm detection by proposing six sarcasm markers and systematically evaluating their linguistic validity and computational performance. The proposed sarcasm markers were constructed from three complementary theoretical perspectives: impoliteness theory, language style theory, and speech act theory. Their conceptual validity was further strengthened through expert interviews and review sessions with a linguistics specialist. This process led to the acceptance of all proposed sarcasm markers. The study used a labelled dataset of 1,970 Malay social media posts and followed three main processes. The proposed sarcasm markers were developed through linguistic mapping of the Malaysian Communications and Multimedia Commission (MCMC) *Bahasa Terlarang* taxonomy and validated using expert review and quantitative measures. They were subsequently evaluated in six classifications including conventional machine learning and transformers-based models. The result showed that the best-performing machine learning configuration (Logistic Regression, M4) achieves F1(Sarc) = 0.659 and ROC-AUC = 0.742, surpassing the prior baseline ROC-AUC by +0.015, while the best transformer configuration (DistilmBERT with tag injection, T1) achieves the highest F1(Sarc) = 0.674 across all evaluated configurations. This study establishes a validated computational baseline for linguistically grounded Malay sarcasm detection, offering a replicable evaluation framework for low-resource figurative language research.

Keywords: Malay Sarcasm Detection · Sarcasm Detection · Natural Language Processing · Machine Learning · Transformer · Low-Resource Language

N. Mohd Sharef et al. (Eds.): CAMP 2026, CCIS 3093, pp. 25–38, 2027.
https://doi.org/10.1007/978-981-92-4302-0_3

1 Introduction

Automated sarcasm detection in low-resource languages remains one of the least-addressed challenges in figurative language Natural Language Processing (NLP), with the absence of both linguistically principled marker schemes and established computational baselines constituting the primary bottlenecks [1–4]. For the Malay language, a low-resource language spoken by an estimated 290 million people, this research issue is particularly noticeable [4]. Existing studies on sarcasm detection have largely relied on non-systematic lexical features, with limited systematic evaluation of linguistically grounded markers and no established computational baseline for pragmatically informed Malay sarcasm detection [5–10]. This evaluation issue hinders scientific progress, as it prevents cumulative comparison across studies and obscures which aspects of linguistic marker design contribute most to detection performance.

This study addresses this gap through a focused baseline experiments evaluation of the construction of new proposed Malay sarcasm markers, comprising six sarcasm markers grounded in language impoliteness (*ketidaksantunan bahasa*) [11], language style (*ragam bahasa*) [12], and speech act theory (*lakuan bahasa*) [13]. The baseline evaluation systematically assesses each marker's individual statistical validity by using coverage, precision, and lift metrics. The study also evaluates their baseline experiments across eight machine learning and two transformer-based classification configurations on a dataset of 1,970 Malay social media text. This study makes three main contributions. First, it provides the systematic statistical validity assessment of pragmatically linguistic Malay sarcasm markers. It also establishes a reproducible baseline using eight machine learning and transformer-based classification models, and presents an incremental feature ablation analysis to identify the feature combinations that lead to the largest performance improvements.

2 Proposed Malay Sarcasm Markers

The proposed of the new Malay sarcasm markers comprises six grounded sarcasm markers constructed through linguistically systematic mapping of *Suruhanjaya Komunikasi dan Multimedia Malaysia*, Malaysian Communications and Multimedia Commission ((MCMC, 2017)) *Bahasa Terlarang* [14] taxonomy onto three theoretical frameworks. Each marker is constructed through a mechanism that explains the communicative act expressed through sarcasm. It is also linked to its corresponding theoretical grounding. In addition, each marker is defined through Malay lexical cue patterns to support computational identification.

The use of *Bahasa Terlarang* [14] as the main reference is justified on five converging ground, domain expert validation, institutional and legal authority, cultural-pragmatic alignment with Malay communicative norms, semantic comprehensiveness, and the bridging of regulatory and computational frameworks. Together, these grounds establish it as the most theoretically sound and practically relevant foundation for constructing new Malay-specific sarcasm markers. Furthermore, to date, no dedicated or more recent taxonomy exists that systematically classifies sarcastic and offensive language specifically within the Malay-language context for the purpose of sarcasm marker construction. The absence of an alternative authoritative resource further reinforces the continued relevance and suitability of the *Bahasa Terlarang* [14] taxonomy as the primary reference in this study. Given this gap in the existing literature, adopting a regulatory document that is officially recognized, culturally grounded, and comprehensively structured remains the most valid and defensible methodological choice for developing linguistically informed sarcasm markers in the Malay language.

The conceptual validity of the proposed sarcasm markers was established through a two-stage expert validation. A linguistics specialist in Malay pragmatics and impoliteness first judged, in a semi-structured interview, whether each marker corresponded to a recognizable sarcastic mechanism with an appropriate theoretical grounding. The expert then independently rated a stratified sample of 240 dataset instances by all the proposed 6 sarcasm markers, were compared against the markers appear by the lexical-cue rules. Inter-rater agreement was quantified using Cohen's kappa, yielding $\kappa = 0.72$, indicating substantial agreement. Disagreements were resolved through discussion, and the resulting consensus assignments were used to finalize the lexical cues of each marker. This two-stage process led to the acceptance of all six proposed sarcasm markers and provides a documented, reproducible basis for their construction.

Table 1 presents the six proposed Malay sarcasm markers together with their theory, mechanism, lexical cues, and functional category. The table shows that the marker scheme is grounded in three main linguistic theories, namely language impoliteness, language style, and speech act theory, reflecting the view that sarcasm is better explained through speaker intention, communicative context, and shared social knowledge than through lexical form alone [15, 16]. Sarcasm markers M1, M2, and M4 are constructed from language impoliteness through different types of face-threatening acts, M3 and M6 are grounded in speech act theory through indirect and provocative linguistic meanings, and Marker M5 is independently grounded in language style theory via stylistic hyperbole [17]. Each marker is specified with its definition and the corresponding grounding mechanism, establishing a theoretically principled distinction from purely lexical or sentiment-based marker approaches [18, 19]. In addition, the lexical cues listed in the table provide the operational basis for identifying each marker computationally in Malay social media text.

The markers are not a direct subset of any single source theory but a purposed six markers maps the offence categories of the *Bahasa Terlarang* [14] taxonomy onto the three theories, so each marker inherits a theoretical mechanism rather than only a word list. Lexical cues were selected on three criteria, where the cue must appear in the taxonomy under the offence category the marker operationalizes. After that, the proposed markers be attested in the dataset, and be confirmed for that marker by the linguistics expert. Only cues meeting all three were retained, excluding incidental or purely sentiment-bearing words.

The sarcasm marker process follows two explicit rules. First, a marker exists only when its lexical cue co-occurs with an evaluative or face-threatening context, so a cue in a neutral, literal sentence is not counted and this operationalizes the requirement that sarcasm arises from a mismatch between literal form and intended meaning rather than from a keyword alone. Second, where an instance satisfies more than one marker, priority goes to the marker with the more specific offence category, preventing double counting.

Table 1. Proposed Malay sarcasm markers

ID	Marker	Pragmatic Theory	Pragmatic Mechanism	Lexical Cues
M1	Individual Sarcasm	Language Impoliteness	Face-threatening act: directly insulting someone by using offensive or demeaning names	*korang, padan muka*
M2	Organisational Sarcasm	Language Impoliteness	Institutional delegitimisation through implicature: surface complaint + implied accusation of institutional betrayal	*cukai, GST, rakyat*
M3	Dehumanisation-objectification Sarcasm	Speech Act Theory	Indirect speech act of contempt: semantic incongruity between human referent and non-human predicate	*babi, anjing, lembu*
M4	Human physical Sarcasm	Language Impoliteness	Embodied face-threatening act: using crude or vulgar body-related language to shame or humiliate someone	*berat badan, anggota*
M5	Character & Behavioural Sarcasm	Language Style	Pragmatic meaning using exaggeration or false praise to say the opposite of what is really meant	*bodoh, tipu, celaka*
M6	Religious & Racial Sarcasm	Speech Act + Impoliteness	Religiously/racially charged indirect provocation: Using religious words combined with negative or critical statements about politics	*islam, kafir, murtad*

These rules distinguish the scheme from keyword matching, where a keyword list flags any occurrence of a term, whereas these markers flag a term only when embedded in an impolite, stylistic, or indirect speech-act context.

The six sarcasm markers are constructed to be mutually distinct at the level of their grounding mechanism. M1, M2, and M4 are separated by the target of the face-threatening act (the individual, the organisational, and the human-physical sarcasm marker). Meanwhile, M3 and M6 are separated by the type of indirect speech act (the dehumanisation-objectification and the religious racial sarcasm marker), and M5 is the only marker grounded in language style through hyperbole. The operational distinction is enforced by the priority rule above, and the residual overlap between markers is reported empirically through the combined any-marker signal in Table 2, which quantifies the extent to which instances are captured by more than one marker.

3 Baseline Experimental

The baseline experiments conducted in this study presents the dataset, evaluation methods, and classification configurations used to establish the experimental baseline for Malay sarcasm detection.

3.1 Dataset

The evaluation dataset was adopted from [20], comprising 1,970 Malay social media comments collected and labelled from Facebook public pages. The dataset contains 969 sarcastic comments (49.2%) and 1,001 non-sarcastic comments (50.8%), representing a near-balanced class distribution. The dataset covers political commentary, economic critique, and social satire. It also contains frequent code-switching, informal slang, phonetic abbreviations, and non-standard spelling, which makes it suitable for developing linguistically markers rather than relying only on general lexical features. This dataset was used because it remains one of the few labelled Malay social media sarcasm dataset available and still provides linguistically relevant sarcastic patterns for baseline evaluation.

The dataset was labelled at the instance level by human annotators, with each comment assigned a binary label as sarcastic or non-sarcastic according to its overall intended meaning rather than the presence of any predefined keyword. The labelling was therefore not produced by a keyword, lexical, or n-gram rule where the lexical and n-gram features were used only afterwards as model inputs, not as the source of the ground-truth labels. This separation ensures that the marker and n-gram features evaluated in this study are tested against an independently human-judged sarcasm label, avoiding circularity between the labelling method and the features under evaluation.

3.2 Evaluation Methods

The research contains three evaluation methods. The first evaluation method assessed the statistical validity of the proposed sarcasm markers using three measures, which are coverage, precision, and lift. Coverage refers to the proportion of dataset in which a marker appears [21], precision measures the proportion of marker occurrences that are correctly associated with sarcastic instances [22], and lift indicates how strongly a marker is associated with sarcasm relative to the baseline sarcasm rate of 0.492, where positive values reflect better performance predictive value [23]. Classification performance was measured using seven metrics: Accuracy, Balanced Accuracy, Macro-F1, F1(sarc), ROC-AUC, PR-AUC, and Matthews Correlation Coefficient (MCC). This evaluation methods was adopted to ensure a comprehensive and reliable evaluation of classifier behavior across both sarcastic and non-sarcastic classes in a near balanced binary classification setting [24–27]. Incremental improvement was computed by comparing performance across feature configurations to measure the contribution of each added feature group to overall classification performance.

Coverage is a marker level frequency measure, not a classification outcome, and is therefore distinct from true-positive and false-positive counts. Coverage is the proportion of all instances in which a marker appears, regardless of whether the instance is sarcastic. True and false positives are defined only when a marker is treated as a classifier, where a marker occurrence on a sarcastic instance is a true positive, and on a non-sarcastic instance, a false positive. The precision reported alongside coverage is therefore the true-positive rate among marker occurrences. The two are complementary which a marker may have high coverage yet low precision if it appear often but on many non-sarcastic instances, as observed for M1 in Table 2.

3.3 Classification Configuration

There are four machine learning feature configurations were evaluated in this study, namely M1, M2, M3, and M4, each representing an incrementally enriched set of features applied to sarcasm detection. Each configuration was tested across seven classification models, which include Logistic Regression, Linear SVM, BernoulliNB, CatBoost, Random Forest, LibSVM, and k-NN, in order to identify the most effective combination of features and classifier for Malay sarcasm detection. These classification configurations were constructed incrementally to systematically examine the contribution of each feature set to overall sarcasm detection performance. Specifically, M1 included only the Malay sarcasm markers, M2 added punctuation patterns and TF-IDF unigram-bigram (1–2) features, M3 further incorporated pragmatic context features, and M4 extended the feature space with emoji boost features. This stepwise design enabled a systematic assessment of the added value of each feature group. Two transformer-based classification were evaluated using DistilmBERT [28] to assess the effect of explicit pragmatic information on sarcasm detection performance. Transformer (T0) represented the standard baseline, in which dataset instances were encoded without marker annotation. In contrast, Transformer (T1) employed pragmatic tag injection, whereby each instance was pre-annotated with marker tags such as *[M2_ORGANISATIONAL]* and *[M6_RELIGIOUS-RACIAL]* prior to transformer encoding. By embedding these tags directly into the input sequence, the model was able to attend to pragmatically salient marker positions during self-attention [29].

4 Baseline Results

Table 2 presents the full statistical validation results for all six proposed sarcasm markers and the combined any-marker signal against the evaluation dataset.

The results show a clear and consistent pattern in which the discriminative strength of each marker reflects its underlying linguistic basis. M2 (Organisational Sarcasm Marker) achieves the strongest overall statistical profile, recording the highest coverage (0.364), the highest precision (0.516), and—critically—the only unambiguously positive lift (+0.024) among the six individual markers. This result is expected, as M2 targets organisational sarcasm marker which is a form of sarcastic expression that is particularly common in Malaysian political social media, and this explains its high coverage and strong association with the sarcasm class. The support count of 717 that representing 36.4% of the entire dataset and further confirms that politically charged organisational sarcasm constitutes the dominant form of Malay social media sarcasm in this dataset.

Meanwhile, M4 (Human Physical Sarcasm Marker) presents a contrasting but theoretically significant profile: the highest precision among all individual sarcasm markers (0.500) despite very low coverage (0.022), with a marginally positive lift (+0.008). This precision-coverage inversion reflects the specificity of embodied face-threatening acts as a pragmatic sarcasm mechanism when body targeting coarse language co-occurs with ironic framing, it is an exceptionally reliable sarcasm signal, but such occurrences are infrequent. M1 (Individual Sarcasm Marker) shows moderate coverage (0.141) but low precision (0.394) and a negative lift (−0.098), suggesting that while personal attacks are

Table 2. Statistical validity of Malay sarcasm markers

ID	Marker	Coverage	Precision	Lift	Total
M1	Individual Sarcasm	0.141	0.394	-0.098	277
M2	Organisational Sarcasm	**0.364**	**0.516**	**0.024**	**717**
M3	Dehumanisation-objectification Sarcasm	0.015	0.133	-0.359	30
M4	Human physical Sarcasm	0.022	0.5	0.008	44
M5	Character & Behavioural Sarcasm	0.065	0.398	-0.093	128
M6	Religious & Racial Sarcasm	0.039	0.368	-0.123	76
≥1 marker	Combined signal	0.512	0.483	-0.009	1,008

common in social media text, they do not always indicate sarcasm unless supported by additional contextual signals.

M5 (Character and Behavioural Sarcasm Marker) achieves moderate coverage (0.065) with near-base precision (0.398), suggesting that exaggerated language used to attack a person's character or behaviour does not consistently indicate sarcasm on its own and requires additional contextual support for reliable detection. M6 (Religious and Racial Sarcasm Marker) and M3 (Dehumanisation-Objectification Sarcasm Marker) record the lowest coverage values (0.039 and 0.015, respectively), with precision levels indicating limited individual sarcasm marker discriminative utility as isolated lexical markers. However, their retention in proposed sarcasm markers is justified on linguistic, pragmatic, and legal grounds. M6 is retained for a legal reason. Religious and racial sarcasm marker carries direct legal implications under Malaysian law [8], so the marker remains important even though it appears rarely. M3, in contrast, is retained for a data reason. Its low coverage occurs because dehumanizing language is genuinely uncommon in the dataset, not because the marker is poorly defined. The combined any marker signal captures 51.2% of the dataset (support = 1,008) with near-base precision (0.483), confirming that no individual sarcasm marker alone is sufficient for robust detection and that the detection pattern that requiring co-occurrence of primary markers with supplementary signals, is an architecturally necessary feature of the proposed sarcasm markers.

Table 3 presents the classification performance across all six evaluated configurations, encompassing seven machine learning and two transformer-based models. Among the machine learning configurations, M3 and M4 are both employing Logistic Regression with markers, punctuation, and n-gram(1–2) features that achieve the highest overall Accuracy of 0.673 and an identical Balanced Accuracy of 0.672, indicating stable and consistent performance across both the sarcastic and non-sarcastic classes. In terms of sarcasm class detection specifically, M2 records a marginally higher F1(Sarc) of 0.660 compared to 0.659 for M3 and M4, despite achieving a slightly lower Accuracy of 0.672.

Table 3. Baseline classification results for sarcasm detection

Config	Feature Set	Model	Acc	Bal.Acc	Mac-F1	F1 (Sarc)	ROC-AUC	PR-AUC	MCC
M1	Markers only	BernoulliNB	0.539	0.543	0.507	0.633	0.56	0.528	0.102
M2	Markers + Punct + ngram (1–2)	Logistic Regression	0.672	0.672	0.672	0.66	0.743	0.728	0.344
M3	Markers + Punct + ngram (1–2)	Logistic Regression	0.673	0.672	0.672	0.659	0.742	0.728	0.345
M4	Markers + Punct + ngram (1–2)	**Logistic Regression**	**0.673**	**0.672**	**0.672**	**0.659**	**0.742**	**0.728**	**0.345**
T0	Markers + Punct + ngram (1–2)	Distil-mBERT	0.584	0.587	0.567	0.653	0.63	0.593	0.191
T1	Markers + Punct + ngram (1–2)	**Distil-mBERT**	**0.602**	**0.605**	**0.581**	**0.674**	**0.646**	**0.611**	**0.236**

This outcome suggests that combining sarcasm markers with punctuation patterns and TF-IDF n-gram(1–2) features constitutes the most effective feature combination for identifying sarcastic instances, as M2 achieves a better balance between precision and recall in detecting the target sarcasm class. The further addition of linguistic-pragmatic context features in M3, and emoji boost features in M4 introduces only marginal gains in overall Accuracy without any improvement in F1(Sarc), indicating diminishing returns beyond the M2 feature configuration. This finding suggests that combining linguistically grounded sarcasm markers with basic text features such as punctuation and word patterns is sufficient to detect Malay sarcasm effectively, and that adding more features beyond this combination does not noticeably improve the model's ability to identify sarcastic instances.

Regarding the ROC-AUC metric, M2 achieves the highest value among all machine learning configurations at 0.743, with M3 and M4 recording 0.742, a negligible difference that further supports the conclusion that the M2 feature set represents the optimal operating point for machine learning-based detection in this study. The PR-AUC values remain consistent across M2, M3, and M4 at 0.728, while MCC improves marginally from 0.344 (M2) to 0.345 (M3 and M4), confirming that the incremental feature additions beyond M2 provide no practically meaningful improvement across any of the seven evaluated metrics.

For the transformer-based configurations, T1-DistilmBERT with pragmatic tag injection is consistently outperforms the standard baseline T0 across all seven metrics, improving Accuracy from 0.584 to 0.602, Balanced Accuracy from 0.587 to 0.605, Macro-F1 from 0.567 to 0.581, and MCC from 0.191 to 0.236. Most notably, T1 records an F1(Sarc) of 0.674, the highest sarcasm class F1 value across all six configurations in the study, confirming that the integration of pragmatic tag injection into transformer encoding meaningfully enhances DistilmBERT's sensitivity to sarcastic instances beyond what standard contextual self-attention achieves without explicit marker guidance. The ROC-AUC for T1 improves from 0.630 (T0) to 0.646, and the PR-AUC increases from 0.593 to 0.611, collectively demonstrating that marker-aware transformer encoding strengthens discriminative performance across both threshold-dependent and threshold-independent evaluation perspectives.

It is notable that the best transformer configuration still trails the best machine learning configuration on overall Accuracy and ROC-AUC, with Logistic Regression under M4 reaching ROC-AUC 0.742 against 0.646 for T1. Three factors explain why DistilmBERT underperforms Logistic Regression in this setting. First, the dataset is small for transformer fine-tuning at 1,970 instances, whereas Logistic Regression with sparse TF-IDF and marker features is far less data-hungry and is well suited to a near-balanced binary task of this size. Second, DistilmBERT is a distilled multi-lingual model that is not specialized for Malay, and the dataset contains heavy code-switching, informal slang, and non-standard spelling that fall outside its pre-training distribution, which weakens its contextual representations. Third, the explicit marker and n-gram features give the linear model a direct, high-precision signal, while the transformer must infer comparable cues from limited fine-tuning data. Consistent with this, tag injection narrows the gap by supplying the transformer with the same explicit pragmatic signal, which is why T1 attains the highest sarcasm-class F1 even though its threshold-independent ranking remains below the linear baseline.

In this study, Accuracy is adopted as the primary overall performance indicator given the near-balanced class distribution of the dataset evaluation (49.2% sarcastic; 50.8% non-sarcastic), where it provides a reliable and interpretable measure of overall classification correctness without systematic bias toward either class. F1(Sarc) and ROC-AUC are simultaneously prioritized as the principal sarcasm detection metrics for two complementary reasons. First, F1(Sarc) directly captures the model's ability to balance precision and recall in detecting the target sarcasm class, a property that overall Accuracy alone cannot adequately reflect when the analytical focus is on the performance of a specific class. Second, and critically for this study, both F1(Sarc) and ROC-AUC are selected to facilitate a direct and meaningful comparison with the primary baseline study by [14], which reported sarcasm detection performance using F-measure and ROC-AUC as its principal evaluation metrics. Adopting the same metric framework ensures that the comparative analysis between the proposed sarcasm markers and the prior baseline is methodologically consistent and scientifically valid, enabling a fair assessment of performance advancement across both studies.

Table 4. Baseline results reported by prior baseline [20] for Malay sarcasm detection

Config	Feature Set	Model	F-m	ROC-AUC
F1	Punctuations + ngram (1–2)	k-NN	**0.651**	**0.658**
F2	Punctuations + ngram (1–2)	LibSVM	**0.726**	**0.727**

Table 4 summarizes the sarcasm detection results reported by [20], which serve as the primary baseline for comparison in this study. They proposed a sarcasm detection and classification approach for Malay social media text using two feature types, namely n-gram and punctuation marks, and evaluated the performance using k-NN and LibSVM classifiers. The LibSVM evaluation achieved the strongest performance, recording an F-measure of 0.726 and a ROC-AUC of 0.727, while k-NN produced a lower F-measure of 0.651 and ROC-AUC of 0.658. These results demonstrate that LibSVM with surface-level punctuation and n-gram features constitutes the strongest baseline for Malay sarcasm detection in prior work. However, both configurations relied exclusively on orthographic and lexical features without any linguistically principled or pragmatically-grounded sarcasm markers, representing a methodological limitation that the proposed *Malay* sarcasm markers directly addresses through linguistically theoretically-grounded marker construction and dual-level expert validation.

5 Result Analysis

Tables 3 and 4 present a direct comparison between the best classification results obtained in this study and those reported by [20], which represents the primary prior work on Malay sarcasm detection. The comparison focuses on two key evaluation metrics, namely F1(Sarc) and ROC-AUC, which were selected to enable a consistent and fair performance assessment across both studies.

As shown in Table 4, the results reported by [20] indicate that the highest-performing models were k-NN, which achieved an F-measure of 0.651 and ROC-AUC of 0.658, and LibSVM, which achieved an F-measure of 0.726 and ROC-AUC of 0.727. Both models relied solely on punctuation and n-gram(1–2) features without incorporating any linguistically grounded sarcasm markers. In the proposed study, the best-performing machine learning model is Logistic Regression under configuration M4, which achieves F1(Sarc) = 0.659 and ROC-AUC = 0.742. The best-performing transformer model is DistilmBERT with tag injection under configuration T1, which achieves F1(Sarc) = 0.674 and ROC-AUC = 0.646.

Comparing results directly in Table 3, the proposed Logistic Regression configuration surpasses [20] best ROC-AUC by +0.015 (0.742 vs. 0.727), indicating stronger threshold-independent discriminative capacity for sarcasm detection. The Distil-mBERT configuration achieves F1(Sarc) = 0.674, exceeding [20] k-NN F-measure of 0.651 by +0.023, though marginally below LibSVM's F-measure of 0.726. Critically, the proposed results are achieved through linguistically-grounded thematic markers derived from language impoliteness, language style, and speech act theory, a linguistically principled approach that extends beyond the surface-level n-gram and punctuation features employed in prior work. The superior ROC-AUC of the proposed machine learning configuration confirms that linguistic marker features provide stronger overall discriminative power, establishing a theoretically interpretable and methodologically replicable advancement over existing Malay sarcasm detection baselines.

This study reports performance differences using standard metrics but does not include statistical significance testing. A paired test such as McNemar's test [31] requires per-instance predictions from both compared systems on a shared test set where these are available for the configurations developed here but not for the prior baseline [20], so a direct significance comparison against [20] was not feasible. Full significance testing across configurations is identified as a priority for future work, alongside the larger-scale evaluation discussed above.

Table 5 presents constructed Malay examples illustrating how the markers identify sarcasm, together with one false-positive and one false-negative case. The markers target a pragmatic mismatch rather than surface keywords. In the M2 example, the praise term "bagus" is literally positive yet becomes sarcastic when set against rising prices, so the organisational target triggers M2. In the M1 example, the personal praise "pandai" is sarcastic only through its evaluative framing against failure, which triggers M1. The false-positive case shows the same "pandai" cue appearing in sincere praise of a child, where no pragmatic mismatch exists yet the marker is still triggered, illustrating the limitation of rule-based matching. The false-negative case shows culturally implicit sarcasm carrying no explicit lexical cue, which the rule-based markers miss, motivating the marker-aware transformer.

Table 5. Illustrative marker-specific examples (Constructed)

Sarcasm Marker/ Case	Malay instance (illustrative)	Pragmatic basis
M2 (correct)	"Bagus betul kerajaan kita, sampai harga barang pun naik tiap-tiap bulan."	Literal praise contradicts the negative situation; triggers M2.
M1 (correct)	"Pandai sangat dia ni sampai kerja simple pun tak boleh buat."	Personal praise framed against failure; evaluative mismatch triggers M1.
False positive	"Pandai betul anak saya, dapat nombor satu kelas."	M1 cue present but sincere; no mismatch, yet marker triggers.
False negative	"Terima kasih banyak-banyak sebab tolong." (said after being let down)	Culturally implicit sarcasm, no explicit cue; missed by markers.

Figure 1 summarizes the comparison between the configurations evaluated in this study and the prior baseline [20] on the two principal metrics, F1(Sarc) and ROC-AUC. The figure makes the deeper pattern explicit. where the proposed machine learning configurations achieve the strongest ROC-AUC, the marker-aware transformer T1 achieves the strongest sarcasm-class F1 among all this-study configurations, and the prior LibSVM retains the highest F-measure. This visual comparison confirms that the contribution of the proposed sarcasm markers is strongest on threshold-independent discrimination (ROC-AUC) while remaining competitive on sarcasm-class F1.

The approach can be situated against techniques for other languages. Indonesian sarcasm detection relies on pre-trained embedding and benchmark language models, treating sarcasm as a representation-learning problem [3, 30]. Besides, Urdu language work adopts cognitive, relationship-based features [19], and recent Turkish language research compares transformers with ensemble learning [29]. The common thread is reliance on distributional features or general transformer encoders, with limited use of explicit, theoretically grounded markers tailored to the language. This study differs by foregrounding language-specific pragmatic markers and combining them with both

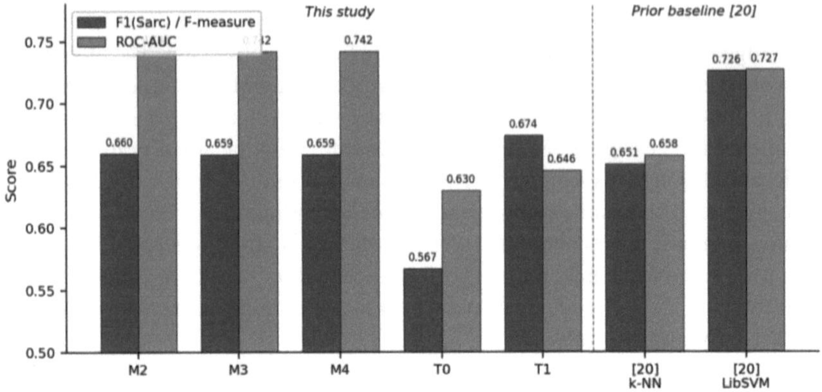

Fig. 1. F1(Sarc) and ROC-AUC across this-study configurations and the prior baseline [20].

linear and transformer models, a marker-centered strategy that yields the interpretability and reproducibility emphasized here while remaining compatible with transformer techniques used elsewhere.

6 Conclusion

This study proposed and evaluated six linguistically-grounded Malay sarcasm markers, constructed by mapping the MCMC *Bahasa Terlarang taxonomy* [14] onto three linguistic theoretical frameworks and adapted from the pragmatic framework of [17], addressing the absence of a culturally situated sarcasm detection resource for Malay. The markers cover the individual, organisational, dehumanisation-objectification, human-physical, character and behavioural, and religious and racial sarcasm, reflecting culturally embedded forms of criticism rather than universal phenomena [30]. Statistical validation on a near-balanced dataset of 1,970 instances identified M2 (Organisational) and M4 (Human physical) as the most discriminative markers, while the combined any-marker signal (51.2% coverage) confirmed that multi-marker co-occurrence is essential for robust detection. Combining markers with TF-IDF unigram-bigram features gave the best machine learning performance (Accuracy 0.673, ROC-AUC 0.742, MCC 0.345), and pragmatic tag injection into DistilmBERT produced the highest sarcasm-class F1 (0.674), confirming that embedding explicit markers enhances transformer-based detection. These results establish the proposed markers as a theoretically sound and computationally effective foundation for Malay sarcasm detection. Future work will expand the dataset to support more stable transformer fine-tuning, incorporate conversational and thread-level context to recover culturally implicit sarcasm, and adopt marker-level annotation to enable finer evaluation of individual markers and Malay-specific pre-trained models.

Acknowledgment. The authors would like to express sincere gratitude to the Faculty of Information Science and Technology (FTSM), Universiti Kebangsaan Malaysia (UKM), Bangi, Selangor, Malaysia, for the support and facilities provided throughout the process of this research. The authors also wish to thank the supervisor and domain experts who contributed valuable insights and guidance in the development of the proposed Malay sarcasm markers for this study.

Declaration of Generative AI and AI-Assisted Technologies in the Writing Process.. During the preparation of this manuscript, we used an AI tool to edit the language. After using this tool/service, the authors reviewed and edited the content as needed and took full responsibility for the content of the published article.

References

1. Chen, W., Lin, F., Li, G., Liu, B.: A survey of automatic sarcasm detection: fundamental theories, formulation, datasets, detection methods, and opportunities. Neurocomputing. **578**, 127428 (2024)
2. Băroiu, A.C., Trăuşan-Matu, Ş.: Automatic sarcasm detection: systematic literature review. Information. **13**(8), 399 (2022)
3. Rosid, M.A., Siahaan, D., Saikhu, A.: Pre-trained word embeddings for sarcasm detection in Indonesian tweets: a comparative study. In: 2022 9th International Conference on Information Technology, Computer, and Electrical Engineering (ICITACEE), pp. 281–286. IEEE (2022, August)
4. Khan, S. et al.: An automated approach to identify sarcasm in low-resource language. PLoS One. **19**(12), e0307186 (2024)
5. Suhaimin, M.S.M., Hijazi, M.H.A., Moung, E.G.: Annotated dataset for sentiment analysis and sarcasm detection: bilingual code-mixed English-Malay social media data in the public security domain. Data Brief. **55**, 110663 (2024)
6. Suhaimin, M.S.M., Hijazi, M.H.A., Alfred, R., Coenen, F.: Modified framework for sarcasm detection and classification in sentiment analysis. Indones. J. Electr. Eng. Comput. Sci. **13**(3), 1175–1183 (2019)
7. Suhaimi, S.H., Bakar, N.A.A., Azmi, N.F.M.: Proposing Malay sarcasm detection on social media services: a machine learning approach. Open Int. J. Inf. **9**(Special Issue 2), 1–10 (2021)
8. Suhaimi, S.H., Bakar, N.A.A., Azmi, N.F.M.: Malay sarcasm detection on social media: a review, taxonomy, and future directions. In: 2022 IEEE 7th International Conference on Information Technology and Digital Applications (ICITDA), pp. 1–8. IEEE (2022, November)
9. Suhaimi, S.H., Bakar, N.A. A., Azmi, N.F.M.: Constructing and analysing the MalaySarc Dataset: a resource for detecting and understanding sarcasm in malay language (2023)
10. Suhaimin, M.S.M., Hijazi, M.H.A., Moung, E.G.: MSSThreD: multitask social media sentiment analysis with sarcasm for public security threat detection. In: 2024 11th International Conference on Information Technology, Computer, and Electrical Engineering (ICITACEE), pp. 25–30. IEEE (2024, August)
11. Zulhairi, N.I., Kasdan, J.: Ungkapan makian dalam wacana politik dan kepemimpinan di Twitter: Analisis ketidaksantunan berbahasa. J. Wacana Sarjana. **6**(1), 1–16 (2022)
12. Saimon, A., et al.: Penggunaan Ragam Bahasa Sindiran Dalam Kalangan Pengguna Twitter. ISI KANDUNGAN PRAKATA KERTAS KERJA, 104 (2022)
13. Ibrahim, N.A., Yusof, M.: Sindiran dalam media sosial: Perspektif lakuan bahasa. J. Komunikasi. **36**(2), 269–288 (2020)
14. SKMM: Bahasa Terlarang dalam Media: Panduan Kesantunan Bahasa. MCMC, Cyberjaya (2018) https://www.mcmc.gov.my
15. Yao, B., Zhang, Y., Li, Q., Qin, J.: Is sarcasm detection a step-by-step reasoning process in large language models? Proc. AAAI Conf. Artif. Intell. **39**(24), 25651–25659 (2025, April)

16. Zhang, Y., Zou, C., Lian, Z., Tiwari, P., Qin, J.: Sarcasmbench: towards evaluating large language models on sarcasm understanding. IEEE Trans. Affect. Comput. **16**, 2560–2578 (2025)
17. Burgers, C., Van Mulken, M., Schellens, P.J.: Verbal irony: differences in usage across written genres. J. Lang. Soc. Psychol. **31**(3), 290–310 (2012)
18. Deshmukh, M.R., Joshi, N., Bharati, M.: Context-aware sarcasm detection in text using NLP. In: 2025 IEEE Pune Section International Conference (PuneCon), pp. 1–3. IEEE (2025, December)
19. Khan, M.Y., Ahmed, T., Siddiqui, M.S., Wasi, S.: Cognitive relationship-based approach for urdu sarcasm and sentiment classification. IEEE Access. **11**, 126661–126690 (2023)
20. Suhaimin, M.S.M., Hijazi, M.H.A., Alfred, R., Coenen, F.: Mechanism for sarcasm detection and classification in malay social media. Adv. Sci. Lett. **24**(2), 1388–1392 (2018)
21. Rizinski, M., Peshov, H., Mishev, K., Jovanovik, M., Trajanov, D.: Sentiment analysis in finance: from transformers back to explainable lexicons (xlex). IEEE Access. **12**, 7170–7198 (2024)
22. Zhang, L., Faseeh, M., Naqvi, S.S.A., Hu, L., Ghani, A.: Enhancing sarcasm detection on social media: a comprehensive study using LLMs and BERT with multi-headed attention on SARC. PLoS One. **20**(11), e0334120 (2025)
23. Diaz-Garcia, J.A., Ruiz, M.D., Martin-Bautista, M.J.: A survey on the use of association rules mining techniques in textual social media. Artif. Intell. Rev. **56**(2), 1175–1200 (2023)
24. Dowd, P.A.: Accuracy and precision. In: Encyclopedia of Mathematical Geosciences, pp. 1–4. Springer International Publishing, Cham (2023)
25. Yacouby, R., Axman, D.: Probabilistic extension of precision, recall, and f1 score for more thorough evaluation of classification models. In: Proceedings of the First Workshop on Evaluation and Comparison of NLP Systems, pp. 79–91 (2020, November)
26. Islam, M.S., Chowdhury, M.J.U., Tareq, M.M.U., Ahmed, T., Aktarujjaman, M., et al.: SarcasmSense: a novel multitask learning framework for vlog-based political sarcasm and irony detection. IEEE Access. (2026)
27. Chicco, D., Jurman, G.: The advantages of the Matthews correlation coefficient (MCC) over F1 score and accuracy in binary classification evaluation. BMC Genomics. **21**(1), 6 (2020)
28. Vakili, Y.Z., Fallah, A., Sajedi, H.: Distilled BERT model in natural language processing. In: 2024 14th International Conference on Computer and Knowledge Engineering (ICCKE), pp. 243–250. IEEE (2024, November)
29. Eser, M., Bilgin, M.: Irony and sarcasm detection in Turkish texts: a comparative study of transformer-based models and ensemble learning. Appl. Sci. **15**(23), 12498 (2025)
30. Suhartono, D., Wongso, W., Handoyo, A.T.: Idsarcasm: benchmarking and evaluating language models for Indonesian sarcasm detection. IEEE Access. **12**, 87323–87332 (2024) last accessed 2023/10/25
31. Dietterich, T.G.: Approximate statistical tests for comparing supervised classification learning algorithms. Neural Comput. **10**(7), 1895–1923 (1998)

Data Science

An Agricultural Digital Twin Platform for Integrated Monitoring, Prediction, and Decision Support: Design and Usability Evaluation

Ghodratollah Solimanian[1], Nurfadhlina Mohd Sharef[1,2(✉)], Syaifulnizam Abd Manaf[2], and Muhamad Sufri Muhammad[2]

[1] Institute of Mathematical Research, Universiti Putra Malaysia, Serdang, Selangor, Malaysia
nurfadhlina@upm.edu.my

[2] Faculty of Computer Science and Information Technology, Universiti Putra Malaysia, Serdang, Selangor, Malaysia

Abstract. Although there are many systems that monitor and predict to help farmers, decision support system that manage to translate these into meaningful intervention advises is still limited. Therefore, this paper introduces AgroPulse Twin Hub, a knowledge-driven agricultural digital twin platform based on human in-the-loop-design that integrates sensing, predictive analytics, and simulation within a unified decision environment. The system positions AI as a decision partner for the farmers instead of a replacement and full automation. Farming data is structured as a continuous pipeline from observation and diagnosis to prediction, prescription, and counterfactual exploration, and allows users to run "what-if" scenarios and make informed decisions. We tested the system with a group of users to perceive how well the system could support them in making decisions. The results show that knowledge-driven, user-centric decision support had users reporting positive experiences and recognizing its value in supporting decision-making.

Keyword: Agriculture intelligence · digital twin · decision support system · usability testing

1 Introduction

In the agricultural sector, the use of digital twins systems that assist in making decisions based on data and models that generate predictions using machine learning has increased as a result of technological advancements. These systems bring together types of data and provide useful information to help manage crops and make agriculture more productive (Pylianidis et al., 2021; Verdouw et al., 2021).

Despite advances in agricultural sensing, predictive analytics, and digital twin technologies, many existing solutions remain fragmented. Monitoring systems often focus on real-time data visualization, predictive models are commonly developed as standalone

N. Mohd Sharef et al. (Eds.): CAMP 2026, CCIS 3093, pp. 41–51, 2027.
https://doi.org/10.1007/978-981-92-4302-0_4

analytical components, and decision-support functionalities are implemented separately. As a result, users are required to interpret outputs across multiple tools and translate analytical results into management decisions themselves. Furthermore, limited attention has been given to understanding how users interact with integrated agricultural digital twin environments and whether such systems effectively support decision-making in practice.

Therefore, this study proposes a knowledge-driven agricultural digital twin platform called Agro Pulse Twin Hub. This platform is based on human-in-the-loop approach. It is designed to transform raw agricultural data into structured knowledge, enabling users to explore insights, predict outcomes, and evaluate management strategies interactively through simulation module. This paper makes the following contributions:

The contributions this paper makes are threefold:

(i) It presents the design and implementation of a knowledge-driven agricultural digital twin platform that integrates monitoring, predictive analytics, simulation, and decision-support capabilities within a unified environment.
(ii) It shows how heterogeneous agricultural data and predictive models can be operationalized to support continuous monitoring, crop assessment, prediction, and intervention planning.
(iii) It evaluates the usability and perceived usefulness of the platform through System Usability Scale (SUS) assessment and qualitative user feedback, providing insights into user interaction with integrated agricultural decision-support systems.

The paper is constructed as follows. The first section provides the background of the paper, followed by a review of related works in section two. The details on the knowledge-driven agriculture platform are presented in the third section while the next section elaborates on the results. Finally, the conclusion and future work are presented in section five.

2 Related Work

2.1 Digital Twin in Agriculture

One of the key strengths of digital twin technology is its ability to create a virtual representation of real farming environments. This allows users to continuously monitor conditions, generate predictions, and explore different management strategies without affecting actual farm operations (Verdouw et al., 2021; Pylianidis et al., 2021; Métwalli et al., 2025). In practice, many existing solutions focus on integrating real-time sensor data with visualization dashboards, providing users with situational awareness of environmental factors such as temperature, humidity, and nutrient levels which are particularly useful in controlled environments such as greenhouses (Ayaz et al., 2022; Hosny et al., 2025).

Even with all the progress so far, a key limitation remains because most digital twin systems today keep their monitoring, prediction, and simulation tools as separate solutions rather than one connected decision-support system. This could be the reason on the limitations of digital twins in providing ongoing reasoning, exploring different scenarios, or making concrete decisions as conditions change. This issue can be solved by redesigning digital twins in agriculture as not merely digital copies of the real world, and as smart environments that blend data, models, and user input so decision-making becomes a seamless, continuous process instead.

2.2 Advances in Agricultural Monitoring and Predictive Analytics

Agricultural monitoring technologies development today been largely driven by the convergence of IoT-based sensing, remote sensing, and machine learning (Talaviya et al., 2020; Zhang et al., 2019). These technologies complement each other well. For example, IoT enables fine-grained, real-time observation of soil and environmental conditions, while remote sensing provides a broader view of crop variability at scale. The ability to move beyond monitoring is made possible by machine learning models such as Random Forest, XGBoost, and Long Short-Term Memory (LSTM) networks that could generate predictions on crop growth, yield, and environmental responses to support future farm insights (Jeong et al., 2016; Khaki & Wang, 2019).

In terms of precision agriculture, emphasis has been growing on data-driven modelling to better understand spatial variability and crop performance. For example, geospatial and statistical approaches have been applied in oil palm to relate soil fertility and nutrient variability to yield outcomes (Balasundram et al., 2006; Balasundram, 2015). At the same time, work in crop and soil modelling has explored nutrient cycling and growth dynamics across different conditions, including nitrogen behaviour in flooded soils and modelling at the watershed scale (Khairudin et al., 2017, 2022).

There are also many recent studies that have combined sensing and machine learning for crop condition assessment, such as estimating chlorophyll from image-based features and predicting crop quality using thermal imaging. There are also works that push predictive digital twin approaches further, such as for nutrient management where relationships between nutrients are modelled to estimate their concentrations and interactions (Hassanijalilian et al., 2020; Nagy et al., 2024; Mohd Ali et al., 2022; Ghazvini et al., 2024).

But even with these improvements, most predictive capabilities are still built in isolation, either as separate models or as part of different analytical pipelines. This makes it hard to connect them smoothly with real-time monitoring and the bigger picture of the system. Because of this, many systems have trouble moving from fragmented insights to a decision-making process that is more continuous and coherent.

2.3 Agricultural Decision Support Systems

The purpose of agricultural decision support systems (DSS) is to connect data and decision-making, including data sources, analytical models, and domain knowledge (Gonzalez et al., 2024; Petraki et al., 2025). Modern DSSs consist of modules for data acquisition, data management, predictive analytics, and visualization, and some systems offer recommendation capabilities such as irrigation scheduling, fertilization planning, and crop intervention strategies (Gonzalez et al., 2024; Petraki et al., 2025).

Yet, DSS implementations are still often limited by fragmented system design, with monitoring systems, predictive models, and recommendation components that are developed independently. Furthermore, they often require users to interpret outputs through multiple interfaces. The consequences of this little attention to user interaction and usability further limit their effectiveness in real-world applications, specifically because users have difficulty interpreting analytical outputs and translating them into actionable decisions (Rose et al., 2016; Zhai et al., 2020; Gonzalez et al., 2024; Petraki et al., 2025).

2.4 User-Centric Evaluation and Adoption in Agricultural Systems

One of the factors that has not gained enough attention in ensuring the success of digital agriculture systems is user interaction. Current research shows that many agricultural systems are built by emphasising analytical performance, with little consideration given to user experience and interaction design, making it difficult for users to comprehend model outputs, understand system interfaces, and translate analytical insights into action (Rose et al., 2016; Petraki et al., 2025).

Usability evaluation methods like the System Usability Scale (SUS) help measure how satisfied users are, how easy the system is to learn, and how efficient interactions feel (Brooke, 1996). Researchers have started looking at usability in agricultural systems, but usually, those studies only focus on specific features like visualization, interpretability, or workload (Htun et al., 2022; Brenes et al., 2024). Although those insights are useful, but they do not really show what happens when users work with more connected environments especially digital twin systems that bring together monitoring, prediction, and decision-making all in one place. Furthermore, the users interaction with predictive or scenario-driven outputs has not really gotten much attention which are in fact important in advisory-based decision support system.

2.5 Research Gap

These limitations highlight a critical gap in current digital agriculture systems. While sensing technologies, predictive analytics, and modelling approaches have advanced significantly, their integration into a unified, knowledge-driven digital twin framework remains limited. Furthermore, insufficient attention has been given to user interaction and usability, particularly in understanding how agricultural trainees and practitioners engage with such systems.

There is a need for systems that not only combine monitoring and prediction, but also structure these components into a coherent knowledge pipeline while supporting intuitive interaction, interpretation, and decision-making. This study addresses this gap by proposing a knowledge-driven agricultural digital twin platform that integrates heterogeneous data sources, predictive analytics, and decision-support functionalities within a unified and user-centered environment.

3 Knowledge-Driven Agricultural Digital Twin Platform

AgroPulse Twin Hub is aimed at providing an integrated setting for decision supporting agricultural monitoring and support application that integrates monitoring, assessment, prediction, advisory generation, and simulation, within a unified digital twin framework. Instead of viewing these functionalities as separate analytical packages, the platform allows users to constantly transition from monitoring farm dynamics to assessing future results and considering possible interventions. The overall architecture of the platform is illustrated in Fig. 1.

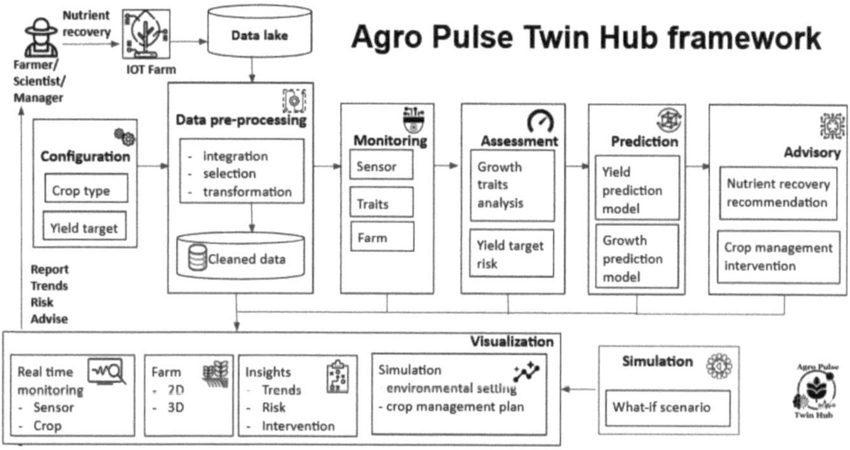

Fig. 1 Framework and workflow of Agro Pulse Twin Hub

The system begins with data acquisition and configuration, where users (e.g., farmers, researchers, and managers) define crop types and yield targets while the system collect and visualize data analytics from IoT-enabled farm environments (Figs. 2 and 3). These include environmental parameters such as pH, electrical conductivity (EC), and temperature, as well as crop traits such as plant height, leaf count, and leaf length. All incoming data are stored in a centralized data repository, enabling unified access to both real-time and historical datasets.

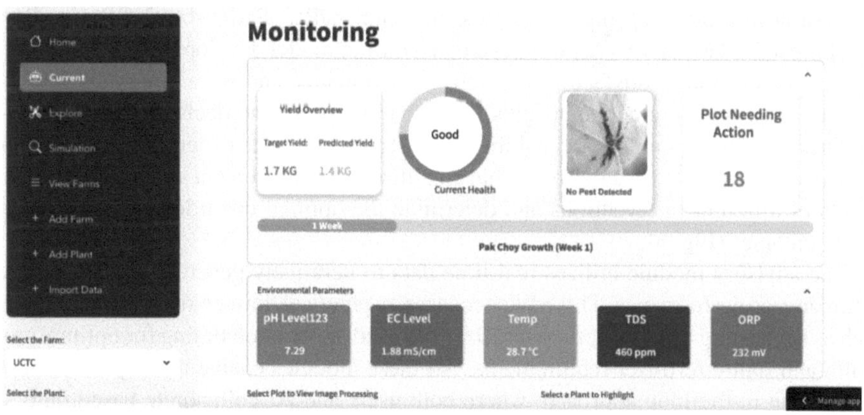

Fig. 2 Overview page with real-time monitoring and summarised intelligence about the farm

There are two sequential stages in the modelling workflow. In the first stage, environmental parameters (EP) and crop indicators (CI) are predicted using historical observations and temporal patterns. Environmental parameters include variables such as pH, EC, and temperature, while crop indicators include plant height, leaf count, and longest

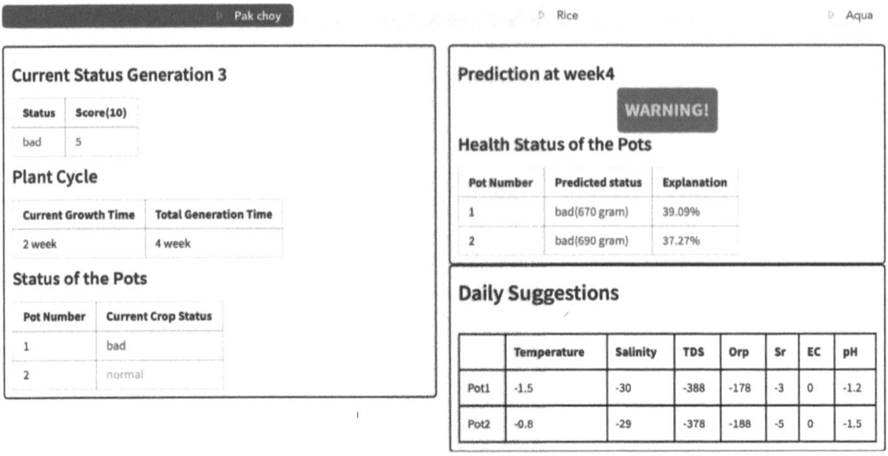

Fig. 3 Application of prediction models to forecast farm condition

leaf measurements. Such predictions provide an estimate of future crop conditions and crop development.

In the second stage, the predicted crop indicators are used as inputs to the yield prediction model to estimate the expected harvest outcomes. This sequential prediction workflow makes it possible for the platform to estimate future crop development before converting those projected conditions into yield forecasts. The derived predictions are provided through the digital twin interface to support monitoring, planning, and intervention decisions.

Several machine learning algorithms, including GRU, SARIMA, XGBoost, Random Forest, LSTM, and SVM, were evaluated during model development. The selected models demonstrated satisfactory predictive performance and were integrated into the platform to provide predictive insights for crop monitoring and decision support.

The digital twin simulation capability (Fig. 4) allows users to experiment with various what-if scenarios, altering environmental conditions or management strategies to test the effects of different interventions and determine the optimal conditions to reach target yield outcomes (Fig. 5).

The advisory module utilizes real-time data to iteratively generate predictions and optimize crop performance. This advice focuses on practical, farmer-friendly parameters, such as suggesting the appropriate fertilizer type and application timing for optimal crop health and future fertilizer requirements. All these modules enable the implementation of a human-in-the-loop approach, where both users and AI collaborate based on both system insights and domain expertise.

The uniqueness of AgroPulse Twin Hub is the aggregation of monitoring, prediction, advisory creation, and simulation in one user interface. The users can see the current status of the farm, view the forecasted future prospects, get intervention recommendations,

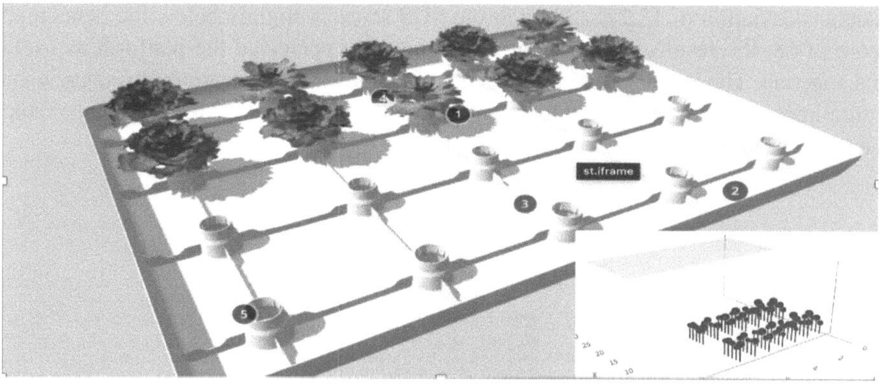

Fig. 4 Digital twin of the current hydroponics farm status (left) and predicted farm condition based on the current situation (right, bottom)

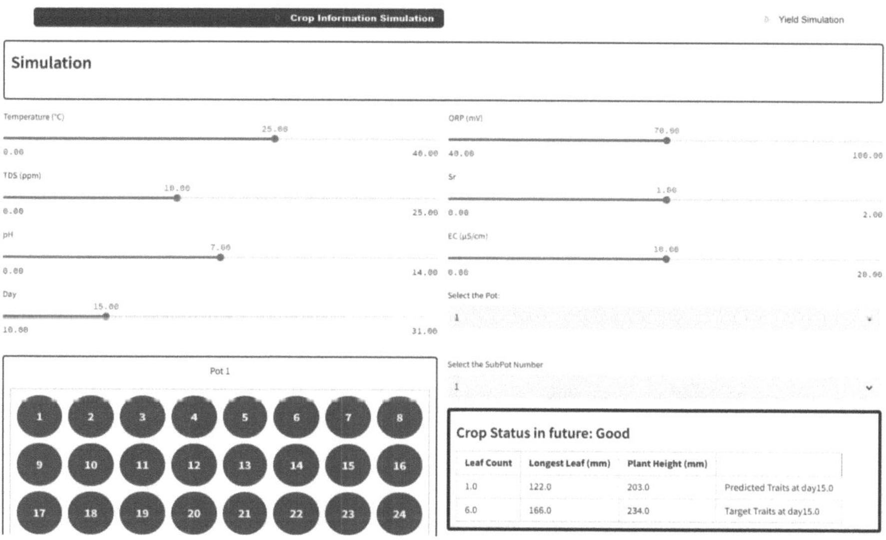

Fig. 5 Simulation module for users to gauge predicted farm condition based on any input scenario

and study alternative management options through what-if simulations without switching from one system to another. This combined functionality enables a more continuous decision-making process as opposed to traditional monitoring- or prediction-based farming systems.

4 Results and Evaluation

Following the system demonstration, the usability of the system has been evaluated by 58 respondents comprising of agricultural officers and students. The standard scoring procedure was used to calculate the SUS score, yielding a mean score of 64.83 with a standard deviation of 13.27. Although the SUS score is slightly below the benchmark score of 68, the results indicate that users generally perceived the platform as useful and coherent. The lower scores associated with complexity and learnability suggest that future improvements should focus on interface simplification and user guidance rather than analytical functionality.

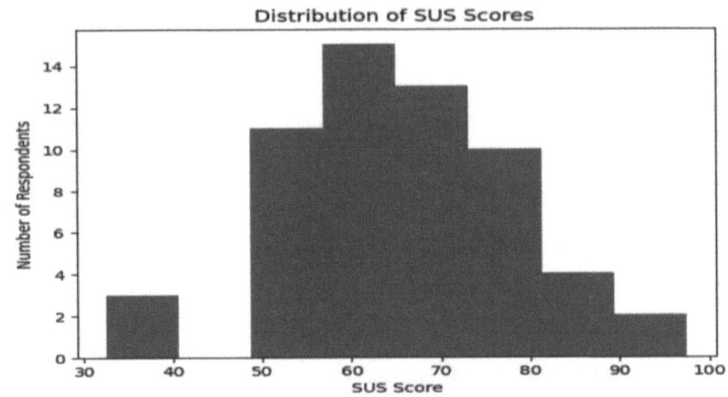

Fig. 6 Distribution of SUS score by the participants

Figure 6 lays out how participants scored the system where some people found it straightforward, but others had a tougher time probably comes down to how comfortable each person is with digital platforms and analysis tools. Note that none of the respondent has experienced using a digital twin system for supporting farm management, although they have previously been exposed to a dashboard that visualises sensors reading.

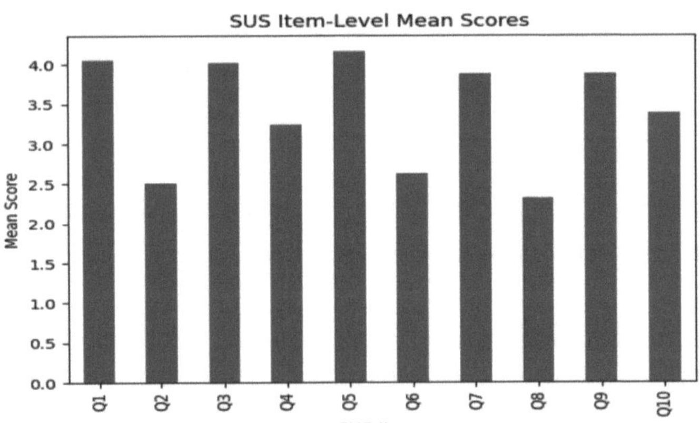

Fig. 7 Mean SUS score by each question

Figure 7 breaks down the SUS responses by item which summarises that users generally think the system is useful and fits together well. They like its functions and what it can do. On the flip side, the scores drop on the negative items, which shows people find it a bit complicated and need to put in extra effort to learn how to use it. This could be trigged by tricky and new interface design, as well as how the interactions are laid out.

The qualitative feedback is mostly inclined towards the positive responses where participants described the system as interactive and informative, saying the features are helpful for understanding crop conditions and forecasts. This indicates that the mix of monitoring and predicting features seems to land well with users. However, they pointed out that the using the system for supporting farm management is complicated and they commented that the system can be better if the explanations are simpler and visuals are clearer.

From a system design perspective this input is consistent with our principles that although predictive analytics allows for the creation of future-oriented insights, the quality of such insights relies on how they are presented and understood by users. The data also shows that the AgroPulse Twin Hub's knowledge-based setup helps with interpretation and learning. Still, it needs some tuning to make things easier for folks who are not tech-savvy so that the information produced is not only accurate but clear, accessible, and practical for making decisions.

5 Discussion

This study shows the need to integrate data, predictive models, and decision support mechanisms in a cohesive framework, as opposed to the isolated functionalities of conventional agricultural systems, and to provide structured pipelines that help users move from passive monitoring to active decision exploration with predictive insights and simulation capabilities, all with a human-in-the-loop design to maintain transparency and trust in AI-driven recommendations in the context of complex agricultural decisions based on a combination of data, experience, and contextual knowledge.

Unlike many agricultural studies that primarily evaluate predictive performance, this work additionally examines user interaction with the proposed platform through usability assessment. The findings highlight that successful agricultural digital twin deployment depends not only on analytical capability but also on the extent to which users can understand, interpret, and act upon system-generated insights. This reinforces the importance of user-centred evaluation in future agricultural decision-support research.

In addition, the amount of information provided by users during the study indicates that usability must be taken into account to aid in the adoption of agricultural technologies. An advanced level of data analytics will not be useful when the data is presented to the user in a way that is confusing, lacks intuitive design, or that provides outputs that do not relate to what the user needs in order to be actionable. Feedback provided by users indicates many opportunities for improvement, particularly with regard to simplifying how users interact with the system, and by providing guidance on what practical decision variables should be used.

6 Conclusion

The AgroPulse Twin Hub is an innovative platform based on data-driven knowledge that leverages predictive analytics and simulation for agricultural decision support. It creates a unified, structured data-to-information pipeline, which addresses the limitations of current systems by transforming raw agricultural and environmental data into actionable management decisions.

By combining predictive analytics, digital twin simulation, and advisory capabilities within a unified platform, users can evaluate potential outcomes of alternative agricultural management strategies and explore what-if scenarios before implementing interventions in the physical farm.

Looking ahead, AgroPulse Twin Hub plans to bring in real-time IoT data streams, use transfer learning across farms, and harness explainable AI to make everything more scalable and easier to understand. By blending data, models, and hands-on human interaction, AgroPulse Twin Hub will push the development of smarter decision-making tools for agriculture.

References

Ayaz, M., Ammad-Uddin, M., Sharif, Z., Mansour, A., Aggoune, E.H.M.: Internet-of-things (IoT)-based smart agriculture: toward making the fields talk. IEEE Access. **7**, 129551–129583 (2022)

Balasundram, S.K.: Precision agriculture in oil palm: current status and future perspectives. J. Oil Palm Res. **27**(3), 239–246 (2015)

Balasundram, S.K., Robert, P.C., Mulla, D.J., Allan, D.L.: Relationship between oil palm yield and soil fertility as determined by spatial variability. Precis. Agric. **7**(2), 85–100 (2006)

Brenes, J.A., López, G., Ferrández-Pastor, F.J., Marín-Raventós, G.: Usability assessment of a greenhouse context-aware alert system for small-scale farmers. Front. Comput. Sci. **6**, 1412913 (2024)

Brooke, J.: SUS: a "quick and dirty" usability scale. In: Jordan, P.W., Thomas, B., McClelland, I.L., Weerdmeester, B. (eds.) Usability Evaluation in Industry, pp. 189–194. Taylor & Francis (1996)

Ghazvini, A., Sharef, N.M., Balasundram, S.K., Lee, L.S.: A concentration prediction-based crop digital twin using nutrient co-existence and composition in regression algorithms. Appl. Sci. **14**(8), 3383 (2024)

Gonzalez, P., Azpiroz, I., Zabala, M., Paolini, G., Maiza, M.: White paper on user Centred design to address challenges in the design and creation of agricultural data visualisation interfaces. Agriculture. **14**(10), 1808 (2024)

Hassanijalilian, O., Igathinathane, C., Bajwa, S.G.: Chlorophyll estimation in soybean leaves in-field with smartphone digital imaging and machine learning. Comput. Electron. Agric. **174**, 105433 (2020)

Hosny, K.M., El-Hady, W.M., Samy, F.M.: Technologies, protocols, and applications of internet of things in greenhouse farming: a survey of recent advances. Inf. Process. Agric. **12**(1), 91–111 (2025)

Htun, N.-N., Rojo, D., Ooge, J., De Croon, R., Kasimati, A., Verbert, K.: Developing visual-assisted decision support systems across diverse agricultural use cases. Agriculture. **12**(7), 1027 (2022)

Jeong, J.H. et al.: Random forests for global and regional crop yield predictions. PLoS One. **11**(6), e0156571 (2016)

Khaki, S., Wang, L.: Crop yield prediction using deep neural networks. Front. Plant Sci. **10**, 621 (2019)

Métwalli, A. et al.: Enhancing hydroponic farming with machine learning: growth prediction and anomaly detection. Eng. Appl. Artif. Intell. **157**, 111214 (2025)

Mohd Ali, M., Hashim, N., Abd Aziz, S., Lasekan, O.: Quality prediction of different pineapple (Ananas comosus) varieties during storage using infrared thermal imaging technique. Food Control. **138**, 108988 (2022)

Nagy, A., Szabó, A., Elbeltagi, A., Nxumalo, G.S., Bódi, E.B.a.T.J.: Hyperspectral indices data fusion-based machine learning enhanced by MRMR algorithm for estimating maize chlorophyll content. Front. Plant Sci. **15**, 1419316 (2024)

Nurulhuda, K., Gaydon, D.S., Jing, Q., Zakaria, M.P., Struik, P.C., Keesman, K.J.: Nitrogen dynamics in flooded soil systems: an overview on concepts and performance of models. J. Sci. Food Agric. **98**, 865–871 (2018)

Raffar, N., Zulkafli, Z., Yiwen, M., Muharam, F.M., Rehan, B.M., Nurulhuda, K.: Watershed-scale modelling of the irrigated rice farming system at Muda, Malaysia, using the soil water assessment tool. Hydrol. Sci. J. **67**(3), 462–476 (2022)

Petraki, D. et al.: Digital tools and decision support Systems in Agroecology: benefits, challenges, and practical implementations. Agronomy. **15**(1), 236 (2025)

Pylianidis, C., Osinga, S., Athanasiadis, I.N.: Introducing digital twins to agriculture. Comput. Electron. Agric. **184**, 105942 (2021)

Rao, N., et al.: Decision support systems in agriculture: a review. Agric. Syst. **187**, 103001 (2021)

Rose, D.C. et al.: Decision support tools for agriculture: towards effective design and delivery. Agric. Syst. **149**, 165–174 (2016)

Talaviya, T., Shah, D., Patel, N., Yagnik, H., Shah, M.: Implementation of artificial intelligence in agriculture for optimization of irrigation and application of pesticides and herbicides. Artificial Intelligence in Agriculture. **4**, 58–73 (2020)

Verdouw, C., Tekinerdogan, B., Beulens, A., Wolfert, S.: Digital twins in smart farming. Agric. Syst. **189**, 103046 (2021)

Zhai, Z., Martínez, J.F., Beltran, V., Martínez, N.L.: Decision support systems for agriculture 4.0: survey and challenges. Comput. Electron. Agric. **170**, 105256 (2020)

Zhang, C., Kovacs, J.M.: The application of small unmanned aerial systems for precision agriculture: a review. Precis. Agric. **20**(4), 693–712 (2019)

A Temporally-Adaptive Multimodal Misinformation Detection Framework with Cross-Modal Alignment and Gated Fusion

Omar Bashaddadh[1,2](✉), Nazlia Omar[1], and Masnizah Mohd[3]

[1] Faculty of Information Science and Technology, Centre for Artificial Intelligence Technology, Universiti Kebangsaan Malaysia, Bangi 43600, Malaysia
P131514@siswa.ukm.edu.my, aboshdad09@gmail.com

[2] Department of Computer Science, College of Computers, Seiyun University, Hadhramout, Yemen

[3] Faculty of Information Science and Technology, Centre for Cyber Security, Universiti Kebangsaan Malaysia, Bangi 43600, Malaysia
masnizah.mohd@ukm.edu.my

Abstract. The rapid spread of multimodal misinformation on social media poses a critical challenge to information credibility and automated verification systems. Conventional fake news detection methods that rely solely on textual features are often inadequate for capturing the complex semantic, visual, and temporal characteristics of modern misinformation. To address this limitation, this study proposes a Temporally-Adaptive Multimodal Misinformation Detection Framework (TAMMDF) that integrates RoBERTa textual representations, ResNet-50 visual features, and graph neural network-based temporal propagation embeddings within a unified architecture. The framework employs bidirectional cross-modal alignment and transformer encoders to model rich interdependencies across modalities, and an adaptive gated fusion mechanism to dynamically weight each modality's contribution. Experiments on the Fakeddit dataset (34,704 samples; 80%/20% split) demonstrate an accuracy of 93.40%, precision of 93.41%, recall of 93.40%, and F1-score of 93.39%. Learned fusion weights were distributed as 30.20% text, 31.56% image, and 38.24% temporal, reflecting balanced tri-modal integration. Ablation studies confirm that removing the temporal modality reduces accuracy to 91.87%, while removing cross-modal alignment collapses text-image consistency detection to 55.54% demonstrating that accuracy alone is an insufficient metric for evaluating alignment quality. These results establish that temporally grounded adaptive fusion substantially improves the robustness of fake news detection systems for real-world social media environments.

Keywords: Multimodal fake news detection · misinformation detection · temporal modelling · cross-modal attention · adaptive fusion · graph neural networks · transformers

N. Mohd Sharef et al. (Eds.): CAMP 2026, CCIS 3093, pp. 52–66, 2027.
https://doi.org/10.1007/978-981-92-4302-0_5

1 Introduction

Fake news, defined as the deliberate dissemination of false or misleading information, constitutes a major global challenge [1, 2]. Its rapid spread has been intensified by social media and digital news platforms, including Facebook, X (formerly Twitter), Reddit, Instagram, Snapchat, YouTube, and TikTok. that support multimodal content involving text, images, videos, social interactions, and temporal signals, which collectively amplify the reach and impact of misinformation. Traditional single-modal methods, mainly based on textual analysis, are limited in their inability to capture the complex multimodal interactions of modern misinformation [3, 4].

Recent multimodal fake news detection methods have introduced advanced technique such as dynamic fusion, graph neural networks, and cross-modal attention. Graph-based models, including GS2F (Graph Structure and Guided Semantic Fusion), exploit semantic relationships across modalities to improve detection accuracy, but they often neglect temporal factors essential for real-time news assessment [5]. Similarly, dynamic fusion methods such as MDF assign adaptive weights to textual and visual modalities but still lack comprehensive temporal analysis [6].

Temporal models such as TEMGNN aggregate multimodal data over time to capture evolving propagation patterns, but their contrastive learning utilisation and adaptive fusion capabilities remain limited [4].

Despite these advances, existing methods have significant gaps, particularly in the joint integration of temporal dynamics, explicit cross-modal alignment, and adaptive modality-aware fusion within a single end-to-end trainable architecture [7–10]. Therefore, robust multimodal models are still needed to address increasingly complex misinformation strategies.

This study addresses these gaps by proposing TAMMDF, which combines RoBERTa text representations, ResNet-50 visual features, and GNN-based temporal propagation features with bidirectional cross-modal alignment, parallel transformer encoders, and an adaptive gated fusion mechanism. The remainder of this paper is structured as follows: Sect. 2 reviews related works; Sect. 3 presents the proposed framework and experimental results; and Sect. 4 concludes the study.

2 Background and Related Works

This section presents a review of related studies, categorized according to the information source utilized for detection, with a particular focus on both single-modal and multimodal detection methods and Fusion Mechanisms.

2.1 Single-Modal Fake News Detection

Single-modal fake news detection predominantly relies on textual content analysed with natural language processing (NLP) techniques. Traditional methods employ lexical, syntactic, and semantic analyses to detect misinformation patterns. Sentiment and linguistic style analyses effectively utilise emotional tone, vocabulary, and syntactic structure to assess credibility [11, 12]. However, these approaches neglect the crucial visual and temporal contexts that characterise modern misinformation scenarios [3, 13]. Advanced text-based models such as BERT and LSTM capture deep semantic representations but continue to struggle with misinformation involving manipulated visual content and temporal dynamics [14].

2.2 Multi-Modal Fake News Detection

Multimodal fake news detection overcomes single-modal limitations by integrating text, images, temporal data, and social contexts, to capture complementary evidence and improve accuracy, generalization, and robustness [4–6, 15–18]. As illustrated in Fig. 1, research in this area follows three interrelated directions: (1) multimodal data study collecting, verifying, and benchmarking misinformation datasets; (2) multimodal feature study examining cross-modal relations and fusion techniques; and (3) multimodal model study developing machine learning models that effectively exploit fused features. These stages form an integrated framework for scalable multimodal misinformation detection [19].

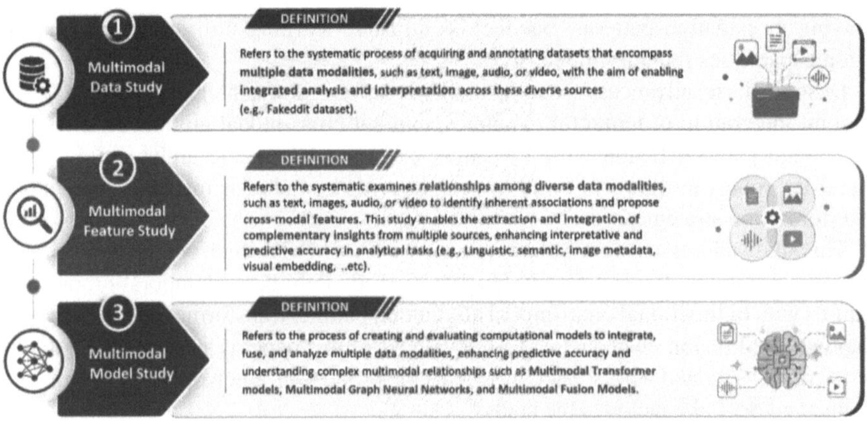

Fig. 1. Overview of the Multimodal Misinformation Detection Direction

2.3 Comparison of Single-Modal and Multimodal Approaches

Single-modal approaches excel at detecting misinformation with clear linguistic anomalies but fail when misinformation involves manipulated visual content or rapidly evolving temporal contexts. Multimodal approaches address these limitations by integrating multiple data streams. For instance, models such as MCAN and CAF explicitly address cross-modal inconsistencies and semantic-physical discrepancies that single-modal frameworks inherently overlook [20, 21]. Transformer-based and graph-based fusion methods further enhance detection accuracy by modelling complex interdependencies across modalities.

2.4 Temporal Information Integration

Temporal dynamics are critical for understanding misinformation dissemination yet are often overlooked in static multimodal models. TEMGNN significantly outperforms static models by integrating temporal GNNs that track misinformation diffusion over time [4]. Hybrid fusion models such as HF-TIM improve classification accuracy by capturing temporal inter-modal relationships [22], further improve classification accuracy by capturing complex temporal interrelationships across modalities [14, 20, 23–25]. Nevertheless, fine-grained temporal modelling and long-horizon trajectory prediction remain underexplored areas.

2.5 Fusion Mechanisms

Data fusion integrates information from multiple modalities to exploit their complementary characteristics. As shown in Fig. 2, fusion strategies are broadly classified into four categories: **early fusion** (feature-level, pre-classification concatenation or pooling); **joint fusion** (intermediate-layer integration with backpropagation); **late fusion** (decision-level aggregation of unimodal outputs); and **hybrid fusion** (combining early and late strategies for balanced performance) [19, 22, 25–27].

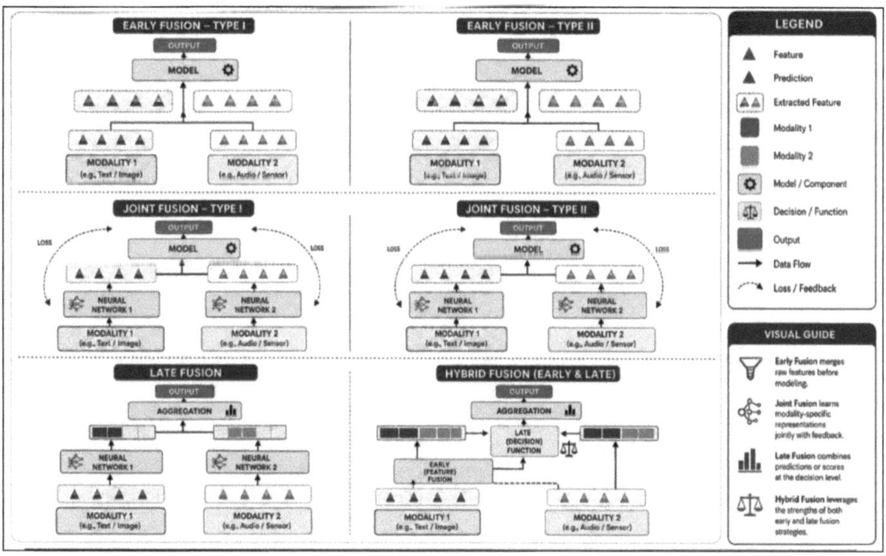

Fig. 2. Fusion Multimodal data mechanisms using Deep Learning

2.6 Advanced Fusion Techniques

Advanced multimodal fusion techniques improve fake news detection by integrating heterogeneous signals. Cross-modal attention remains central: CAF uses bidirectional attention to preserve fine-grained text-image correlations [25]; MMCAN enhances modality-specific representations through co-attention and mutual knowledge distillation [9]; and MIAN exposes semantic contradictions between modalities using inverse attention [8]. Graph-based approaches such as GS2F apply graph contrastive learning for structured text-image reasoning [5], while TEMGNN extends this with temporal and external knowledge [4]. User-centred methods such as MFUIE incorporate user behaviour through GAT-based fusion [28]. Transformer-based fusion models capture complex text-image dependencies but often underutilise temporal diffusion signals [15]. Table 1 provides a comparative overview of these methodologies.

Table 1. Comparative overview of evolving multimodal fusion strategies and limitations

Ref.	Methodology	Modalities	Fusion Technique	Key Strengths	Limitations/Gaps
[5]	GS2F	Text, Images	Graph Contrastive Learning	Robust semantic integration	Lacks temporal dimension
[6]	MDF (Dynamic Weighted)	Text, Images	Dynamic attention	Handles modality uncertainty	No temporal integration
[16]	Mixture-of-Experts	Text, Images	Modality interaction gating	Strong modality-specific feature	Minimal temporal/ external knowledge
[10]	Contrastive + Optimal Transport	Text, Images	Optimal transport alignment	Precise multimodal alignment	No temporal integration
[8]	Inverse Attention (MIAN)	Text, Images	Inverse attention	Highlights cross-modal inconsistencies	Limited temporal analysis
[9]	Matching-aware Co-Attention Networks (MMCAN)	Text, Images	Co-attention + distillation	Enhanced modality refinement	Insufficient temporal integration
[15]	Transformer-based Fusion	Text, Images	Transformer self-attention	Effective semantic interaction	Limited temporal dynamics
[4]	TEMGNN	Text, Images, Temporal, External	Graph attention & convolution	Explicit temporal integration	Limited contrastive learning
[28]	MFUIE	Text, Images, User Posts	GAT + Visual/Text Fusion	Incorporates user behavior	Limited inter-modal reasoning
[29]	MIGCL	Text, Images	Intra-/Intermodal Graph CL	Deep cross-modal dependencies	Lacks user context
[30]	MRAN	Text, Images	Multi-level Encoding + Cross Attention	Superior multimodal feature	Complex; lacks temporal modelling

(*continued*)

Table 1. (*continued*)

Ref.	Methodology	Modalities	Fusion Technique	Key Strengths	Limitations/Gaps
[31]	MDPF	Text, Images, Sentiment	Joint Perception + Routing	Sentiment-informed learning	Focused on sentiment, not credibility

2.7 Performance Overview of Prior Models (Table 2)

Table 2. Performance of previous fake news detection models across multiple datasets.

Ref.	Model	Modality	Dataset	Accuracy	Precision	Recall	F1-score
[5]	GS2F	Text, Images	Fakeddit	0.896	0.889	0.893	0.891
[6]	MDF	Text, Images	weibo	0.810	0.850	0.890	0.870
[16]	MIMoE-FND	Text, Images	GossipCop	0.895	0.841	0.799	0.817
[10]	MCOT	Text, Images	wiebo	0.901	0.900	0.900	0.900
[8]	MAIN	Text, Images	Fakeddit	0.870	–	–	–
[9]	MCAN	Text, Images	wiebo	0.911	0.911	0.911	0.910
[15]	TGA	Text, Images	Twitter	0.918	0.912	0.854	0.918
[28]	MFUIE	Text, Images, User	weibo	0.926	0.9265	0.926	0.927
[29]	MIGCL	Text, Images	Twitter	0.906	0.908	0.907	0.906
[30]	MRAN	Text, Images	Pheme	0.870	0.871	0.868	0.874
[31]	MDPF+CLIP	Text, Images, Ext	Twitter	0.894	0.903	0.885	0.894
[19]	BERT_ResNet50	Text, Image	Fakeddit	0.8909	–	–	–

2.8 Discussion of Research Gaps

A comparative analysis reveales that MFUIE [28] achieved the highest accuracy (0.926) on Weibo by incorporating user features, and Transformer Fusion (TGA) [15] performs strongly on Twitter (0.918). However, performance is consistently dataset dependent. GS2F and BERT_ResNet50 [5, 19], despite modern architectures, achieve only modest accuracy (~0.89) on Fakeddit, highlighting that architectural sophistication alone does not guarantee superior performance without effective modality fusion. Crucially, temporal modelling remains absent or peripheral in most existing models, and no prior work jointly addresses temporal integration, bidirectional cross-modal alignment, and adaptive modality-aware fusion within a single end-to-end trainable architecture. These gaps motivate the proposed TAMMDF framework.

3 Proposed Temporally-Adaptive Multimodal Misinformation Detection Framework

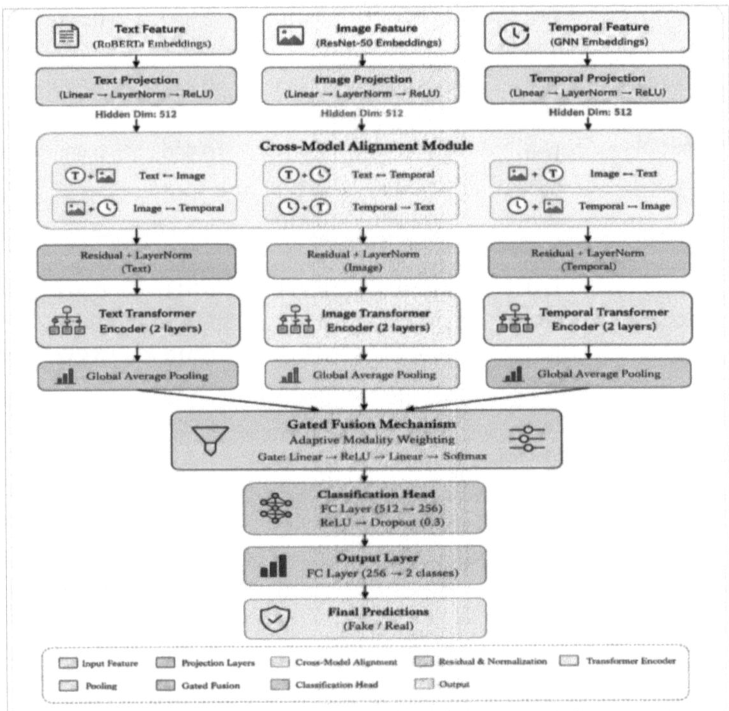

Fig. 3. Proposed TAMMDF Model Architecture

The proposed TAMMDF, illustrated in Fig. 3, processes three complementary modalities: (1) textual representations extracted using RoBERTa embeddings; (2) visual representations obtained from ResNet-50; and (3) temporal propagation patterns encoded by GNNs. Each modality is projected into a shared 512-dimensional latent space via modality-specific linear projection, layer normalization, and ReLU activation. A bidirectional cross-modal alignment module explicitly models text-image, text-temporal, and image-temporal interactions, capturing complementary and inconsistent signals across modalities. Residual connections and layer normalization preserve modality-specific characteristics while incorporating cross-modal information.

Three parallel two-layer transformer encoder branches model intra-modal contextual dependencies for text, image, and temporal features. Each branch applies global average pooling to produce compact modality-aware representations, which are then combined via an adaptive gated fusion mechanism (Linear–ReLU–Linear–Softmax). The fused representation is passed to a lightweight classification head for binary fake/real prediction.

3.1 Temporal Feature Construction

To construct temporal propagation features, user-news interaction graphs are built from Fakeddit metadata. Each graph is a bipartite structure in which nodes represent individual posts (news items) and users, and directed edges encode temporal interactions (comments, reposts, and shares) ordered by timestamp. Node features are initialised with the projected text and image embeddings. Edge features capture temporal attributes, including propagation velocity (interactions per unit time) and cascade depth (number of propagation hops from the source post). A two-layer Graph Convolutional Network (GCN) aggregates neighbourhood information to produce 128-dimensional temporal node embeddings [32], which are subsequently projected to 512 dimensions to match the text and image feature spaces. Although Fakeddit does not provide explicit retweet graph data, temporal interaction sequences were reconstructed from post metadata (submission timestamps and comment threads), following a protocol analogous to that used in TEMGNN [4]

3.2 Bidirectional Cross-Modal Alignment

Six parallel cross-attention modules bidirectionally align all modality pairs: Text ↔ Image, Text ↔ Temporal, and Image ↔ Temporal, as defined in Eqs. (1)–(6).

Text – Image Bidirectional Alignment

$$h_{text}^{aligned} = LayerNorm\left(h_{text} + CrossAttn\left(h_{text},\ h_{image}\right)\right) \tag{1}$$

$$h_{image}^{aligned} = LayerNorm\left(h_{image} + CrossAttn\left(h_{image},\ h_{text}\right)\right) \tag{2}$$

Text – Temporal Bidirectional Alignment

$$h_{text}^{aligned} = LayerNorm\left(h_{text}^{aligned} + CrossAttn\left(h_{text},\ h_{temporal}\right)\right) \tag{3}$$

$$h_{temporal}^{aligned} = LayerNorm\left(h_{temporal} + CrossAttn\left(h_{temporal},\ h_{text}\right)\right) \tag{4}$$

Image – Temporal Bidirectional Alignment

$$h_{image}^{aligned} = LayerNorm\left(h_{image}^{aligned} + CrossAttn\left(h_{image},\ h_{temporal}\right)\right) \quad (5)$$

$$h_{temporal}^{aligned} = LayerNorm\left(h_{temporal}^{aligned} + CrossAttn\left(h_{temporal},\ h_{image}\right)\right) \quad (6)$$

where h denotes the hidden representation for each modality (learned feature vector/embedding for each modality), and CrossAtten is defined as:

$$CrossAttn(Q, K) = Softmax\left(\frac{QK_Q(KW_k)^T}{\sqrt{d_k}}\right)(KW_v) \quad (7)$$

Where Q, K, and V denoted to Query, Key, and Value components of the attention mechanism; W_Q, W_K, and W_V are learned projection matrices; and d_k is the key dimension. Unlike existing methods using simple concatenation [13, 33, 34] or unidirectional attention [33, 35], our bidirectional mechanism ensures mutual modality refinement and extends temporal alignment beyond the text-image paradigm.

3.3 Adaptive Modality-Aware Fusion

Following cross-modal alignment and transformer encoding, an adaptive gated fusion mechanism, learns modality-specific importance weights via a softmax-normalized gate network, as defined in Eqs. (8) and (9).

$$W = Softmax\left(W_2.ReLU\left(W_1.\left[h_{text}^{enc}; h_{image}^{enc}; h_{temporal}^{enc}\right]\right)\right) \quad (8)$$

$$h_{fused} = \sqsupseteq_{text}.h_{text}^{enc} + \sqsupseteq_{image}.h_{image}^{enc} + \sqsupseteq_{temporal}.h_{temporal}^{enc} \quad (9)$$

Where h_{text}^{enc}, h_{image}^{enc}, $h_{temporal}^{enc}$ are the transformer-encoded representations after alignment, $W = [w_{text}, w_{image}, w_{temporal}]^T$, $\sigma_i w_i = 1$, and [.;.;.] represents concatenation, and W_1 and W_2 are learned gate network parameters. Whereas existing multimodal fake news detectors use fixed fusion strategies (e.g., concatenation and averaging) that cannot adapt to varying levels of modality reliability [36]. This mechanism dynamically adjusts modality contribution per sample, automatically emphasising reliable modalities (e.g., privileging temporal signals for coordinated misinformation campaigns with strong propagation patterns but weak text-image inconsistency).

3.4 End-to-End Training and Loss Functions

All components, cross-modal attention, transformer encoders, gated fusion, and temporal GNN encoders are jointly optimised through a combined loss:

$$\mathcal{L}_{total} = \alpha\mathcal{L}_{cls} + \beta\mathcal{L}_{contr} + \gamma\mathcal{L}_{bal} \quad (10)$$

Where the component losses are:

$$\mathcal{L}_{cls} = -\sum_{i=1}^{N}\sum_{c=1}^{c} y_{i,c}\log(\hat{y}_{i,c}) \tag{11}$$

$$\mathcal{L}_{contr} = -\frac{1}{N}\sum_{i=1}^{N}\left(sim\left(h^{i}_{text}, h^{i}_{image}\right) - S^{i}_{target}\right)^{2} \tag{12}$$

$$\mathcal{L}_{bal} = \left\| -\frac{1}{N}\sum_{i=1}^{N} W^{i} - u \right\|^{2} \tag{13}$$

With sim(.,.) as cosine similarity, $S^{i}_{target} = 0.3 + 0.5.\, y_i$ (expected similarity based on label), u = [1/3, 1/3, 1/3]^{T} (uniform weight target), and α = 0.65, β = 0.30, ϒ = 0.05, N = batch size, C = 2 (binary classification). The contrastive loss L_contr drives text-image representations to be semantically coherent, which is the quantity measured by the consistency detection metric. Consistency detection is therefore defined as the proportion of test samples for which the cosine similarity between the aligned text and image representations falls within a threshold of ±0.15 from the label-conditioned target S^{i}_{target}. This metric captures semantic cross-modal coherence that goes beyond surface classification accuracy.

3.5 Baseline Models

The proposed framework is compared against three baselines evaluated on Fakeddit: **Multimodal fusion (BERT, ResNET-50)** [37], integrates text and images using early, joint, and late fusion; **GS2F** [5]**,** exploits graph-based semantic relationships across modalities; and **BERT_ResNet50** [19], a standard text-image multimodal classifier serving as direct modality baseline.

3.6 Experimental Setup

Table 3 summarises all experimental settings to ensure full reproducibility. The Fakeddit dataset was split into 80%/20%. Early stopping (patience = 15 on validation F1-score) was applied to prevent overfitting. All experiments were conducted on an Intel® Core™ i7-6600U CPU @ 2.60 GHz, 12 GB RAM, and Intel HD Graphics 520, running Windows 11 Pro, with a fixed random seed of 42.

Table 3. Experimental configuration and hyperparameter settings

Hyperparameter / Setting	Value / Description
Dataset	Fakeddit (34,704 samples; 2-class: Fake/Real)
Training / Test Split	80%/20%
Class Distribution (Train)	Fake: 52.3%, Real: 47.7%
Text Encoder	RoBERTa (pre-trained, fine-tuned)

(continued)

Table 3. (*continued*)

Hyperparameter / Setting	Value / Description
Image Encoder	ResNet-50 (pre-trained on ImageNet, fine-tuned)
Temporal Encoder	2-layer GCN; node = post, edge = retweet/reply; temporal feature = propagation velocity + cascade depth
Graph Construction	User–news bipartite graph built from Fakeddit metadata; edges encode temporal order of user interactions; nodes carry textual/visual content embeddings
Projection Dimension	512-D shared latent space (Linear → LayerNorm → ReLU)
Transformer Encoder	2-layer per modality; 8 attention heads; d_model = 512
Optimizer	AdamW
Learning Rate	0.0001, Weight_Decay = 1e-4
Batch Size	32
Epochs	100 (early stopping; patience = 15 on validation F1)
Dropout	0.3 in classification head
Loss Weights	$\alpha = 0.70$ (classification), $\beta = 0.25$ (contrastive), $\gamma = 0.05$ (balance)
Random Seed	42 (fixed for reproducibility)
Hardware	Intel® Core™ i7-6600U CPU @ 2.60 GHz, 12 GB RAM, and Intel HD Graphics 520

3.7 Results and Comparison

Table 4 reports overall performance on the Fakeddit test set. The proposed TAMMDF achieves 93.40% accuracy, 93.41% precision, 93.40% recall, and 93.39% F1-score — outperforming all baselines. The improvement over the strongest baseline, Multimodal Fusion BERT+ResNet-50 (90.0%), demonstrates the added value of temporal propagation modelling and adaptive gated fusion. The learned fusion weights (Text: 30.20%; Image: 31.56%; Temporal: 38.24%) confirm balanced tri-modal integration, with temporal features contributing most distinctly.

Table 4. Overall performance comparison on fakeddit dataset

Model	Accuracy (%)	Precision (%)	Recall (%)	F1-Score (%)
GS2F (Graph Semantic Fusion) [5]	0.896	0.889	0.893	0.891
Multimodal fusion (BERT,ResNET-50) [37]	0.900			0.880
BERT_ResNet50 [19]	0.8909			

(*continued*)

Table 4. (*continued*)

Model	Accuracy (%)	Precision (%)	Recall (%)	F1-Score (%)
Proposed Model	**0.9340**	0.9341	0.9340	0.9339

3.8 Ablation Study

Table 5 reports three ablation configurations evaluated on the same Fakeddit test split. Removing the temporal modality (w/o Temporal) reduces accuracy to 91.87%, with fusion weights redistributing to a text-image split of 57%/43%, confirming that temporal GNN features capture propagation patterns irreplaceable by text or image cues alone. Removing cross-modal alignment (w/o Alignment) leaves surface accuracy effectively unchanged at 93.47%; however, the consistency detection metric collapses to 55.54% (near-chance level) while temporal weights inflate to 51.55%, indicating degenerate pseudo-unimodal behaviour. This demonstrates a critical point: accuracy alone is insufficient to evaluate alignment quality. The full TAMMDF model achieves 93.40% accuracy with balanced trimodal fusion weights and 93.00% consistency detection, establishing that cross-modal alignment is architecturally fundamental for semantically grounded multimodal fake news detection. The marginal difference between the accuracy of the full model (93.40%) and the w/o Alignment variant (93.47%) is within expected statistical variation across evaluation runs, and does not indicate that removing alignment is beneficial; the consistency metric explicitly reveals the degradation in semantic coherence.

Table 5. Results of ablation study on Fakeddit dataset

Model	Accuracy %	Text weight	Image weight	Temporal weight	Consistency %
T-GNN	0.6178	–	–	100%	–
Proposed model (w/o) Temporal	0.9187-(0.153)	0.5700	0.4300	–	
Proposed model (w/o) alignment	93.47	0.2318	0.2518	0.5155	0.5554
Proposed Model	0.9340	0.320	0.3156	0.3824	0.9300

4 Conclusion

This study proposes TAMMDF, a temporally adaptive multimodal misinformation detection framework that addresses persistent limitations of existing fake news detection systems, specifically inadequate temporal modelling, weak cross-modal interaction learning, and inflexible modality fusion. The framework integrates RoBERTa text embeddings, fine-tuned ResNet-50 image features, and GNN-based temporal propagation features within a unified architecture utilising bidirectional cross-modal alignment, parallel transformer encoders, and adaptive gated fusion. Experimental evaluation on the Fakeddit dataset achieved 93.40% accuracy, 93.41% precision, 93.40% recall, and 93.39% F1-score. The learned fusion weights confirmed balanced tri-modal integration, with temporal propagation patterns contributing most distinctly (38.24%) to the final prediction.

Ablation experiments validated the architectural necessity of each component: removing the temporal modality reduced accuracy to 91.87%, while removing cross-modal alignment caused consistency detection to collapse to 55.54% despite a marginal accuracy change, underscoring the alignment module's critical role in preserving semantic coherence beyond raw classification performance. These findings collectively demonstrate that integrating temporal dissemination dynamics with semantically grounded adaptive fusion substantially enhances the robustness and interpretability of multimodal misinformation detection, offering a scalable and practically deployable framework for real-world social media environments. Future work will investigate statistical significance across multiple runs, extended temporal graph architectures, and cross-dataset generalisation.

Acknowledgments. This study was supported by the Universiti Kebangsaan Malaysia under Grant TAP-K011456.

Disclosure of Interest. The authors declare no known financial or personal relationships that could have influenced this work and confirm that no conflicts of interest or competing interests are associated with this manuscript.

Declaration of Generative AI and AI-Assisted Technologies in the Writing Process. During the preparation of this manuscript, the authors used an AI-assisted tool to edit the language and enhanced the resolution of figures. After using this tool/service, the authors carfully reviewed and edited the manuscript as needed and took full responsibility for the content of the published article.

References

1. Bashaddadh, O., Omar, N., Mohd, M., Khalid, M.N.A.: Machine learning and deep learning approaches for fake news detection: a systematic review of techniques, challenges, and advancements. IEEE Access. **13**, 90433–90466 (2025). https://doi.org/10.1109/ACCESS.2025.3572051
2. Raza, N., et al.: Enhancing fake news detection with transformer-based deep learning: a multidisciplinary approach. PLoS One. **20**, 1–32 (2025). https://doi.org/10.1371/journal.pone.0330954

3. Guo, H., Huang, T., Huang, H., Fan, M., Friedland, G.: A systematic review of multimodal approaches to online misinformation detection. In: Proceeding of the 5th International Conference on Multimedia Information Processing and Retrieval, MIPR 2022, pp. 312–317 (2022). https://doi.org/10.1109/MIPR54900.2022.00062
4. Qu, Z., Zhou, F., Song, X., Ding, R., Yuan, L., Wu, Q.: Temporal enhanced multimodal graph neural networks for fake news detection. IEEE Trans. Comput. Soc. Syst. **11**, 7286–7298 (2024). https://doi.org/10.1109/TCSS.2024.3404921
5. Zhou, D., Ouyang, Q., Lin, N., Zhou, Y., Yang, A.: GS2F: multimodal fake news detection utilizing graph. **24** (2025). https://doi.org/10.1145/3708536
6. Lv, H., Yang, W., Wei, F., Peng, J., Geng, H.: MDF: a dynamic fusion model for multi-modal fake news detection ⋆. **17** (2024)
7. Jing, C., Gao, H., Zhang, X., Gao, T., Zhou, C.: DPSG: dynamic propagation social graphs for multi-modal fake news detection. Inf. Fusion. **113**, 102595 (2025). https://doi.org/10.1016/j.inffus.2024.102595
8. Zhang, T., Yu, E., Shao, Y., Li, S., Hou, S., Sun, J.: Multimodal inverse attention network with intrinsic discriminant feature exploitation for fake news detection. **1** (2025)
9. Hu, L., Zhao, Z., Qi, W., Song, X., Nie, L.: Multimodal matching-aware co-attention networks with mutual knowledge distillation for fake news detection. Inf. Sci. (Ny). **664**, 120310 (2024). https://doi.org/10.1016/j.ins.2024.120310
10. Shen, X., Huang, M., Hu, Z., Cai, S., Zhou, T.: Multimodal fake news detection with contrastive learning and optimal transport. Front. Comput. Sci. **6**, 1473457 (2024). https://doi.org/10.3389/fcomp.2024.1473457
11. Hamed, S.K., Ab Aziz, M.J., Yaakub, M.R.: Fake news detection model on social media by leveraging sentiment analysis of news content and emotion analysis of users' comments. Sensors. **23**, 1748 (2023). https://doi.org/10.3390/s23041748
12. Chen, X., Chen, Y., Liu, Y., Pan, J.: TMEF-BI: trusted multimodal evidential fusion considering behavior information for fake news detection. IEEE Trans. Comput. Soc. Syst., 1–14 (2025). https://doi.org/10.1109/TCSS.2025.3530002
13. Uppada, S.K., Patel, P., Sivaselvan, B.: An image and text-based multimodal model for detecting fake news in OSN's. J. Intell. Inf. Syst. **61**, 367–393 (2023). https://doi.org/10.1007/s10844-022-00764-y
14. Yu, K., Jiao, S., Ma, Z.: Fake news detection based on BERT multi-domain and multi-modal fusion network. Comput. Vis. Image Underst. **252**, 104301 (2025). https://doi.org/10.1016/j.cviu.2025.104301
15. Yang, P., Ma, J., Liu, Y., Liu, M.: Multi-modal transformer for fake news detection. Math. Biosci. Eng. **20**, 14699–14717 (2023). https://doi.org/10.3934/mbe.2023657
16. Liu, Y., Liu, Y., Li, Z., Yao, R., Zhang, Y., Wang, D.: Modality interactive mixture-of-experts for fake news detection. (2025). https://doi.org/10.1145/3696410.3714522.
17. Guo, Y. et al.: Consistency-heterogenity balanced fake news detection via cross-modal matching. IEEE Trans. Artif. Intell. **00**, 1–10 (2025). https://doi.org/10.1109/TAI.2025.3527921
18. Wu, Y., Zhan, P., Zhang, Y., Wang, L., Xu, Z.: Multimodal fusion with co-attention networks for fake news detection. Find. Assoc. Comput. Linguist. ACL-IJCNLP. **2021**, 2560–2569 (2021). https://doi.org/10.18653/v1/2021.findings-acl.226
19. Abdali, S., Shaham, S., Krishnamachari, B.: Multi-modal misinformation detection: approaches, challenges and opportunities. ACM Comput. Surv. **57**, 1–29 (2024). https://doi.org/10.1145/3697349
20. Luvembe, A.M., Li, W., Li, S., Liu, F., Wu, X.: CAF-ODNN: complementary attention fusion with optimized deep neural network for multimodal fake news detection. Inf. Process. Manag. **61**, 103653 (2024). https://doi.org/10.1016/j.ipm.2024.103653

21. Zhang, Y., Ma, J., Jia, Y.: MCAN: Multimodal Cross-Aware Network for Fake News Detection by Extracting Semantic-Physical Feature Consistency. Springer US (2025). https://doi.org/10.1007/s11227-024-06815-1
22. Hamed, S.K., Aziz, M.J.A., Yaakub, M.R.: Improving data fusion for fake news detection: a hybrid fusion approach for unimodal and multimodal data. IEEE Access. **12**, 112412–112425 (2024). https://doi.org/10.1109/ACCESS.2024.3443092
23. Hamed, S.K., Ab Aziz, M.J., Yaakub, M.R.: A review of fake news detection approaches: a critical analysis of relevant studies and highlighting key challenges associated with the dataset, feature representation, and data fusion. Heliyon. **9**, e20382 (2023). https://doi.org/10.1016/j.heliyon.2023.e20382
24. Hamed, S.K., Ab Aziz, M.J., Yaakub, M.R.: A review of fake news detection models: highlighting the factors affecting model performance and the prominent techniques used. Int. J. Adv. Comput. Sci. Appl. **14**, 379–390 (2023). https://doi.org/10.14569/IJACSA.2023.0140742
25. Abduljaleel, I.Q., Ali, I.H.: Deep learning and fusion mechanism-based multimodal fake news detection methodologies: a review. Eng. Technol. Appl. Sci. Res. **14**, 15665–15675 (2024). https://doi.org/10.48084/etasr.7907
26. Njoku, J.N., Caliwag, A.C., Lim, W., Kim, S., Hwang, H.J., Jeong, J.W.: Deep learning based data fusion methods for multimodal emotion recognition. J. Korean Inst. Commun. Inf. Sci. **47**, 79–87 (2022). https://doi.org/10.7840/kics.2022.47.1.79
27. Huang, S.C., Pareek, A., Seyyedi, S., Banerjee, I., Lungren, M.P.: Fusion of medical imaging and electronic health records using deep learning: a systematic review and implementation guidelines. npj Digit. Med. **3**, 136 (2020). https://doi.org/10.1038/s41746-020-00341-z
28. Hao, X., Xu, W., Huang, X., Sheng, Z., Yan, H.: MFUIE: a fake news detection model based on multimodal features and user information enhancement. EAI Endorsed Trans. Scalable Inf. Syst. **12**, 1–13 (2025). https://doi.org/10.4108/eetsis.7517
29. Cui, W., Shang, M.: MIGCL: fake news detection with multimodal interaction and graph contrastive learning networks. Appl. Intell. **55**, 1–23 (2025). https://doi.org/10.1007/s10489-024-05883-3
30. Yang, H., Zhang, J., Zhang, L., Cheng, X., Hu, Z.: MRAN: multimodal relationship-aware attention network for fake news detection. Comput. Stand. Interfaces. **89** (2024). https://doi.org/10.1016/j.csi.2023.103822
31. Lu, Q. et al.: Multimodal dual perception fusion framework for multimodal affective analysis. Inf. Fusion. **115**, 102747 (2025). https://doi.org/10.1016/j.inffus.2024.102747
32. Stepanova, N., Ross, B.: Temporal generalizability in multimodal misinformation detection. In: GenBench 2023 - GenBench 1st Work. Gen. NLP, Proc, pp. 173–184 (2023). https://doi.org/10.18653/v1/2023.genbench-1.6
33. Raza, A., Usmani, R.S.A., Khan, S.U.R.: A hybrid deep learning based fake news detection system using temporal features. Asian Bull. Big Data Manag. **4**, 264–273 (2024). https://doi.org/10.62019/abbdm.v4i02.179
34. Jin, Z., Cao, J., Zhang, Y., Zhou, J., Tian, Q.: Novel visual and statistical image features for microblogs news verification. IEEE Trans. Multimed. **19**, 598–608 (2017). https://doi.org/10.1109/TMM.2016.2617078
35. Wang, J., Mao, H., Li, H.: FMFN: fine-grained multimodal fusion networks for fake news detection. Appl. Sci. **12**, 1093 (2022). https://doi.org/10.3390/app12031093
36. Mu, G., Ju, X., Yan, H., Li, J., Gao, H., Li, X.: An enhanced misinformation detection model based on an improved beluga whale optimization algorithm and cross-modal feature fusion. Biomimetics. **10**, 128 (2025). https://doi.org/10.3390/biomimetics10030128
37. Lin, S.Y., Chen, Y.C., Chang, Y.H., Lo, S.H., Chao, K.M.: Text-image multimodal fusion model for enhanced fake news detection. Sci. Prog. **107**, 1–31 (2024). https://doi.org/10.1177/00368504241292685

Beyond Ratings: A Multi-Source Human-Centered Perspective on Restaurant Selection

Viktor Handrianus Pranatawijaya[1,2](✉), Aslina Baharum[3], Azizzeanna Hassan[1], and Yunli Lee[1,4]

[1] Department of Smart Computing and Cyber Resilience, School of Computing and Artificial Intelligence, Faculty of Engineering and Technology, Sunway University, Petaling Jaya, Selangor, Malaysia
viktorhp@it.upr.ac.id, viktor.p@imail.sunway.edu.my

[2] Department of Informatics Engineering, Faculty of Engineering, University of Palangka Raya, Palangka Raya, Indonesia

[3] The Design School, Faculty of Innovation & Technology, Taylor's University, Subang Jaya, Selangor, Malaysia

[4] Research Centre for Human-Machine Collaboration (HUMAC), Faculty of Engineering and Technology, Sunway University, Petaling Jaya, Selangor, Malaysia

Abstract. Restaurant recommender systems increasingly support dining decisions through ratings, reviews, and contextual information. However, existing approaches remain performance-oriented, prioritizing predictive accuracy while giving limited attention to human-centered factors shaping restaurant selection. This study adopts a multi-source exploratory approach integrating Google Maps restaurant intelligence from 131,007 review records in Bandar Sunway, Malaysia, cross-country pilot surveys from Malaysia and Indonesia, and stakeholder interviews. Descriptive analysis, lightweight review intelligence, reliability assessment, sentiment overview, and thematic synthesis were conducted to examine restaurant-related signals and selection preferences. Findings indicate that restaurant selection extends beyond ratings and geographic proximity, emphasizing contextual, experiential, and trust-related considerations. The study contributes a requirements-grounded human-centered perspective showing how heterogeneous restaurant-related signals may inform restaurant intelligence and future recommendation design. Findings provide insights for more explainable, context-aware, and human-centered restaurant selection support. Future work may operationalize the proposed perspective through advanced explainable and context-aware recommendation mechanisms to support adaptive restaurant intelligence.

Keywords: Restaurant Recommender Systems · Explainable Recommendation · Context-Aware Recommendation · Human-Centered AI · Dining Experience · Review Intelligence

N. Mohd Sharef et al. (Eds.): CAMP 2026, CCIS 3093, pp. 67–81, 2027.
https://doi.org/10.1007/978-981-92-4302-0_6

1 Introduction

Restaurant recommender systems increasingly support dining decisions through ratings, reviews, and contextual information [1, 2]. However, restaurant selection remains highly subjective and is shaped by contextual, emotional, and experiential considerations beyond ratings or geographic proximity [3, 4]. Despite the growing availability of restaurant information, transforming large-scale user-generated content into meaningful restaurant intelligence for human-centered and contextually relevant dining decisions remains challenging.

Recent advances in restaurant recommender systems have improved recommendation quality through machine learning, sentiment analysis, context-aware recommendation, and Location-Based Services (LBS) [5–8]. Online reviews provide restaurant-related intelligence through customer feedback and ratings [9, 10]. Despite these advances, existing approaches remain performance-oriented, prioritizing predictive accuracy while giving limited attention to human-centered factors such as contextual suitability, trust, explainability, and dining experience [3, 11, 12].

This limitation reflects fragmentation in restaurant recommendation research, where review intelligence, contextual recommendation, explainability, and user experience are often examined separately rather than integrated into restaurant selection understanding [13, 14]. Consequently, existing systems may insufficiently capture how individuals evaluate restaurants during dining decisions [15, 16]. Therefore, the objective of this study is to explore and synthesize restaurant-related intelligence, user preferences, and stakeholder perspectives to develop a preliminary human-centered understanding of restaurant selection beyond ratings, geographic proximity, and predictive relevance.

To address this gap, this study adopts a multi-source exploratory approach integrating Google Maps restaurant intelligence in Bandar Sunway, cross-country pilot surveys from Malaysia and Indonesia, and stakeholder interviews. The study contributes a requirements-grounded human-centered perspective integrating review intelligence, contextual suitability, explainability, and experiential dining considerations. Findings provide a foundation for explainable, context-aware, and human-adaptive recommendation design beyond predictive performance evaluation.

2 Related Work

2.1 Restaurant Recommendation and Review Intelligence

Restaurant recommender systems (RRS) support restaurant selection through user preferences and contextual factors. Traditional approaches commonly employ content-based and collaborative filtering to estimate relevance from historical interactions [1, 2], although these methods often face cold-start, sparsity, and contextual limitations [2, 16]. Recent studies increasingly adopt context-aware recommendation, Location-Based Services (LBS), and machine learning to improve personalization through contextual and spatial signals [5, 17–19].

Online restaurant reviews provide restaurant intelligence through customer feedback and ratings [4, 12]. Review analytics and sentiment-based approaches are used to examine factors such as food quality, service, atmosphere, affordability, and customer satisfaction [8, 9, 20]. However, existing approaches remain prediction-oriented, giving limited attention to contextual, experiential, emotional, and trust-related factors shaping restaurant selection [10, 13, 15, 21], highlighting the need for human-centered and contextually grounded restaurant intelligence.

2.2 Human-Centered Recommendation and Restaurant Selection

Recommendation quality increasingly extends beyond predictive accuracy toward transparency, trust, and user experience, particularly in subjective domains such as restaurant selection. Explainability and contextual intelligence improve recommendation effectiveness by supporting user understanding while incorporating situational conditions and preferences [7, 10, 13, 22]. Recent studies increasingly recognize user experience as an integral recommendation component rather than solely a post hoc evaluation outcome [15, 23].

Restaurant selection further involves contextual, experiential, emotional, and social considerations beyond algorithmic relevance [3, 12]. However, explainability, contextual recommendation, review intelligence, and user experience are often examined separately rather than integrated into restaurant selection understanding [15, 16]. Despite these advances, prior studies remain fragmented across recommendation accuracy, review analytics, explainability, contextual recommendation, and user experience. Limited research has integrated heterogeneous restaurant-related signals into a human-centered understanding of restaurant selection. This study therefore adopts a multi-source exploratory approach combining review intelligence, cross-country user preferences, and stakeholder perspectives to develop a preliminary human-centered perspective.

3 Methodology

3.1 Research Design

This study adopted an exploratory multi-source design integrating Google Maps restaurant intelligence from Bandar Sunway, Malaysia, a cross-country pilot survey involving respondents from Malaysia and Indonesia, and stakeholder interviews to better understand restaurant selection through review intelligence, user preferences, and contextual stakeholder insights.

A multi-source exploratory design was adopted because restaurant selection represents a complex human-centered phenomenon involving behavioral, contextual, experiential, and social dimensions that may not be adequately captured through a single source of evidence. Restaurant reviews provide large-scale behavioral signals and customer perceptions, pilot surveys capture user preferences and recommendation priorities, while stakeholder interviews contribute contextual and experiential insights from real-world dining environments. The integration of these complementary sources enables a more comprehensive understanding of restaurant selection beyond rating-centric evaluation [24, 25].

3.2 Restaurant Review Intelligence and Data Preparation

Restaurant data were collected from publicly accessible Google Maps through automated web scraping of dining establishments in Bandar Sunway, Malaysia. Data collection was completed between April and May 2026 prior to manuscript preparation and analysis. Retrieved data included restaurant categories, ratings, review counts, and customer reviews, followed by cleaning and duplicate filtering. Exploratory Data Analysis (EDA) and descriptive analysis examined restaurant characteristics, rating distributions, review activity, and restaurant-related signals [21].

For lightweight review intelligence analysis, dining-related themes (e.g., food quality, service, atmosphere, affordability, convenience, and waiting time) were identified through keyword grouping informed by prior literature and stakeholder interview insights [3, 8, 9, 20]. Sentiment classification was performed using the Valence Aware Dictionary and Sentiment Reasoner (VADER), with reviews classified using standard compound score thresholds ($\geq$0.05 positive, ≤ -0.05 negative, and values between -0.05 and 0.05 neutral). Findings were interpreted as indicative signals given the exploratory nature of the analysis. As VADER is optimized for English text, sentiment interpretation should be approached cautiously for localized or mixed-language reviews common in Malaysia. Future work may incorporate transformer-based multilingual approaches, including aspect-based sentiment analysis.

3.3 Cross-Country Restaurant Selection Preferences

A cross-country pilot survey examined restaurant selection preferences among respondents from Malaysia and Indonesia. Eligible respondents were aged 18 years or older with prior experience using digital restaurant platforms. Following data screening, 57 valid responses were retained.

Five constructs were examined: (1) Emotional Considerations [3, 13], (2) Contextual and Location-Based Factors [7, 13], (3) Explainability and Transparency [10, 11], (4) User Experience (UX) [15, 23], and (5) Sustainability Considerations [12, 15]. Descriptive country-level comparisons were conducted to identify preliminary recommendation preferences. Given the exploratory scope and unequal country distribution, findings were not intended to be statistically generalizable. Survey quality was assessed using Cronbach's alpha and Corrected Item-Total Correlation (CITC) [26, 27]. Harman's single-factor test indicated no substantial common method bias, with the first extracted factor below the 50% threshold [28].

3.4 Stakeholder Perspectives on Restaurant Selection

To complement quantitative findings, exploratory stakeholder interviews were conducted with restaurant stakeholders from Malaysia and Indonesia. Participants were purposively selected to provide diverse operational perspectives across restaurant and coffee shop settings, including owners, managers, supervisors, and food and beverage practitioners.

Semi-structured interviews were conducted to provide contextual understanding beyond survey findings by exploring emotional, contextual, experiential, and sustainability-related influences on restaurant selection, alongside perceptions of online reviews, affordability, trust, recommendation transparency, and customer expectations. The six stakeholders included restaurant owners, café operators, and food and beverage practitioners from Malaysia and Indonesia.

Interviews lasted approximately 20–45 min and were audio-recorded with participant consent. Interview data were analyzed thematically to identify recurring patterns related to emotional, contextual, experiential, and trust-related considerations in restaurant decision-making. Findings were interpreted as exploratory stakeholder insights intended to contextualize the proposed recommendation perspective rather than serve as definitive qualitative validation.

4 Findings and Discussion

4.1 Restaurant Review Intelligence in Bandar Sunway

This section presents exploratory findings from Google Maps restaurant data in Bandar Sunway, Malaysia, to examine restaurant characteristics and customer engagement relevant to recommendation. Following preprocessing, 326 valid restaurants were retained, including 323 with detailed customer reviews. As summarized in Table 1, the dataset comprised 80 restaurant categories and 131,007 review records, with an average rating of 4.31 and a median rating of 4.40.

Figure 1 illustrates restaurant rating and review activity patterns. Most restaurants were concentrated within the high-rating range (4.0–5.0), suggesting limited discriminatory value from ratings alone. Review activity exhibited a right-skewed distribution, indicating uneven customer engagement and the complementary role of review volume in restaurant evaluation. Although restaurants represented the most frequent category, shopping malls generated the highest review activity.

Table 1. Descriptive Statistics of Restaurants in Bandar Sunway

Indicator	Value
Number of valid restaurant records	326
Restaurants with detailed reviews	323
Number of restaurant categories	80
Average restaurant rating	4.31
Median restaurant rating	4.40
Total review records	131,007
Median review count per restaurant	282
Most frequent category	Restaurant
Highest review-volume category	Shopping mall

The concentration of highly rated restaurants may partially reflect platform-related positivity bias, where highly reviewed or visible establishments attract greater customer

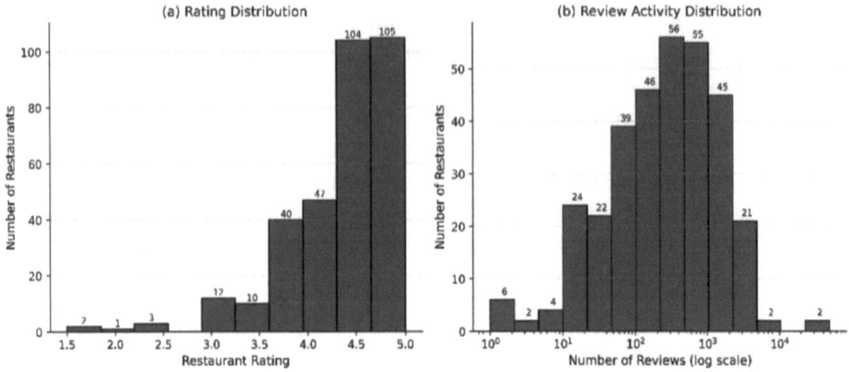

Fig. 1. (a) Restaurant Rating Distribution and (b) Review Activity Distribution.

engagement. This observation aligns with prior restaurant review studies suggesting that online review platforms exhibit positivity bias, where highly visible or popular establishments accumulate disproportionately favorable ratings and customer engagement [3, 4].

4.1.1 Dining-Related Theme and Sentiment Intelligence

Lightweight review intelligence analysis identified recurring dining-related themes in Google Maps reviews. As shown in Fig. 2, food quality emerged as the most frequent theme (74.60%), followed by service (64.80%) and atmosphere (55.01%). Affordability (41.23%), waiting time (17.98%), and convenience-related factors (6.85%) appeared less frequently yet remained relevant, suggesting that dining evaluation extends beyond ratings toward experiential and contextual considerations. These findings align with prior studies identifying food quality, service, and atmosphere as dominant factors influencing customer evaluation [8, 9, 20].

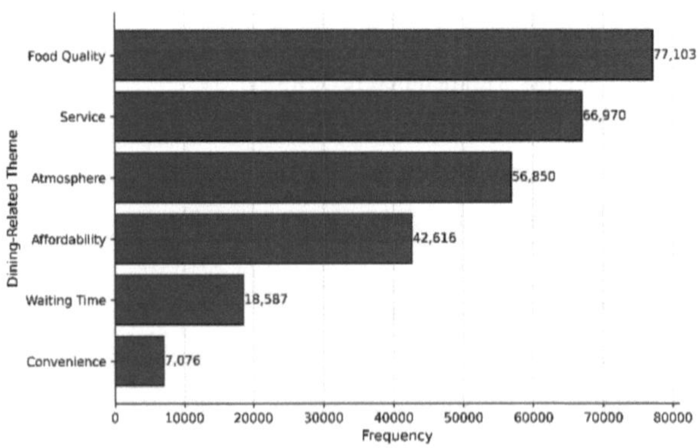

Fig. 2. Most Frequent Dining-Related Themes Extracted from Google Maps Reviews

As summarized in Table 2, sentiment analysis indicated positive perceptions (84.84% positive reviews), with atmosphere (89.24%) and service (87.99%) showing the highest positive sentiment, whereas waiting time showed the highest negative sentiment (10.94%), highlighting service delays as dissatisfaction.

Table 2. Positive and Negative Sentiment Distribution Across Dining-Related Aspects

Aspect	Positive (%)	Negative (%)
Atmosphere	89.24	6.60
Service	87.99	7.99
Food Quality	87.01	8.64
Affordability	86.87	8.80
Waiting Time	85.22	10.94

4.2 Cross-Country Restaurant Selection Insights

This section presents preliminary user insights from cross-country pilot surveys in Malaysia (n = 19) and Indonesia (n = 38). Following expert validation and refinement, 57 valid responses were retained, predominantly Indonesian students (66.7%; aged 18–24 = 87.7%; students = 77.2%). Given the exploratory nature and unequal sample sizes, findings should be interpreted as hypothesis-generating rather than statistically generalizable. Findings align with prior recommendation studies emphasizing transparency, contextual suitability, and user experience as influences on recommendation acceptance beyond predictive accuracy [10, 14, 23].

4.2.1 Restaurant Decision-Making Preferences

As summarized in Table 3, contextual and location-based factors received the highest evaluation across both countries, followed by explainability and user experience dimensions. These findings suggest that restaurant decision-making extends beyond numerical ratings alone. Indonesian respondents reported slightly higher construct means, although country-level differences should be interpreted cautiously given the exploratory nature of the study.

Table 3. Cross-Country Comparison of Restaurant Decision-Making Preferences

Construct	Malaysia Mean	Indonesia Mean	Overall Mean	Overall Interpretation
Emotional Considerations	4.01	4.20	4.13	High

(continued)

Table 3. (*continued*)

Construct	Malaysia Mean	Indonesia Mean	Overall Mean	Overall Interpretation
Contextual and Location-Based Factors	4.24	4.42	4.36	Very High
Explainability and Transparency	4.02	4.45	4.31	Very High
User Experience (UX)	4.09	4.43	4.32	Very High
Sustainability Considerations	3.99	4.15	4.10	High

4.2.2 Recommendation Importance Factors

The second pilot survey examined respondents' perceived importance of recommendation-related factors during restaurant selection. As summarized in Table 4, explainability factors received the highest overall evaluation (M = 4.38), followed by contextual factors (M = 4.33) and user experience factors (M = 4.32). Emotional (M = 4.00) and sustainability factors (M = 4.15) remained positively perceived but comparatively secondary.

Table 4. Cross-Country Comparison of Recommendation Importance Factors

Factor	Malaysia Mean	Indonesia Mean	Overall Mean	Overall Interpretation
Emotional Factors	3.74	4.13	4.00	High
Contextual Factors	4.22	4.38	4.33	Very High
Explainability Factors	4.32	4.41	4.38	Very High
User Experience (UX) Factors	4.22	4.37	4.32	Very High
Sustainability Factors	4.07	4.19	4.15	High

These findings suggest that respondents value recommendation systems that are understandable, contextually relevant, and trustworthy, particularly in unfamiliar dining situations. Across both countries, transparency, contextual suitability, and experiential considerations consistently emerged as important factors beyond predictive relevance alone.

4.2.3 Reliability and Common Method Bias Assessment

Pilot reliability was assessed using Cronbach's alpha and Corrected Item-Total Correlation (CITC). As summarized in Table 5, the restaurant decision-making survey demonstrated acceptable to good reliability (α = 0.749–0.891), while the recommendation importance survey showed good to excellent reliability (α = 0.839–0.933), with final retained items demonstrated acceptable reliability.

Table 5. Reliability Assessment of Cross-Country Pilot Surveys

Survey	Construct	Final Items	Cronbach's Alpha	Reliability
Decision-Making	Emotion-Based	6	0.749	Acceptable
Decision-Making	Contextual & LBS	7	0.804	Good
Decision-Making	Explainability	8	0.891	Good
Decision-Making	User Experience	8	0.866	Good
Decision-Making	Sustainability	8	0.880	Good
Recommendation Importance	Emotion	8	0.887	Good
Recommendation Importance	Contextual	8	0.866	Good
Recommendation Importance	Explainability	8	0.912	Excellent
Recommendation Importance	User Experience	8	0.839	Good
Recommendation Importance	Sustainability	8	0.933	Excellent

To assess potential common method bias, Harman's single-factor test was conducted for both self-reported surveys. The first extracted factor explained 32.94% of total variance for the restaurant decision-making survey and 44.47% for the recommendation importance survey, both remaining below the 50% threshold, indicating that common method bias was unlikely to substantially influence the findings.

4.3 Exploratory Stakeholder Interview Insights

Exploratory stakeholder interviews provided contextual understanding regarding restaurant decision-making and dining experiences. Across six interviews involving restaurant stakeholders from Malaysia and Indonesia, recurring patterns consistently emphasized emotional experiences, contextual suitability, trust in digital information, and experiential dining considerations.

Restaurant selection was frequently described as extending beyond food quality or ratings alone, involving emotional attachment, comfort, nostalgia, and memorable experiences. Several stakeholders emphasized that repeat visits were often associated with familiarity and long-term emotional relationships with particular places. As one stakeholder explained,

> Customers often return not only because of the food, but also because of memories and emotional attachment developed over time (Restaurant stakeholder, P05, Indonesia).

Restaurant suitability was additionally described as dependent on contextual and situational needs, including work, study, meetings, social gatherings, accessibility, and time constraints. Participants emphasized that restaurant preferences frequently shift according to dining purpose and context. As one stakeholder explained,

> Customers choose different places depending on what they want to do. Some come to work or study, while others prefer a place to relax or gather with friends (Restaurant stakeholder, P06, Malaysia).

Trust in online reviews, customer feedback, restaurant photographs, and digital information sources was consistently highlighted as important for reducing uncertainty during restaurant selection. A stakeholder from Indonesia observed,

> Customers usually check reviews, ratings, and food photos before deciding (P02, Indonesia).

Several participants additionally expressed concerns regarding manipulated reviews and expectation mismatch. One stakeholder noted,

> Online reviews are sometimes misleading because expectations created online may not match the actual dining experience (Restaurant stakeholder, P03, Malaysia),

Such concerns underscore the importance of recommendation transparency and trustworthy review information in reducing expectation mismatch.

Stakeholders further emphasized that restaurant recommendation extends beyond numerical ratings or popularity indicators, involving authenticity, affordability, familiarity, consistency, and long-term customer relationships. As another stakeholder reflected,

> Some customers return after many years because they remember the place, the taste, and the experience they had before (Restaurant stakeholder, P05, Indonesia).

Collectively, these findings suggest that restaurant recommendation may benefit from moving beyond rating-centric approaches toward more human-centered support aligned with emotional experiences, contextual needs, trustworthiness, and authentic dining behavior. These stakeholder insights are broadly consistent with prior human-centered recommendation research emphasizing emotional relevance, contextual suitability, trust, and experiential considerations in subjective decision-making domains such as restaurant selection [3, 12, 13, 15].

4.4 Knowledge Synthesis Across Multi-Source Evidence

To develop the proposed human-centered perspective, findings from restaurant reviews, pilot surveys, and stakeholder interviews were synthesized through an iterative insight integration process. Restaurant reviews contributed large-scale restaurant intelligence, including dining themes and sentiment patterns. Pilot surveys provided evidence regarding user preferences and recommendation priorities. Stakeholder interviews supplied contextual explanations regarding emotional attachment, trust, affordability, and dining experiences. Insights that consistently emerged across the three sources were consolidated into a preliminary human-centered restaurant selection perspective.

4.5 Implications for Human-Centered Restaurant Intelligence

The findings suggest that restaurant selection may benefit from integrating signals beyond ratings and ranking mechanisms alone. Review intelligence contributed evidence regarding dominant dining themes and customer perceptions. Pilot surveys highlighted the importance of contextual suitability, explainability, and user experience. Stakeholder interviews further emphasized emotional attachment, trust, affordability, and situational dining needs. Together, these complementary sources provide converging evidence supporting a more human-centered perspective for restaurant selection and future restaurant recommendation design.

From an implementation perspective, contextual suitability, emotional attachment, and trust-related considerations may be operationalized through review analytics, aspect-based sentiment analysis, contextual preference modeling, and explainability mechanisms to support more human-centered restaurant recommendation design.

Table 6 summarizes how each data source contributed distinct yet complementary insights. Rather than functioning independently, these sources collectively informed the development of the proposed human-centered restaurant selection perspective.

Table 6. Contribution of Each Data Source

Data Source	Main Contribution
Google Maps Reviews	Restaurant intelligence, dining themes, sentiment signals
Cross-Country Pilot Survey	User preferences, recommendation priorities
Stakeholder Interviews	Contextual, experiential, emotional, and trust-related insights

Figure 3 illustrates how insights extracted from restaurant reviews, pilot surveys, and stakeholder interviews were synthesized into a preliminary human-centered restaurant selection perspective. Rather than representing a finalized recommendation architecture, the framework demonstrates how complementary evidence from multiple sources may be integrated to support future restaurant recommendation design.

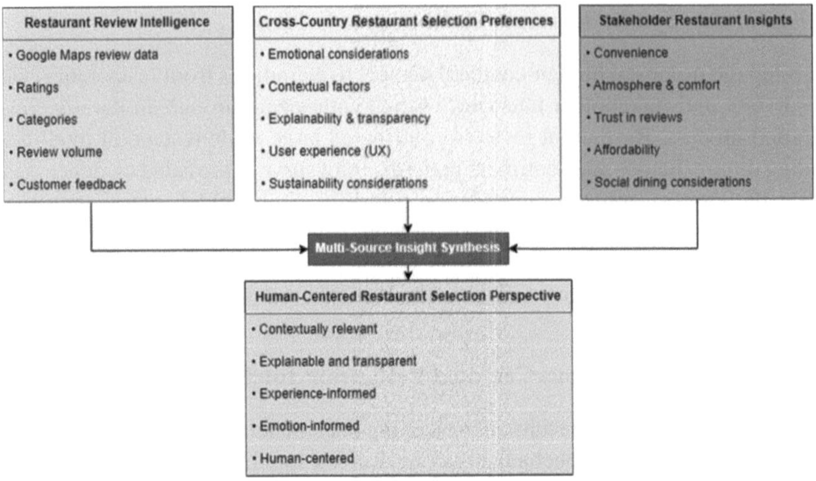

Fig. 3. Multi-source framework for human-centered restaurant selection.

The proposed perspective comprises three interacting layers. The Restaurant Intelligence Layer incorporates restaurant-level signals from Google Maps, including ratings, review activity, and dining-related themes. The User Preference Layer reflects contextual suitability, explainability, and user experience dimensions identified through pilot surveys. The Experiential Dining Layer captures socially and situationally embedded considerations, including convenience, affordability, atmosphere, trust in reviews, and emotional attachment derived from stakeholder interviews. Rather than operating independently, these layers interact during restaurant evaluation and selection, providing an initial requirements-grounded foundation for future restaurant recommendation design integrating contextual, experiential, and explainability-oriented user needs.

5 Conclusion

This study explored restaurant selection through a multi-source exploratory approach integrating Google Maps restaurant intelligence, cross-country pilot surveys, and stakeholder interviews. Findings suggest that restaurant selection extends beyond traditional relevance signals, highlighting contextual suitability, explainability, emotional attachment, affordability, trust in digital information, and socially embedded dining behavior. Review intelligence, user preferences, and stakeholder perspectives consistently emphasized that restaurant evaluation involves experiential and contextual considerations beyond ratings or algorithmic relevance alone.

The study contributes a preliminary human-centered perspective by integrating fragmented restaurant-related signals into a more coherent understanding of restaurant selection. Rather than emphasizing predictive optimization alone, the study provides requirements grounding for more explainable, context-aware, and human-adaptive restaurant recommendation design informed by real-world dining behavior. The contribution emerges from the synthesis of large-scale review intelligence, cross-country user preferences, and stakeholder perspectives, demonstrating how complementary evidence sources may jointly inform restaurant recommendation design.

Several limitations should be acknowledged. The exploratory pilot surveys, unequal country distribution, and relatively small sample size may limit broader generalizability, while stakeholder interviews were intended to provide contextual richness rather than thematic saturation. The study should therefore be interpreted as an exploratory human-centered restaurant intelligence study intended to inform future recommendation design rather than predictive performance evaluation. Future work may operationalize the proposed perspective through richer review understanding, contextual preference modeling, and human-centered recommendation evaluation to support adaptive restaurant intelligence. Future research may further strengthen sentiment interpretation through multilingual transformer-based approaches and selective manual validation.

Acknowledgments. AI-assisted language support was used solely for language refinement and copy editing. Research design, analysis, interpretation, and manuscript decisions remained the independent responsibility of the authors.

Ethics Approval. This study received ethical approval from the Sunway University Research Ethics Committee (Approval No: 2026/REC0024). Participation was voluntary with informed consent, while Google Maps data were collected from publicly accessible sources without retaining personally identifiable information.

Appendix

See Table 7.

Table 7 Example Keywords Used for Lightweight Review Intelligence Analysis.

Theme	Example Keywords
Food Quality	food, taste, tasty, delicious, flavor, menu, dish
Service	staff, service, friendly, waiter, helpful, hospitality
Atmosphere	cozy, ambience, comfortable, environment, relaxing
Affordability	price, affordable, value, cheap, expensive, budget, cost
Convenience	parking, location, nearby, easy access, accessible
Waiting Time	queue, waiting, crowded, line, delay, slow service, long wait

References

1. Adomavicius, G., Tuzhilin, A.: Toward the next generation of recommender systems: a survey of the state-of-the-art and possible extensions. IEEE Trans. Knowl. Data Eng. **17**(6), 734–749 (2005). https://doi.org/10.1109/TKDE.2005.99

2. Burke, R.: Hybrid web recommender systems. In: The Adaptive Web, pp. 377–408. Springer, Berlin, Heidelberg (2007). https://doi.org/10.1007/978-3-540-72079-9_12
3. Baek, J., Choe, Y.: Differential effects of the valence and volume of online reviews on customer share of visits: the case of US casual dining restaurant brands. Sustainability. **12**(13), p. 5408 (2020). https://doi.org/10.3390/su12135408
4. Yalcinkaya, B., Just, D.R.: Comparison of customer reviews for local and chain restaurants: multilevel approach to Google reviews data. Cornell Hosp. Q. **64**(1), 63–73 (2023). https://doi.org/10.1177/19389655221102388
5. Idalisa, N., Hazhar, M.H.M., Muslim, N., Albashah, N.L.S.: Restaurant recommendation system in Malaysia using machine learning approach. In: Tallon-Ballesteros, A.J. (ed.) Frontiers in Artificial Intelligence and Applications, pp. 319–325. IOS Press (2024). https://doi.org/10.3233/FAIA241357
6. Khot, S., Dhumal, R., Singh, T.P.: Location based restaurant recommendation system (RECOSYS) using sentimental analysis. In: Proceeding of International Conference on Integration of Computational Intelligent System (ICICIS), pp. 1–4. IEEE (2023). https://doi.org/10.1109/ICICIS56802.2023.10430251
7. Feremans, L., Verachtert, R., Goethals, B.: A neighbourhood-based location- and time-aware recommender system. In: Proceeding of 5th Workshop on Online Recommender Systems and User Modeling (ORSUM), CEUR Workshop Proceeding, vol. 3303, Seattle, WA, USA (2022)
8. Kandhro, I.A., Ali, F., Uddin, M., Kehar, A., Manickam, S.: Exploring aspect-based sentiment analysis: an in-depth review of current methods and prospects for advancement. Knowl. Inf. Syst. **66**(7), 3639–3669 (2024). https://doi.org/10.1007/s10115-024-02104-8
9. Gupta, V., Rattan, P.: Advancing sentiment analysis in restaurant reviews through unsupervised machine learning algorithms. Int. J. Intell. Eng. Syst. **17**(4), 1108–1121 (2024). https://doi.org/10.22266/IJIES2024.0831.82
10. Zhang, Y., Chen, X.: Explainable recommendation: a survey and new perspectives. Found. Trends Inf. Retr. **14**(1), 1–101 (2020). https://doi.org/10.1561/1500000066
11. Tintarev, N., Masthoff, J.: Explaining recommendations: design and evaluation. In: Recommender Systems Handbook, pp. 353–382. Springer, Boston, MA (2015). https://doi.org/10.1007/978-1-4899-7637-6_10
12. Klouvidaki, M.I., Antonopoulos, N., Styliaras, G.D., Kanavos, A.: Virtual restaurants: customer experience keeps their businesses alive. Information. **14**(7), 406 (2023). https://doi.org/10.3390/info14070406
13. Polignano, M., Narducci, F., de Gemmis, M., Semeraro, G.: Towards emotion-aware recommender systems: an affective coherence model based on emotion-driven behaviors. Expert Syst. Appl. **170**, 114382 (2021). https://doi.org/10.1016/j.eswa.2020.114382
14. Govea, J., Gutierrez, R., Villegas-Ch, W.: Transparency and precision in the age of AI: evaluation of explainability-enhanced recommendation systems. Front. Artif. Intell. **7** (2024). https://doi.org/10.3389/frai.2024.1410790
15. Ho, S.-H., Chang, C.-J.: User experience of restaurant recommendation systems with different service modes. In: Proceeding of IEEE Gaming, Entertainment, and Media Conference (GEM), pp. 1–4. IEEE (2025). https://doi.org/10.1109/GEM66882.2025.11155235
16. Jannach, D., Resnick, P., Tuzhilin, A., Zanker, M.: Recommender systems beyond matrix completion. Commun. ACM. **59**(11), 94–102 (2016). https://doi.org/10.1145/2891406
17. Keya, A.J., Arpona, S.A., Kabir, M.M., Mridha, M.F.: Recurrent ALBERT for recommendation: a hybrid architecture for accurate and lightweight restaurant recommendations. Cogn. Comput. Syst. **5**(4), 265–279 (2023). https://doi.org/10.1049/ccs2.12090
18. Khaki, M., Dehghani, F.: Unveiling latent structures personalized restaurant recommendations via machine learning. Int. J. Artif. Intell. **23**(1), 155–168 (2025)

19. Prakruthi, M.K., Chaitra, M., Shashidhar, A.S., Suchith Kumar, G.M., Vinay, M.V.: Restaurant feasibility study and behavioural based restaurant recommendation. In: Proceeding of the International Conference on Intelligent and Innovative Technologies in Computing, Electrical and Electronics (IITCEE), pp. 1–4. IEEE (2024). https://doi.org/10.1109/IITCEE59897.2024.10467783
20. Kalaivani, K.S., et al.: Improving the restaurant recommendation performance using sentiment analysis. In: Proceeding of the International Conference on Inventive Research in Computing Applications (ICIRCA), pp. 296–300. IEEE (2023). https://doi.org/10.1109/ICIRCA57980.2023.10220869
21. Philip, J., et al.: Smart dining insights using automated food review analysis and visualization on Google maps. In: Proceeding of the International Conference on Computing Methodologies and Communication (ICCMC), pp. 1909–1914. IEEE (2025). https://doi.org/10.1109/ICCMC65190.2025.11140683
22. Kim, D., Li, Q., Jang, D., Kim, J.: AXCF: Aspect-based collaborative filtering for explainable recommendations. Expert Syst. **41**(8) (2024). https://doi.org/10.1111/exsy.13594
23. Rana, A., Sanner, S., Bouadjenek, M.R., Di Carlantonio, R., Farmaner, G.: User experience and the role of personalization in critiquing–based conversational recommendation. ACM Trans. Web. **18**(4), 1–21 (2024). https://doi.org/10.1145/3597499
24. Creswell, J.W., Creswell, J.D.: Research Design: Qualitative, Quantitative, and Mixed Methods Approaches, 5th edn. SAGE Publications, Thousand Oaks, CA (2018)
25. Parija, S.C., Kate, V.: Thesis, dissertation and project. In: Thesis Writing for Master's and Ph.D. Program, pp. 3–10. Springer, Singapore (2018). https://doi.org/10.1007/978-981-13-0890-1_1
26. Hair, J.F., Black, W.C., Babin, B.J., Anderson, R.E.: Multivariate Data Analysis. Cengage, Boston, MA (2019)
27. Pallant, J.: SPSS Survival Manual. Routledge, London (2020). https://doi.org/10.4324/9781003117452
28. Podsakoff, P.M., MacKenzie, S.B., Lee, J.-Y., Podsakoff, N.P.: Common method biases in behavioral research: a critical review of the literature and recommended remedies. J. Appl. Psychol. **88**(5), 879–903 (2003). https://doi.org/10.1037/0021-9010.88.5.879

Digital Audio Poisoning Via MFCC-Space Bilevel Optimization

Haslizatul Mohamed Hanum(✉) and Muhammad Hazim bin Abdul Rahman

Faculty of Computer and Mathematical Sciences, Universiti Teknologi MARA (UiTM), Shah Alam, Malaysia
haslizatul@uitm.edu.my, mdhazimrahman@gmail.com

Abstract. Recent advances in generative and discriminative audio models have intensified concerns about unauthorized use of music in training datasets and the resulting ethical implications for creative work. While unlearnable data and audio poisoning have been proposed as defensive mechanisms, most existing methods target deep end-to-end architectures, while far less is known about how classical MFCC-based music classifiers behave under large-scale feature-space poisoning. This study investigates an MFCC-centric adversarial pipeline against a Linear SVM genre classification victim model trained on 40-dimensional MFCC mean–standard deviation features from the Free Music Archive (Rock vs. Electronic). Empirically, the MFCC with Linear SVM classifier shows surprising global robustness, shown with 25–75% of training MFCC vectors poisoned as the margin-based objective did not reduce test accuracy, but manages to substantially reduce signed margins and flips labels for many targeted points. These results show that MFCC-space bilevel poisoning in a low-dimensional linear setting can be locally active yet globally benign, positioning this MFCC-based pipeline as a lightweight analogue to more complex unlearnable-music frameworks and motivating future work on end-to-end poisoning.

Keywords: Machine Learning · Adverserial Audio · Data Poisoning · Bilevel Optimization

1 Introduction

Recent work shows that modern audio systems are vulnerable to poisoning and adversarial perturbations, but most evidence comes from deep learning and waveform-level attacks. This leaves it unclear how these threats translate to classical models based on Mel-Frequency Cepstral Coefficients (MFCCs), a standard feature representation in audio processing that approximates human auditory perception, and feature-space poisoning [4, 6, 10, 15]. Little is known about how classical MFCC-based music classifiers behave under large-scale feature-space poisoning. This study uses 40-dimensional MFCC mean–standard deviation features to classify music genres in the Free Music Archive (FMA) small dataset [3], focusing on binary classification between the Rock and the Electronic tracks. Each audio track is represented using 20 MFCC coefficients

N. Mohd Sharef et al. (Eds.): CAMP 2026, CCIS 3093, pp. 82–94, 2027.
https://doi.org/10.1007/978-981-92-4302-0_7

per frame, summarized through global mean and standard deviation, then standardized using a training-set scaler. A Linear Support Vector Machine (SVM) is selected as the victim classifier due to its interpretability, efficiency, and common use in MFCC-based pipelines, especially in resource-constrained environments.

In this study, a margin-based outer objective is utilized to probe and perturb the local decision geometry of a Linear SVM in a low-dimensional MFCC feature space by applying dataset-level feature-space poisoning directly in MFCC space using bilevel optimization. Poisoning is performed on 25%, 50%, and 75% of training vectors to test the classifier's robustness at substantial poisoning fractions. The inner optimization retrains the Linear SVM, while the outer optimization uses Simultaneous Perturbation Stochastic Approximation (SPSA) to generate perturbations that reduce classification performance or manipulate decision margins. By analyzing both global model performance and local margin distributions, the study aims to determine whether MFCC-based classifiers are vulnerable to feature-space poisoning and how poisoning-induced margin changes relate to classifier robustness. This thesis addresses that gap by systematically studying MFCC-space bilevel poisoning and its relationship to model performance and waveform-level attacks in a Linear SVM-based genre classification setting.

The main contributions of this study are as follows:

1. The empirical results demonstrate that margin-based poisoning can successfully manipulate local decision geometry and reduce margins of targeted samples, while having minimal impact on overall classification performance.
2. The findings provide evidence that MFCC-based feature representations combined with linear classifiers exhibit strong robustness against large-scale feature-space poisoning, offering insights into the security characteristics of classical audio classification pipelines.

2 Related Work

Mel-Frequency Cepstral Coefficients (MFCCs) remain a standard representation in speech and audio processing. Typically, 13 to 40 MFCCs per frame are retained. For classification, many systems compute global statistics such as mean and standard deviation per coefficient, yielding a fixed-length feature vector per utterance or track. Even with the rise of deep learning, MFCC-based pipelines with classical models continue to be effective in many audio tasks.

Traditional ML, including SVM, Random Forest, and XGBoost, with deep architectures, including CNN, Long Short-Term Memory (LSTM), and Gated Recurrent Unit (GRU), were compared for music genre classification on GTZAN, using STFT-based features, MFCC, and mel-spectrograms [14]. While deep models often achieve the best performance when data are abundant, SVMs and other classical models with MFCCs can be effective on smaller datasets and offer interpretability [14]. Optimized MFCC-SVM combinations, integrated into a majority-voting scheme, can reach to roughly 97% accuracy across several audio classification tasks [13].

MFCCs also play a central role in protective frameworks like WaveFuzz, which perturb audio to manipulate MFCC features [5]. Studying poisoning directly in MFCC space against a model that uses these features can help contrast with deep-model poisoning where most poisoning successes are targeted on deep networks [1, 4, 16]. A classical ML model of Linear SVM trained with MFCCs may provide a complementary perspective on robustness in low-capacity, low-dimensional regimes.

Given the evidences discussed previously, the choice to use a 40-Dimensional MFCC mean-standard deviation features with Linear SVM classifier can be motivated by several factors. Interpretability and analysability can be achieved with a low-dimensional, linear decision boundary that permits detailed analysis of signed margins, margin shifts under poisoning, and visualizations such as Principal Component Analysis (PCA) scatter plots and margin Cumulative Distribution Functions (CDF).

Using the selected MFCC features also enable computational tractability for bilevel optimization. Bilevel poisoning requires many retraining of the victim model. Linear SVMs trained on 40-Dimensional features can be retrained quickly, making SPSA-based poisoning feasible under limited computational resource.

2.1 Bilevel Formulations

Many poisoning and robust-learning problems can be formulated as bilevel optimization where the inner loop trains a model on possibly poisoned data to minimize training loss, and the outer loop adjusts the data or hyperparameters to optimize an attack or defence objective.

This perspective underlies classic clean-label poisoning methods such as the Bulls-eye Polytope or the CLPA [1, 16], as well as certified defences that sanitize data via generative models [8]. Recent advances in bilevel optimization propose fully first-order and functional methods for large-scale ML [9, 11].

In the specific case of a Linear SVM with hinge loss, the signed margin $m_i = y_i f(x_i)$ for an example (x_i, y_i) is a natural quantity to target in a poisoning attack. Classical SVM theory directly links larger margins to better generalization and robustness of the decision boundary [2, 12]. Reducing the signed margin of a point moves it closer to or across the separating hyperplane, increasing its vulnerability to misclassification. Consequently, many poisoning and adversarial formulations implicitly or explicitly focus on feature-space manipulations that alter geometric relationships among points in ways that reduce margins or bring target examples into regions of confusion [1, 16]. In this study, a margin-based outer objective is therefore a principled way to probe and perturb the local decision geometry of a Linear SVM in a low-dimensional MFCC feature space.

In many audio pipelines, especially those using existing libraries for MFCC and SVMs, the end-to-end system is best treated as a black box. In such situations, zeroth-order methods that approximate gradients from function evaluations, rather than analytic derivatives, are more practical. Simultaneous Perturbation Stochastic Approximation (SPSA) is a popular zeroth-order algorithm that perturbs all parameters simultaneously in random directions by evaluating the objective at two perturbed points $(z + c\Delta, z - c\Delta)$, which the directional derivatives are then estimated from these two values. The algorithm then updates parameters using a stochastic gradient-descent-like rule. SPSA has been adopted in black-box adversarial settings where only loss or confidence scores are observable, not gradients [6, 7].

In this study, SPSA is used for MFCC-space bilevel poisoning where the inner problem is SVM training, and the outer objective is evaluated by retraining on modified

MFCC features. SPSA treats the mapping from MFCC vectors to outer objective values as a black box. This choice is justified by its practicality under limited resources and alignment with black-box threat models, where gradients are not available.

3 Methodology

This study employs the Free Music Archive (FMA) dataset as the primary source of music files, metadata, and precomputed feature tables [3]. The FMA full dataset, containing 106,574 untrimmed MP3 tracks across 161 unbalanced genres, is used for exploratory data analysis (EDA), while the FMA small subset, consisting of 8,000 balanced tracks across 8 genres, is selected for the main experimental pipeline due to its manageable size and suitability for repeated model retraining in poisoning experiments. Metadata tables are loaded using the official FMA helper utilities to obtain top-level genre labels, subset indicators, and official train, validation, and test split assignments. Tracks are filtered based on the selected target genres identified during EDA, and separate metadata DataFrames are created for each split. For feature extraction, each track is loaded using librosa with a sampling rate of 22.05 kHz and converted to mono to ensure consistency. MFCC features are then extracted per frame using 20 coefficients, with parameters n_fft = 1024, hop_length = 512, and a Hann window. This produces a 20 × T MFCC matrix for each track, where T represents the number of frames, which is then summarized using global mean and standard deviation to form a fixed-length 40-dimensional MFCC feature vector. These features serve as the input for the baseline Linear SVM classifier and the MFCC-space poisoning experiments. Any corrupted, missing, or incompatible audio files are automatically skipped during preprocessing to ensure robust feature matrix construction without interrupting the overall pipeline.

The pipeline is separated into 3 phases (Fig. 1). The first phase will cover the feature extraction and the baseline classifier (Sect. 3.1), followed by the MFCC-space bilevel formulation for a single target track (Sect. 3.2). The second phase is then repeated for each randomly selected track, and the resulting perturbations are saved into the third phase (Sect. 3.3).

3.1 MFCC Feature Extraction and Baseline Classifier

After identifying the target genres, the small subset of FMA is utilized and filtered to reduce computation load for repeated model retraining and bilevel optimization. Using the metadata provided by FMA, the training, validation, and test splits containing track IDs and metadata for the selected genres are obtained.

Each track ID is resolved to its corresponding file path in the FMA directory structure using a helper function that maps numeric track IDs to folder and file names [3]. Audio is loaded from disk using 'librosa.load' with a sample rate of 22,050 Hz and mono conversion. The MFCC features are then computed using 'librosa.feature.mfcc' with 20 MFCC coefficients per frame, and the STFT configuration of '1024' window size, '512' hop size, and Hann window. This process yields a 20 × T MFCC matrix across T time frames.

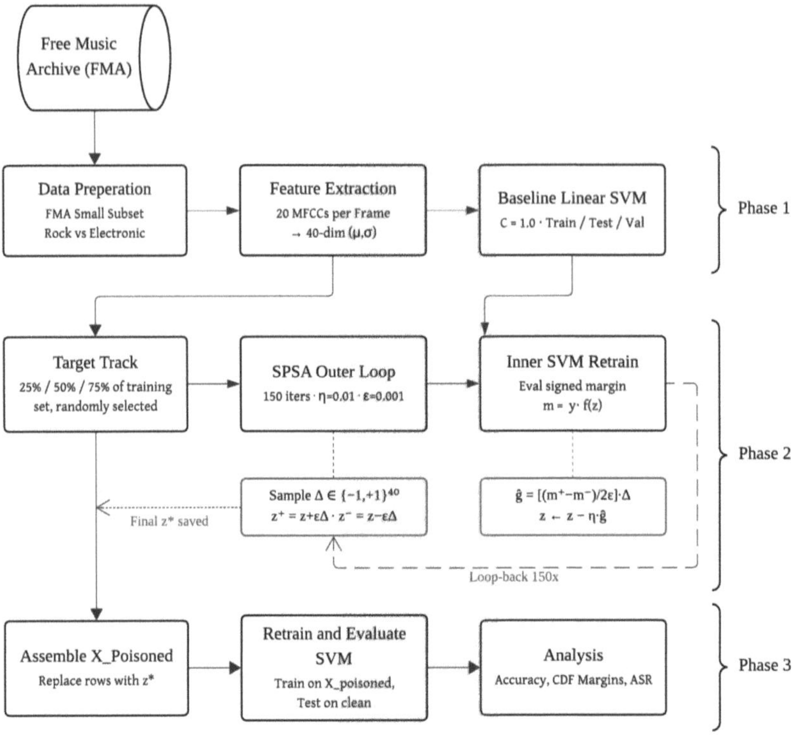

Fig. 1 Experimental pipeline

To obtain a fixed-length representation at the track level, the mean and standard deviation of each MFCC coefficient is computed across time and the statistics concatenated using Eqs. (1), (2), and (3):

$$\mu_k = (1/T)\sum\nolimits_{t=1}^{T} x_{k,t}, k = 1, \ldots, 20 \tag{1}$$

$$\sigma_k = sqrt\left((1/T)\sum\nolimits_{t=1}^{T} \left(x_{k,t} - \mu_k\right)^2\right), k = 1, \ldots, 20 \tag{2}$$

$$z = [\mu_1, \ldots, \mu_{20}, \sigma_1, \ldots, \sigma_{20}]^T \in \mathbb{R}^{40} \tag{3}$$

where $x_{k,\,t}$ is the value of the k-th MFCC coefficient at time frame t, k is the MFCC index, $k = 1, \ldots, 20$, t is the time-frame index, $t = 1, \ldots, T$, T is the total number of time frames, μ_k is the mean of the k-th MFCC across time, σ_k is the standard deviation of the k-th MFCC across time, z is the final 40-dimensional feature vector, and $\mathbb{R}^{40}$ is the 40-dimensional real-valued space.

This results in a 40-dimensional MFCC statistics vector for each track, which forms the basis for both the baseline classifier and the MFCC-space attack. To build the feature matrices, the process is iterated over each split's track list, the corresponding audio loaded, and the 40-dimensional MFCC statistics are computed, and the resulting feature vectors and genre labels are collected as shown below:

$X_{\text{train}} \in \mathbb{R}^{N_{\text{train}} \times 40}$: training feature matrix
$y_{\text{train}} \in \{-1, +1\}^{N_{\text{train}}}$: training labels
$X_{\text{val}} \in \mathbb{R}^{N_{\text{val}} \times 40}$: validation feature matrix
$y_{\text{val}} \in \{-1, +1\}^{N_{\text{val}}}$: validation labels
$X_{\text{test}} \in \mathbb{R}^{N_{\text{test}} \times 40}$: test feature matrix
$y_{\text{test}} \in \{-1, +1\}^{N_{\text{test}}}$: test labels

Given the 40-dimensional MFCC statistics, each feature dimension is then standardised using the training set statistics. A 'StandardScaler' is fit on X_{train} and applied to X_{test} and X_{val} to obtain zero-mean, unit-variance features.

To obtain a baseline classifier, a Linear Support Vector Machine (SVM) is trained with regularization parameter $C = 1.0$ and maximum iteration count of 3,000. The SVM learns a linear decision boundary in the standardized 40-dimensional MFCC space to separate the two chosen genres. The MFCC with Linear SVM serves as the victim model for subsequent adversarial experiment.

3.2 MFCC-Space Single-Target Bilevel Formulation

To study targeted attacks in MFCC feature space, we first consider a single-target bilevel optimization problem. Let $z_{\text{target}} \in \mathbb{R}^{40}$ be the standardized MFCC statistics vector of a chosen target track and y_{target} its label. Given a training feature matrix $X \in \mathbb{R}^{N_{\text{train}} \times 40}$ and corresponding labels y, Linear Support Vector Machine (SVM) is trained and its decision function $f(\cdot)$ is obtained for the inner problem of the bilevel setup. The outer problem focuses on optimizing the feature vector z assigned to the target index to reduce the signed margin $m(z) = y_{\text{target}} f(z)$, where $f(\cdot)$ is the decision function of the SVM retrained on the modified dataset.

Formally, for a target index i and candidate feature vector z, a modified training matrix $X'(z)$ is defined by replacing the i-th row with z. Let f_z denote the SVM decision function trained on $X'(z)$. The outer objective is to minimize $m(z) = y_{\text{target}} f_z(z)$, thereby pushing the target toward or beyond the decision boundary.

The analytic gradients of $m(z)$ are not computed. Instead, Simultaneous Perturbation Stochastic Approximation (SPSA) is employed to estimate gradients with respect to z. At each outer iteration, a random perturbation direction is sampled, the inner problem is evaluated at symmetric positive and negative perturbations around z, and a finite-difference estimate of the gradient is formed. The feature vector z is then updated using a fixed learning rate. The bilevel optimization is run for a fixed number of iterations of 150 iterations to obtain an optimized MFCC feature vector z_* for the target.

For a given target index i and candidate feature vector z, let $z \in R^{40}$ denote its MFCC mean–standard-deviation feature vector. Vector z is treated as a bilevel control variable and Simultaneous Perturbation Stochastic Approximation (SPSA) is applied to minimize its signed margin under a Linear SVM.

At each outer iteration, a random perturbation direction $d \in -1, +1^{40}$ is drawn and the two candidate feature vectors, $z^+ = z + \varepsilon d$, $z^- = z - \varepsilon d$, where ε is a small finite-difference step, were evaluated. For each candidate, a temporary copy of the training feature matrix is created in which the row corresponding to the target index is replaced by (z^+) *or* (z^-). On this modified training set, the inner Linear SVM is trained using scikit-learn's LinearSVC with regularization parameter ($C = 1.0$) and maximum 3,000 iterations, while all other parameters are kept at default.

The signed margin of the target point, $m^+ = y_{target} f(z^+)$, $m^- = y_{target} f(z^-)$, is then computed where $y_{target} \in -1, +1$ is the target label and $f(\cdot)$ is the SVM decision function. A one-dimensional SPSA gradient estimate in direction d is then obtained as shown in Equation (4).

$$g = \left(\frac{m^+ - m^-}{2\varepsilon} \right) d \tag{4}$$

The candidate feature vector z is then updated such that $z < -z - \eta g$ with learning rate η. Repeating this process for the fixed number of 150 outer iterations yields an optimised MFCC feature vector for the target point, along with a trajectory of its margin values. The inner SVMs operate on the same standardised 40-dimensional MFCC feature space (Sect. 3.2) to ensure that the poisoning is performed entirely in the standardised MFCC feature domain.

3.3 Dataset-Level Bilevel Poisoning

The single-target formulation is extended to a dataset-level poisoning scenario by applying bilevel optimization to multiple randomly selected training examples. Given a desired poisoning fraction of 25%, 50%, and 75% of the training set to accommodate for minority, balanced, and majority of the sample to be poisoned. For each randomly selected index, the same SPSA-based bilevel optimization is performed to obtain a modified MFCC feature vector. The resulting poisoned training matrix X_{poisoned} contains perturbed MFCC statistics for the selected tracks and the original statistics for the remaining tracks.

A Linear SVM is then retrained on X_{poisoned} and its performance evaluated on the clean validation and test sets. To extend the single-target bilevel procedure to the dataset level, a random subset of the training points is selected and independently optimised each of their MFCC feature vectors using the margin-based SPSA scheme described previously. Given a desired poisoning fraction $p \in (0, 1]$, $\lfloor pN_{train} \rfloor$ is sampled with distinct indices from the training set without replacement, where N_{train} is the number of training examples. For each selected index, candidate feature vector z were initialised with the corresponding standardised 40-dimensional MFCC feature vector and ran 150 SPSA outer iterations with learning rate $\eta = 10^{-2}$ and finite-difference step $\varepsilon = 10^{-3}$ using the linear inner SVM, with regularisation parameter ($C = 1.0$) and maximum 3,000 iterations, to evaluate signed margins. The optimised vectors replaced the original rows in a copy of the training feature matrix, yielding a poisoned training matrix X_{poisoned}.

After constructing X_{poisoned}, a poisoned Linear SVM were trained using LinearSVC with regularisation parameter ($C = 1.0$) and maximum 3,000 iterations on X_{poisoned} and its original label y_{train}. This model shares the same architecture and hyperparameters as the baseline classifier (Sect. 3.1), but is trained on the poisoned feature space. This poisoned SVM is then evaluated on the original clean training features X_{train}, the poisoned training features X_{poisoned}, and the clean validation and test feature matrices X_{val} and X_{test}, with its train, validation, and test accuracies recorded. This enables to quantify both local changes to the feature space and the global impact of MFCC-space poisoning on generalisation performance.

4 Results and Discussion

4.1 Baseline Classifier Performance

The victim classifier throughout this work is a Linear SVM trained on 40-dimensional MFCC mean-standard deviation features extracted from the FMA small subset restricted to Rock and Electronic tracks as decided on the EDA findings. Using the official FMA training, validation, and test splits, the performance of the baseline Linear SVM (Table 1) with its confusion matrix on the 200-track test set (Fig. 2) were obtained.

Table 1 Baseline classifier performance

Data Split	Accuracy (%)
Training	88.63
Validation	93.50
Test	85.50

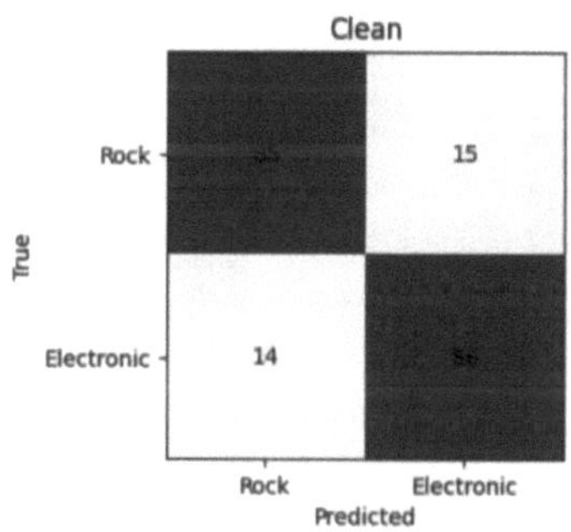

Fig. 2 Confusion matrix of baseline SVM

The baseline SVM correctly predicted 85 tracks as Rock, with 15 incorrectly predicted as Electronic out of 100 Rock tracks. For Electronic tracks, 86 are correctly predicted as Electronic while 14 incorrectly predicted as Rock (Fig. 1). This baseline confirms that the task is non-trivial yet solvable as the error rates are around 14% to 15% per class, and the confusion matrix is relatively balanced. The MFCC with Linear SVM combination provides a clean and interpretable starting point for studying adversarial perturbations.

4.2 Dataset-Level MFCC-Space Bilevel Poisoning

In this experiment, the training set contains 1,600 tracks. Poisoning fractions of 25% (400 tracks), 50% (800 tracks), and 75% (1,200 tracks) are evaluated.

For each poisoning fraction p, training examples $p \cdot N_{\text{train}}$ are randomly selected for the MFCC-space bilevel optimization to be conducted on their feature vectors. A Linear SVM is then retrained on the resulting X_{poisoned} obtained from the Margin-based bilevel objective. The effect on the train, test, and validation accuracy (Table 2) with their confusion matrix (Fig. 3) were obtained.

Table 2 Linear SVM performance on clean vs. MFCC-poisoned training sets for margin-based bilevel poisoning

Data Split	Accuracy (%)			
	Train (Clean)	Train	Test	Validation
25%	87.4	83.7	88.0	94.0
50%	86.3	81.2	88.5	94.0
75%	86.3	79.7	88.0	90.5

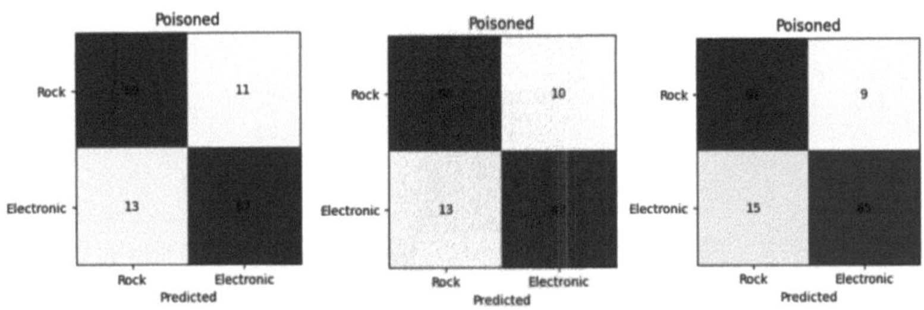

Fig. 3 Confusion matrices of SVM performance at 25%, 50%, and 75% poisoned percentage

At 25%, the model correctly predicted 89 tracks as Rock, with 11 incorrectly predicted as Electronic out of 100 Rock tracks. For Electronic tracks, 87 are correctly predicted as Electronic while 13 incorrectly predicted as Rock, achieving 88.0% accuracy. At 50%, the model correctly predicted 90 tracks as Rock, with 10 incorrectly predicted as Electronic out of 100 Rock tracks. For Electronic tracks, 87 are correctly predicted as Electronic while 13 incorrectly predicted as Rock, achieving 88.5% accuracy. At 75%, the model correctly predicted 91 tracks as Rock, with 9 incorrectly predicted as Electronic out of 100 Rock tracks. For Electronic tracks, 85 are correctly predicted as Electronic while 15 incorrectly predicted as Rock, achieving 88.0% accuracy.

Based on the results, it can be observed that there is no degradation of test accuracy with the accuracy never being lower than the baseline value of 85.5%. At 25% poisoning it reaches 88.5%, and at 50% and 75% it is 88.0%. Thus, for this Linear SVM and MFCC configuration, margin-based MFCC-space poisoning does not break global performance, but improves it instead.

It is also observed that train accuracy on poisoned features decreases. Accuracy on the poisoned feature matrix X_{poisoned} decreases from 83.7% to 81.2% to 79.7% as poisoning fraction increases from 25% to 50% to 75%, reflecting that the model is fitting a more distorted training set.

Validation accuracy also remains high throughout the experiment. The accuracy stays near baseline at 25% and 50% with 94.0% and 93.5% respectively and drops modestly to 90.5% at 75%. This modest drop does not translate into test performance collapse.

Relative to the baseline confusion matrix, Rock misclassifications decrease from 15 for the baseline SVM to 9 to 11 on the poisoned, while Electronic misclassifications remain similar or slightly worse at 75%. Overall, the distribution of errors remains balanced and low.

To inspect local effects on the targeted training points, Cumulative Distribution Functions (CDFs) of margins, and Attack Succes Rate (ASR) for the targeted points are shown for each percentage (Figs. 4 and 5).

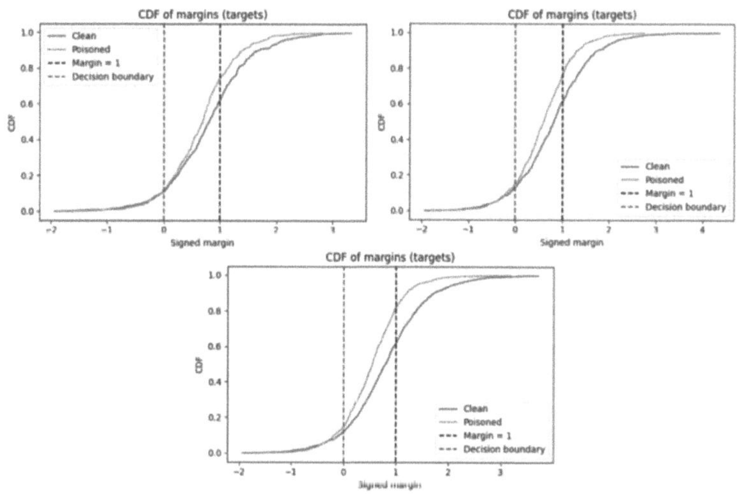

Fig. 4 Cumulative Distribution Functions (CDFs) of margins at 25%, 50%, and 75% poisoned percentage

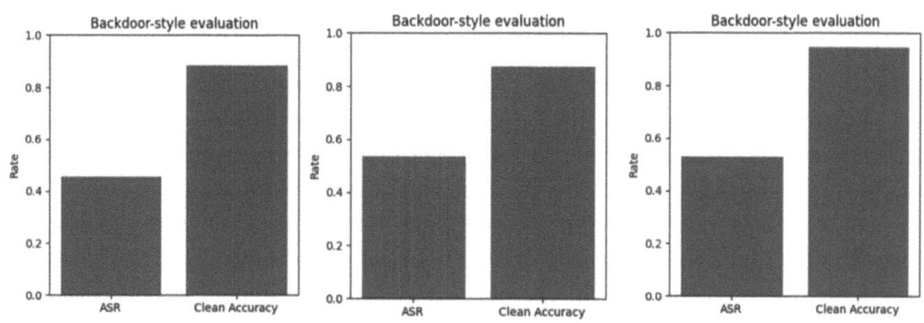

Fig. 5 Attack Succes Rate (ASR) of targeted points at 25%, 50%, and 75% poisoned percentage

Roughly half of the targeted training points change sign of their margins or cross the decision boundary shown by the ASR ≈ 0.5 (Fig. 5) for all poisoning fractions, with 25% slightly lower. This indicates that the attack is locally effective at moving many selected samples across the margin. The CDFs of margins (Fig. 4) indicates that in the interval (0,1), the poisoned curve lies slightly above the clean curve, meaning a larger fraction of targeted samples cluster near the decision boundary after poisoning.

Taken together, these results show that although margin-based poisoning does alter the local geometry around targeted MFCC vectors, these local effects do not accumulate into a global accuracy drop. Instead, the test performance remains stable and instead had slightly improved.

5 Conclusion

This study investigated an MFCC-centric adversarial pipeline against a music genre classifier via bilevel optimization in MFCC feature space. The victim model is a Linear SVM trained on 40-dimensional MFCC mean-standard deviation features extracted from the small subset of the Free Music Archive (FMA) dataset [3], restricted to Rock and Electronic tracks selected via MFCC-based exploratory data analysis. The research evaluates how well these MFCC features support genre classification and how poisoning-induced perturbations alter feature-space structure, decision boundaries, and classification behavior. Special attention is given to SVM margin changes to understand both local and global robustness of the classifier under adversarial conditions.

The empirical findings show that dataset-level MFCC-space poisoning is globally benign in this study, as the margin-based bilevel poisoning objectives did not significantly reduce the test accuracy, even when up to 75% of the training MFCC vectors are perturbed. However, the margin-based poisoning successfully reduces margins and flips a significant portion of targeted training points locally, but these effects do not translate into meaningful global degradation in validation or test performance. This indicates that the 40-dimensional MFCC mean–standard deviation feature space with a Linear SVM classifier is highly robust against large-scale feature-space poisoning.

Several factors likely explain this robustness. The MFCC representation is compact, low-dimensional, and smooth, averaging temporal structure into stable global statistics, which makes small perturbations less likely to create strongly adversarial configurations. Additionally, the Linear SVM uses a single linear decision boundary, so perturbing many points through small MFCC shifts tends to produce only slight boundary adjustments rather than the complex vulnerabilities often seen in deep neural networks. The margin-based objective also focuses on local training-point manipulation rather than directly optimizing test performance, while SPSA updates remain conservative and small. As a result, the poisoning remains stealthy because validation and test accuracy do not visibly decline, making detection difficult if only aggregate performance is monitored. However, this stealth also highlights weak attack effectiveness, suggesting that low-dimensional interpretable features and classical classifiers may naturally offer stronger resistance to poisoning attacks compared to high-capacity deep learning models.

Practical Implications The results suggest that MFCC-based audio classification systems using linear classifiers may possess inherent robustness against large-scale feature-space poisoning. For resource-constrained environments where computational efficiency and interpretability are important, MFCC–SVM pipelines remain a viable solution. The observed stability in validation and test performance indicates that poisoning attacks targeting MFCC statistics alone may be insufficient to substantially degrade the performance of such systems.

Research Implications The findings contribute to the understanding of poisoning attacks in classical machine learning settings and highlight the distinction between local and global attack effectiveness. Although the poisoning process successfully altered decision margins and affected many targeted samples, these local changes did not translate into meaningful degradation of overall model performance. This suggests that future research should investigate more expressive feature representations, nonlinear classifiers, and end-to-end deep learning architectures to better understand the conditions under which audio poisoning attacks become effective.

Acknowledgments. The project is conducted at and supported by the Faculty of Computer and Mathematical Sciences, Universiti Teknologi MARA, Shah Alam, Selangor, Malaysia.

References

1. Aghakhani, H., Meng, D., Wang, Y.-X., Kruegel, C., Vigna, G.: Bullseye polytope: a scalable clean-label poisoning attack with improved transferability. In: Proceeding of the IEEE European Symposium on Security and Privacy (EuroS&P), pp. 159–178 (2021). https://doi.org/10.1109/EuroSP51992.2021.00021
2. Cortes, C., Vapnik, V.: Support-vector networks. Mach. Learn. **20**(3), 273–297 (1995). https://doi.org/10.1007/BF00994018
3. Defferrard, M., Benzi, K., Vandergheynst, P., Bresson, X.: FMA: a dataset for music analysis. arXiv:1612.01840. (2016)
4. Ge, Y., Wang, Q., Yu, J., Shen, C., Li, Q.: Data poisoning and backdoor attacks on audio intelligence systems. IEEE Commun. Mag. **61**(12), 176–182 (2023). https://doi.org/10.1109/MCOM.012.2200596
5. Ge, Y., Wang, Q., Zhang, J., Zhou, J., Zhang, Y., Shen, C.: WaveFuzz: a clean-label poisoning attack to protect your voice. arXiv:2203.13497. (2022)
6. Govindarajulu, G., Shah, P.: Decoding deception: understanding automatic speech recognition vulnerabilities in evasion and poisoning attacks. https://arxiv.org/abs/2509.22060. (2025). https://doi.org/10.48550/arXiv.2509.22060
7. Henry, J., Ergezer, M., Orescanin, M.: Perceptually constrained fast adversarial audio attacks. In: Proceeding of the IEEE International Conference on Machine Learning and Applications (ICMLA), pp. 819–824 (2021). https://doi.org/10.1109/ICMLA52953.2021.00135
8. Hong, S., Carlini, N., Kurakin, A.: Certified robustness to clean-label poisoning using diffusion denoising. https://arxiv.org/abs/2403.11981. (2024)
9. Kwon, J., Kwon, D., Wright, S., Nowak, R.D.: A fully first-order method for stochastic bilevel optimization. In: Proceeding of the 40th International Conference on Machine Learning (ICML), in Proceedings of Machine Learning Research, vol. 202, pp. 18083–18113 (2023) https://proceedings.mlr.press/v202/kwon23c.html

10. Meerza, S.I.A., Sun, L., Liu, J.: HarmonyCloak: Making music unlearnable for generative AI. In: Proceeding of the IEEE Symposium on Security and Privacy (SP), pp. 430–448 (2025). https://doi.org/10.1109/SP61157.2025.00085
11. Petrulionyte, I., Mairal, J., Arbel, M.: Functional bilevel optimization for machine learning. In: Proceeding of the Advances in Neural Information Processing Systems (NeurIPS), vol. 37, pp. 14016–14065, Vancouver, Canada (Dec. 2024). https://doi.org/10.52202/079017-0449
12. Schölkopf, B., Smola, A.J.: Learning with Kernels: Support Vector Machines, Regularization, Optimization, and Beyond. MIT Press, Cambridge, MA, USA (2002)
13. Telembici, L., Rusu, C.: Audio classification for feature-based majority voting optimization and hyperparametric tuning. In: Proceeding of the 33rd European Signal Processing Conference (EUSIPCO), pp. 1412–1416 (2025). https://doi.org/10.23919/EUSIPCO63237.2025.11226494
14. Wu, Y., Cheng, Z., Zhou, K., Chen, B., Tong, G.: Applications and comparative analysis of machine learning and statistical learning signal processing in music genre classification. In: Proceeding of the 9th International Conference on Digital Signal Processing, pp. 15–26 (Feb. 2025)
15. Zhang, Z., et al.: SafeSpeech: robust and universal voice protection against malicious speech synthesis. In: Proceeding of the 34th USENIX Security Symposium (USENIX Security), pp. 4581–4600 (2025)
16. Zhao, B., Lao, Y.: CLPA: Clean-label poisoning availability attacks using generative adversarial nets. Proc. AAAI Conf. Artif. Intell. **36**(8), 9162–9170 (2022)

DroneCare: An Intelligent System for Predicting Drone Failures Using Artificial Intelligence

Noura Yahya Al Hoqani(✉), Thura Ahmed Al Wahaibi, Enas Younis Alshuaili, and Riman Hamed Alhajri

Department of Computing and Information Sciences, University of Technology and Applied Sciences, PO Box 74. Al Khuwair, Muscat, Sultanate of Oman
noura.alhoqani@utas.edu.om

Abstract. The rapid growth in the use of Unmanned Aerial Vehicles (UAVs) across critical sectors such as emergency response, logistics, surveillance, and agriculture has increased the demand for intelligent systems that ensure operational reliability and flight safety. Unexpected in-flight failures caused by battery degradation, sensor malfunctions, adverse environmental conditions, or mechanical faults can result in emergency landings, property damage, and safety risks. To address these challenges, this study presents DroneCare, an intelligent predictive maintenance system designed to anticipate UAV failures before they occur. The proposed system utilizes multiple machine learning techniques, including Logistic Regression, Random Forest, XGBoost, Support Vector Machine (SVM), and Artificial Neural Networks (ANN), to analyse drone telemetry and operational data. Key flight parameters such as battery level, altitude, GPS accuracy, wind speed, payload weight, and flight distance were used to predict flight outcomes. Experimental results demonstrated that the ANN model achieved the best performance, reaching an accuracy of 98.15% and F1-score of 0.9615. The developed model was integrated into a simulation environment using Microsoft AirSim to provide real-time risk assessment and early warning notifications. By enabling proactive maintenance and timely intervention, DroneCare enhances UAV operational safety, reduces the likelihood of unexpected failures, and supports the reliable adoption of drone technology in critical applications.

Keywords: Unmanned Aerial Vehicles · Drone Safety · Predictive Maintenance · Machine Learning · Random Forest · XGBoost · Artificial Neural Networks · Failure Prediction · UAV Reliability

1 Introduction

Unmanned Aerial Vehicles (UAVs), are increasingly used in sectors such as surveillance, logistics, agriculture, and emergency response. Their ability to perform critical tasks efficiently has made them an important technology in modern operations. However, UAVs remain vulnerable to unexpected in-flight failures caused by factors such as battery degradation, sensor malfunctions, adverse environmental conditions, and excessive payload loads. These failures can lead to emergency landings, operational disruptions, property damage, and safety risks.

N. Mohd Sharef et al. (Eds.): CAMP 2026, CCIS 3093, pp. 95–110, 2027.
https://doi.org/10.1007/978-981-92-4302-0_8

Most current drone maintenance approaches are reactive, where faults are addressed only after a failure occurs. As drone deployment continues to expand, there is a growing need for intelligent predictive maintenance systems capable of identifying potential failures before they happen. Recent advances in Artificial Intelligence (AI) and Machine Learning (ML) provide effective solutions for analyzing telemetry data and detecting patterns associated with abnormal flight behaviour. To address this challenge, this study presents DroneCare, an AI-based predictive maintenance system for UAVs. The proposed system analyses key flight parameters, including battery level, altitude, GPS accuracy, wind speed, payload weight, and flight distance, to predict flight outcomes and assess operational risk. Several machine learning models were evaluated, including Logistic Regression, Random Forest, XGBoost, Support Vector Machine (SVM), and Artificial Neural Networks (ANN). Also, The AirSim were used to provide real-time risk assessment and early warning notifications, contributing to safer and more reliable drone operations.

Despite the growing adoption of UAVs, existing maintenance approaches remain largely reactive and often fail to identify potential malfunctions before they occur. Furthermore, there is a lack of comprehensive predictive maintenance systems that analyse multiple flight parameters simultaneously and provide real-time risk assessment. To address this challenge, this study proposes DroneCare, an intelligent UAV predictive maintenance system that utilizes machine learning techniques to predict potential failures, generate early warning notifications, and improve operational safety. The main objectives are to enhance flight reliability, reduce maintenance costs, minimize operational downtime, and support safe drone deployment in critical applications.

2 Literature Review

Recent research has increasingly focused on applying Deep Learning (DL) and Machine Learning (ML) techniques to improve UAV reliability, fault detection, and predictive maintenance. Several studies have demonstrated the effectiveness of data-driven approaches in identifying potential failures and enhancing operational safety.

Table 1 demonstrates the growing adoption of AI and machine learning techniques in UAV predictive maintenance. The reviewed studies successfully addressed specific fault categories such as battery degradation, propeller imbalance, sensor failures, and anomaly detection. However, most approaches focus on individual components or isolated failure scenarios rather than providing a comprehensive assessment of overall UAV operational health.

2.1 Research Gap and Contributions

Despite the significant progress achieved in UAV predictive maintenance, several limitations remain. Most existing studies focus on a single UAV component, such as batteries, propellers, or sensors, rather than considering the combined impact of multiple operational and environmental factors. In addition, many approaches rely on offline analysis and lack real-time monitoring capabilities, limiting their practical deployment during

Table 1 : Literature Review

Table Study	Algorithm / Technique	Results
Iannace et al. (2019)	Artificial Neural Network (ANN)	Achieved 97.63% accuracy in detecting UAV propeller imbalance using acoustic signals.
Ahmad et al. (2022)	Random Forest, SVM, CART	Developed an intelligent framework for automated UAV failure prediction, detection, and classification.
Bell et al. (2022)	Stacked Recurrent Autoencoder (LSTM-AE)	Applied anomaly detection on UAV sensor data and achieved approximately 82.1% detection accuracy.
Ahmad et al. (2024)	Transformer-Based Neural Network	Proposed a framework capable of predicting and classifying six different UAV sensor failure types.
Şahin (2025)	Artificial Neural Network (ANN)	Predicted UAV battery Remaining Useful Life (RUL) with a Mean Absolute Percentage Error (MAPE) of 3.2%.

active flight operations. Furthermore, only a limited number of studies integrate predictive maintenance models within realistic UAV simulation environments for operational validation. o address these limitations, this research proposes DroneCare, an intelligent UAV predictive maintenance system that combines multiple flight telemetry and environmental parameters, including battery level, altitude, GPS accuracy, wind speed, payload weight, and flight distance. The proposed framework evaluates and compares multiple machine learning models, including Logistic Regression, Random Forest, XGBoost, Support Vector Machine (SVM), and Artificial Neural Networks (ANN), to identify the most effective predictive approach. The best-performing model is integrated with Microsoft AirSim to provide real-time risk assessment, early warning notifications, and operational support for safer drone flights.

The main contributions of this research are summarized as follows:

- Development of a comprehensive UAV predictive maintenance framework using multiple flight telemetry parameters.
- Comparative evaluation of five machine learning algorithms for UAV failure prediction.
- Integration of the best-performing predictive model into a Microsoft AirSim simulation environment.
- Real-time risk assessment and warning generation to support proactive maintenance decisions.
- Improvement of UAV operational safety, reliability, and maintenance efficiency through early failure prediction.

3 Methodology

This study adopts a quantitative research approach to develop and evaluate machine learning models for UAV failure prediction. The proposed DroneCare framework utilizes drone telemetry and operational data to identify patterns associated with successful and unsuccessful flight outcomes. The analysis incorporates a combination of operational, environmental, and technical parameters, including battery level, altitude, GPS accuracy, wind speed, payload weight, flight distance, and obstacle encounters.

A quantitative approach was selected because the dataset consists of structured variables that can be processed and analysed using statistical and machine learning techniques. Several classification algorithms were evaluated, including Logistic Regression, Random Forest, XGBoost, Support Vector Machine (SVM), and Artificial Neural Networks (ANN). These models were trained to classify flight outcomes and assess potential failure risks.

The research methodology consists of four main stages: data collection, data preprocessing, model development, and performance evaluation. The best-performing model was subsequently integrated into a Microsoft AirSim simulation environment to provide real-time risk assessment and early warning notifications during drone operations

3.1 Research Design

This study follows a Design and Development research approach to develop and validate an intelligent predictive maintenance framework for Unmanned Aerial Vehicles (UAVs). The proposed DroneCare system is designed to predict potential flight failures by analysing operational, environmental, and technical flight parameters using machine learning techniques.

The research process involves collecting UAV telemetry data, preprocessing the dataset, developing predictive models, and evaluating their performance using standard classification metrics. Multiple machine learning algorithms, including Logistic Regression, Random Forest, XGBoost, Support Vector Machine (SVM), and Artificial Neural Networks (ANN), were implemented and compared to identify the most effective approach for UAV failure prediction.

This research design ensures both theoretical and practical validation of the proposed framework, enabling the development of a reliable predictive maintenance solution capable of supporting safer and more efficient drone operations.

3.2 System Architecture

The proposed DroneCare system follows a multi-stage architecture designed to predict potential UAV failures and support proactive maintenance. The process begins with the collection of drone telemetry data, including battery level, altitude, wind speed, payload weight, GPS accuracy, and other operational flight parameters.

The collected data undergoes a preprocessing stage to improve data quality. This stage includes handling missing values, removing outliers, normalizing numerical features, and encoding categorical variables. After preprocessing, feature selection is performed to identify the most relevant variables that contribute to flight outcome prediction.

The selected features are then used during the model training phase. Multiple machine learning algorithms, including Logistic Regression, Random Forest, XGBoost, and Support Vector Machine (SVM), are evaluated alongside a deep learning approach based on Artificial Neural Networks (ANN). The models learn patterns from historical flight data and classify flight outcomes according to the risk of failure.

Once trained, the predictive model analyses incoming flight parameters and generates failure predictions in real time. If abnormal conditions or potential failures are detected, the system immediately sends warning notifications to the operator, allowing preventive actions to be taken. Otherwise, the drone continues normal operation under continuous monitoring.

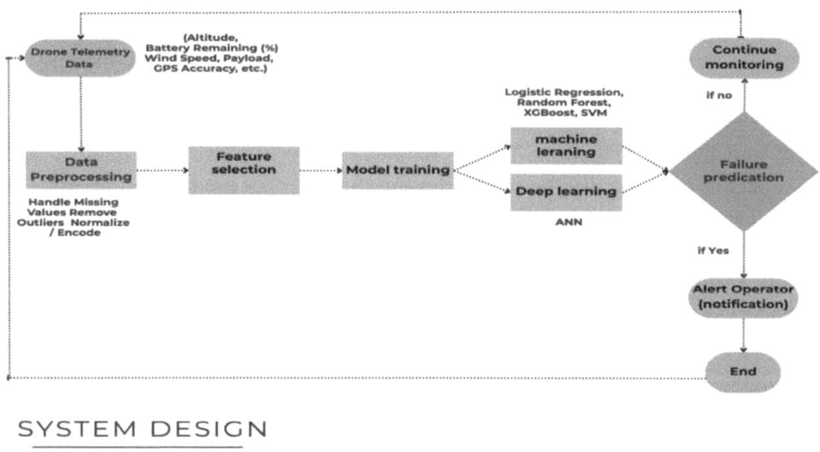

Fig. 1: DroneCare System Architecture

Figure 1 illustrates the overall DroneCare system architecture, showing the workflow from telemetry data acquisition and preprocessing to model training, failure prediction, and real-time alert generation.

4 Data Collection

The dataset for this project was obtained from Kaggle, titled Supplemental Drone Telemetry Data – Drone Operations Log. It includes 557 flight records with features such as Drone ID, Application, Drone Size, Altitude, Flight Duration, Battery Remaining (%), GPS Accuracy (meters), Wind Speed (m/s), Obstacles Encountered, and the target variable Flight Status (Completed or Landed Unexpectedly). The dataset contains a mix of numerical and categorical features.

The data preprocessing process followed a structured workflow to prepare the dataset before model training.

The main steps include:

- Data examination to identify missing values and anomalies

- Handling missing values using appropriate imputation techniques
- Outlier detection and treatment to reduce noise in the data
- Encoding categorical variables into numerical format
- Normalization or standardization for feature scaling
- Handling class imbalance using resampling techniques
- Splitting the dataset into training and testing sets

This process ensures that the dataset is clean, consistent, and suitable for machine learning models, ultimately improving prediction accuracy and overall model performance.

4.1 Data Preprocessing

This section presents the preliminary analysis conducted on the DroneCare dataset to prepare it for model development. The purpose of this stage is to ensure that the data is clean, consistent, and suitable for machine-learning processing. Several preprocessing steps were applied, including data transformation, normalization, descriptive statistical analysis, and feature reduction. These procedures help improve data quality, address inconsistencies, and identify the most relevant attributes for predicting the drone flight situation. The following sections describe each preprocessing step, the techniques used, and the outcomes obtained during the preparation of the dataset.

Figure 2 illustrates the data preprocessing workflow adopted in the DroneCare system. The process began with data examination to identify missing values and anomalies, followed by missing value handling and outlier treatment. Categorical variables were then encoded into numerical form, and feature scaling was applied through normalization and standardization techniques. To improve model performance, class imbalance was addressed using appropriate resampling methods before splitting the dataset into training and testing sets. These preprocessing steps ensured that the dataset was clean, balanced, and suitable for machine learning model development.

4.2 Data Transformation

Data Transformation is a crucial preprocessing step, preparing the dataset for efficient machine learning modelling. In the DroneCare project, several transformation techniques were implemented to ensure data purity, consistency, and suitability for analysis. These transformations included data type conversion and normalization.

4.3 Data Type Conversion

During the initial inspection, some numerical attributes were incorrectly stored as object types, preventing machine learning algorithms from interpreting them as continuous numeric values. These columns were converted to their appropriate numerical formats (float64)

Columns converted from object to float:

- Altitude (meters)
- Battery Remaining (%)

Fig. 2: Data preprocessing workflow used in the DroneCare system

- Distance Flown (km)
- Flight Duration (minutes)

After conversion, the dataset contained the following numerical attributes:

- Propeller Count
- Max Carry Weight (kg).
- Actual Carry Weight (kg)
- Altitude (meters).
- Flight Duration (minutes)
- Distance Flown (km).
- Battery Remaining (%)
- GPS Accuracy (meters).
- Wind Speed (m/s)

Converting these features into numerical types ensures that they can be used in scaling, statistical analysis, and machine learning training.

4.4 Feature Scaling (Standardization)

To enhance data quality and guarantee correct model performance, a number of preprocessing techniques were used. Outliers were found and dealt with to lessen their influence on the model, and missing values were handled using suitable methods such mean or mode imputation.

Additionally, Z-score scaling was used to standardize numerical features such that they are all on a same scale. By rescaling the data to have a mean of 0 and a standard deviation of 1, this transformation enhances model performance and keeps features with greater magnitudes from taking over the learning process.

4.5 Selected Machine Learning Models

Multiple machine learning models and a deep learning model were used to predict drone flight outcomes. These include Logistic Regression, Support Vector Machine (SVM), Random Forest, XGBoost, and Artificial Neural Networks (ANN). Logistic Regression was used as a baseline model, while SVM was applied to find optimal decision boundaries. Random Forest and XGBoost were used as ensemble models to improve prediction accuracy and reduce overfitting. The Artificial Neural Network (ANN) model was designed to capture complex patterns within the dataset and improve overall prediction performance. The models were trained and evaluated to determine the most effective approach for failure prediction during flight.

4.6 Data Analysis and Visualization

Exploratory Data Analysis (EDA) was conducted to better understand the dataset and identify key patterns influencing drone performance and failure prediction. Key features such as battery level and wind speed were analyzed due to their direct impact on drone stability and energy consumption.

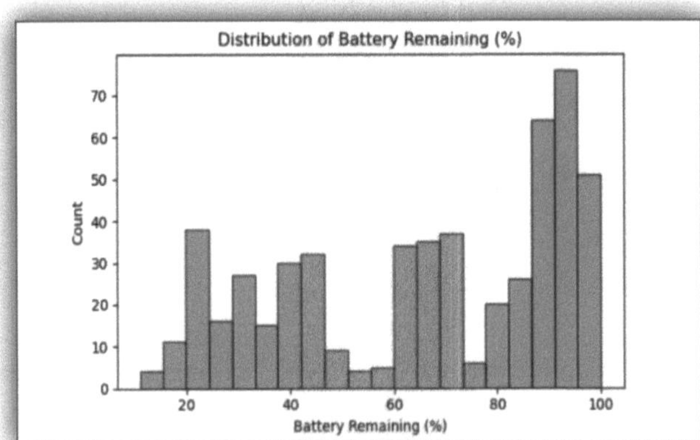

Fig. 3: Distribution of Battery Remaining (%)

The distribution of battery remaining percentage is shown in Fig. 3. The analysis indicates that most flights operate at higher battery levels, typically between 70% and 100%. However, a portion of the data reflects lower battery levels, which are more likely to be associated with unexpected landings and increased failure risk.

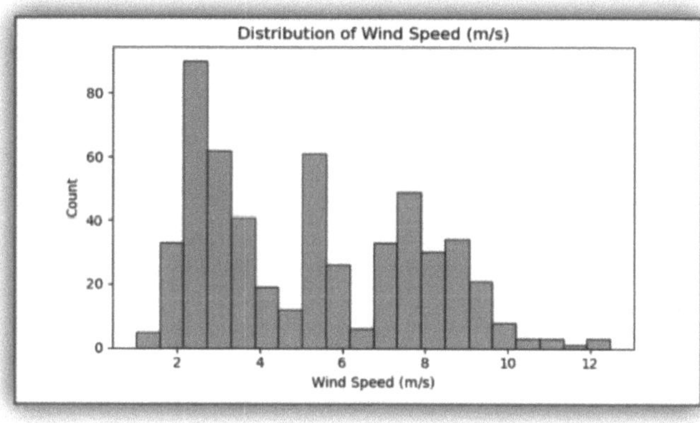

Fig. 4: Distribution of Wind Speed (m/s)

Figure 4 shows the distribution of Wind Speed (m/s) in the dataset. The graph indicates that most drone flights were conducted under low to moderate wind conditions, while fewer flights experienced higher wind speeds. Wind speed is considered an important environmental factor because strong winds can increase energy consumption, reduce flight stability, and contribute to unexpected flight failurcs.

To further explore feature relationships, Fig. 5 illustrates the relationship between wind speed and battery level. A clear negative trend can be observed, indicating that higher wind speeds lead to faster battery depletion. This is due to the increased energy required to maintain stability under strong wind conditions.

These findings highlight the importance of both environmental and operational factors in drone performance. Wind speed plays a critical role in energy consumption, making it a key feature for predicting potential drone failures. These insights were essential in selecting relevant features and improving the overall performance of the proposed model.

To further explore feature relationships, Fig. 6 illustrates the relationship between flight distance and battery remaining percentage. Each point represents an individual drone flight, while the red line indicates the general trend of the data. A clear negative relationship can be observed between the two variables, meaning that as the flight distance increases, the remaining battery level tends to decrease. This result is expected because longer flights require more energy from the drone's battery. The visualization demonstrates how flight distance can significantly affect battery consumption during drone operations.

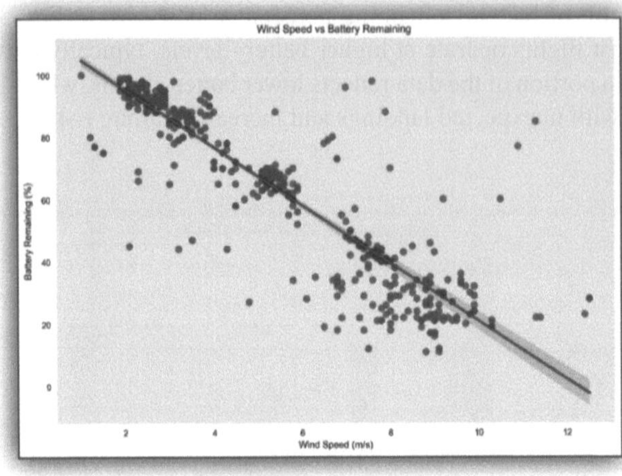

Fig. 5: Relationship between wind speed and battery level

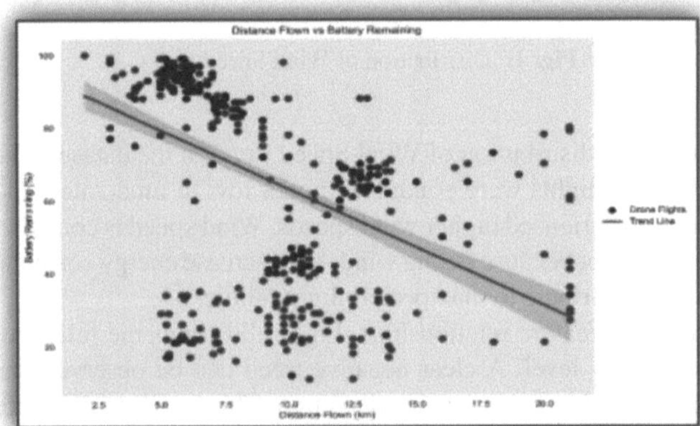

Fig. 6: Relationship between Distance Flown and Battery Remaining

4.7 Feature Importance Analysis

To better understand the model's decision-making process, feature importance analysis was conducted. The results showed that battery level, wind speed, and payload weight are the most influential factors in predicting drone failures. This finding aligns with real-world scenarios, where battery depletion and environmental conditions significantly affect flight stability. To further understand the relationship between features and the target variable, correlation analysis was conducted as shown in Fig. 7.

The correlation analysis shows the strength and direction of the relationship between each feature and the flight status. Features such as obstacles encountered, GPS accuracy, and payload weight show a positive correlation with failure risk. On the other hand, battery remaining percentage shows a strong negative correlation, indicating that lower battery levels are highly associated with unexpected landings. This analysis confirms

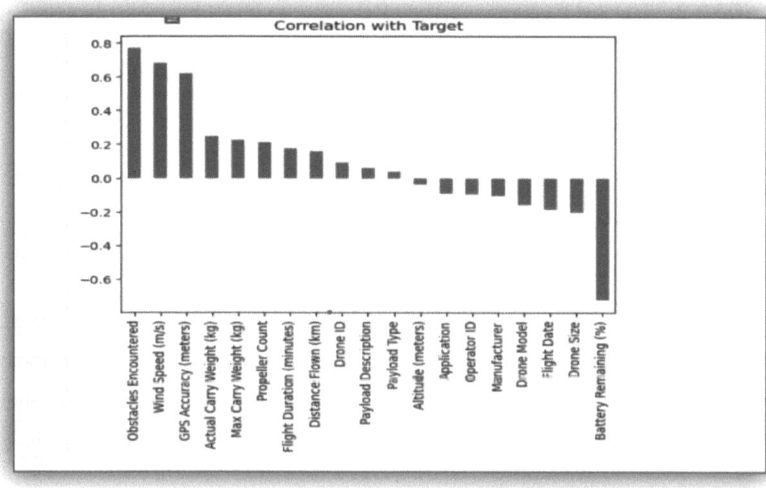

Fig. 7: Correlation between features and flight status

that battery level, environmental conditions, and operational factors play a significant role in predicting drone failures.

4.8 Feature Selection

Thirteen features were selected based on their domain relevance to drone flight safety and their empirical correlation with the target variable (Flight Status)

As shown in Table 2, the selected features were identified based on their domain relevance to UAV flight safety, operational significance, and empirical correlation with the target variable. These features capture key environmental, technical, and mission-related factors that influence drone performance and failure risk. The inclusion of variables such as battery level, wind speed, GPS accuracy, and payload characteristics ensures that the model is trained on a comprehensive and meaningful feature set, enabling more accurate and reliable prediction of flight outcomes.

Table 2: Selected Features and Their Relevance to Flight Status Prediction

Features	Type	Relevance to Flight Status
Battery Remaining (%)	Numeric	Low battery is a primary cause of unexpected landings
Wind Speed (m/s)	Numeric	High wind increases mission risk significantly
GPS Accuracy (meters)	Numeric	Poor GPS accuracy leads to navigation failure
Obstacles Encountered	Binary (0/1)	Direct indicator of flight interruption risk
Actual Carry Weight (kg)	Numeric	Overloading reduces flight performance
Max Carry Weight (kg)	Numeric	Defines operational capacity threshold
Distance Flown (km)	Numeric	Longer distances correlate with higher failure rates
Altitude (meters)	Numeric	Extreme altitudes affect motor and sensor performance
Drone Model	Categorical (Encoded)	Different models have varying failure profiles
Drone Size	Categorical (Encoded)	Size affects stability and payload capacity
Propeller Count	Numeric	Affects drone reliability and flight stability
Payload Type	Categorical (Encoded)	Different payloads impose different risk profiles
Application	Categorical (Encoded)	Mission type correlates with operational conditions

5 Model Performance Comparison

The performance of the evaluated machine learning models is summarized in Table 3.

Table 3 : Performance Comparison of the Evaluated Models

Model	Accuracy	Precision	Recall	F1 Score
Logistic Regression	0.9630	0.8966	0.9630	0.9286
Random Forest	0.9630	0.8966	0.9630	0.9286
XGBoost	0.9630	0.8710	1.0000	0.9310
SVM	0.9444	0.9200	0.8519	0.8846
ANN	0.9815	1.0000	0.9259	0.9615

As shown in Table 3, the Artificial Neural Network (ANN) achieved the highest performance among all evaluated models, obtaining an accuracy of 98.15%, a precision of 1.0000, a recall of 0.9259, and an F1-score of 0.9615. Although Random Forest and XGBoost also demonstrated strong predictive capabilities, the ANN model outperformed all other approaches due to its ability to learn complex nonlinear relationships among UAV operational and environmental parameters. These results indicate that deep learning techniques can provide more effective and reliable UAV failure prediction compared to traditional machine learning methods.

5.1 Evaluation Metrics

The performance of the models was evaluated using standard classification metrics, including accuracy, precision, recall, and F1-score. These metrics provide a comprehensive assessment of model performance. In addition, confusion matrices were used to visualize classification results and identify correct and incorrect predictions. These evaluation techniques help ensure that the model can effectively distinguish between safe flights and potential failures.

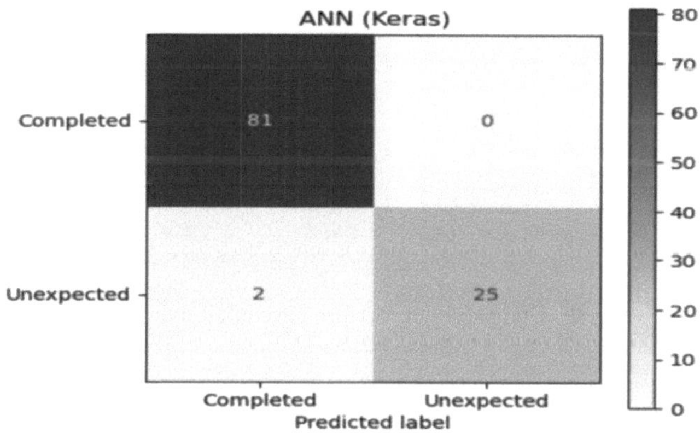

Fig. 8: Confusion matrix of the ANN model for drone flight classification

As shown in Fig. 8, the model correctly predicted 81 completed flights and 25 unexpected landings, with only 2 misclassifications. No incorrect predictions were observed for completed flights. These results indicate that the ANN model achieves high accuracy and is highly effective in distinguishing between successful flights and potential failures.

5.2 Comparative Analysis of Applied Algorithms

A comparative analysis of the evaluated algorithms indicates that ensemble-based methods, particularly Random Forest, achieved strong overall performance, attaining an accuracy of 96.30% and an F1-score of 0.9286.

The strong performance of Random Forest can be attributed to its ability to capture nonlinear relationships and handle complex feature interactions within the dataset. In contrast, Logistic Regression, being a linear model, showed slightly lower predictive performance (accuracy 96.30%, F1-score 0.9286).

SVM and XGBoost also demonstrated good results, achieving accuracy values of 94.44% and 96.30%, with F1-scores of 0.8846 and 0.9310, respectively. The ANN (Keras) model provided the highest competitive performance (accuracy 98.15%, F1-score 0.9615), effectively capturing complex patterns and successfully detecting most failure cases.

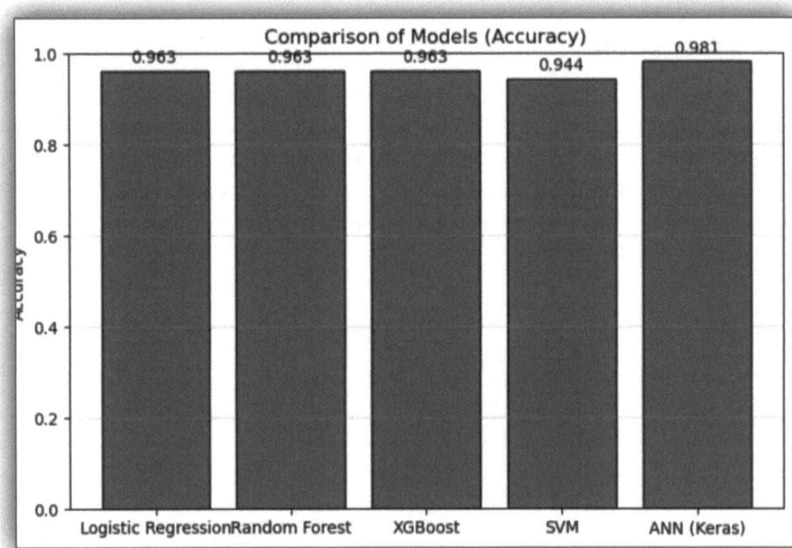

Fig. 9: Comparing the performance of all five models

Figure **9** presents a bar chart comparing the performance of all five models for drone failure prediction, based on the Accuracy metric. The analysis shows that the ANN model achieved the highest accuracy (0.9815), followed by Random Forest (0.9630). The remaining models, including Logistic Regression, XGBoost, and SVM, showed slightly lower but comparable accuracies of 0.9630, 0.9630, and 0.9444, respectively

6 Comparison with Related Studies

According to Gino Iannace, Giuseppe Ciaburro, and Amelia Trematerra (2019), the study presents a fault diagnosis approach for UAV propellers using an Artificial Neural Network (ANN) model. The research focuses on detecting blade imbalance, which is considered a critical factor that may lead to drone failure or unsafe operation. The proposed system relies on acoustic measurements, where the noise emitted by the UAV under different operating conditions (balanced and unbalanced blades) is recorded and analyzed.

In comparison, the referenced study achieved an accuracy of 97.63% in classifying UAV blade conditions. The proposed DroneCare system, on the other hand, achieved a slightly higher accuracy of 98.15%. While the referenced study focuses on detecting faults in a single component (propeller), the DroneCare system analyzes multiple flight parameters such as battery level, wind speed, GPS accuracy, and payload weight. This enables the model to provide a more comprehensive and reliable prediction of drone failures.

Table 4 : Comparison between the previous study and the proposed DroneCare system.

Table Feature	Previous Study (Blade Fault Detection)	Proposed System (DroneCare)
Model	Artificial Neural Network (ANN)	Artificial Neural Network (ANN)
Task Type	Classification	Classification
Scope	Single Component (Propeller)	Full Drone System
Accuracy	97.63%	98.15%
Performance	High Accuracy	Higher & More Comprehensive

Table 4 shows, a comparison between the ANN-based UAV blade fault detection study and the proposed DroneCare system, highlighting the improved performance and broader scope of the proposed model

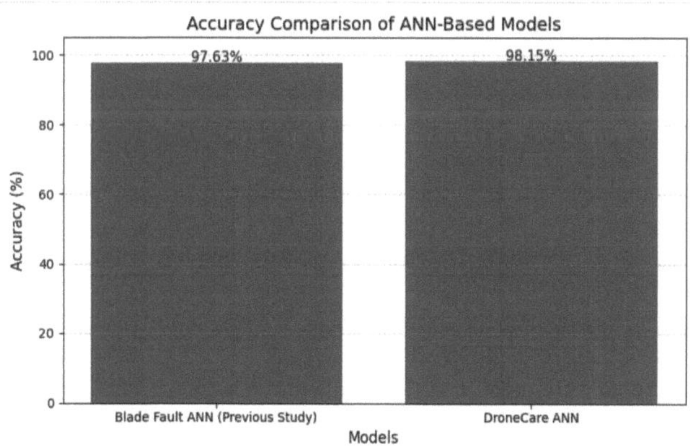

Fig. 10: Accuracy comparison of ANN-based models

Figure 10 show the Accuracy comparison of ANN-based models between the previous UAV blade fault detection study and the proposed DroneCare system. The comparison shows that although both models use ANN, the DroneCare system achieves better performance due to the use of multiple input features.

7 Conclusion

This project created DroneCare, an intelligent system that uses telemetry data to forecast drone malfunctions. With an accuracy of 98.15%, the Artificial Neural Network (ANN) model outperformed other models like Logistic Regression, SVM, Random Forest, and XGBoost, according to the results.

By facilitating early failure identification, the technology shows how predictive analytics may enhance drone dependability and operational safety. This improves productivity, lowers maintenance costs, and lessens crashes, particularly in vital applications like emergency response and logistics.

To further enhance prediction performance, future work can concentrate on adding real-time data and growing the dataset.

Acknowledgments. The authors would like to express their sincere gratitude to their project supervisor, Dr. Noora Yahya Al Hoqani, for her continuous guidance, support, and valuable feedback throughout the development of this project. The authors also extend their appreciation to all faculty members and colleagues who provided constructive suggestions and support during this research.

References

Ahmad, M.W., Akram, M.U., Ahmad, R., Hameed, K., Hassan, A.: Intelligent framework for automated failure prediction, detection, and classification of mission critical autonomous flights. ISA Trans. **129**, 355–371 (2022)

Arafat, M.Y., Alam, M.M., Moh, S.: Vision-based navigation techniques for unmanned aerial vehicles: review and challenges. Drones. **7**(2), 89 (2023)

Babichev, S.A., Ries, J., Lvovsky, A.I.: Quantum scissors: teleportation of single-mode optical states by means of a nonlocal single photon. arXiv:quant-ph/0208066. (2002)

Bell, V., Arce, I.M., Mase, J.M., Rengasamy, D., Rothwell, B., Figueredo, G.P.: Anomaly detection for unmanned aerial vehicle sensor data using a stacked recurrent autoencoder method with dynamic thresholding. SAE Int. J. Aerosp. **15**(2), 219–229 (2022)

Cabahug, J., Eslamiat, H.: Failure detection in quadcopter UAVs using K-means clustering. Sensors. **22**(16), 6037 (2022)

Hepsiba, G., et al.: Autonomous UAV forced landing site prediction using machine learning. Int. J. Sci. Res. Arch. **12**(1), 1010–1016 (2024)

Iannace, G., Ciaburro, G., Trematerra, A.: Fault diagnosis for UAV blades using artificial neural networks. Robotics. **8**(3), 59 (2019)

Kaggle: Supplemental drone telemetry data – Drone operations log. https://www.kaggle.com/ (2025). Accessed 2025

Petritoli, E., Leccese, F., Ciani, L.: Reliability and maintenance analysis of unmanned aerial vehicles. Sensors. **18**(9), 3171 (2018)

Roshanski, I., Roshanski, M., Kalech, M.: Real-time sensor fault detection in drones: a correlation-based algorithmic approach. OASIcs. **125**, 17 (2024).:1–17:20

Sadhu, V., Anjum, K., Pompili, D.: On-board deep-learning-based unmanned aerial vehicle fault cause detection and classification via FPGAs. IEEE Trans. Robot. **39**(4), 3319–3331 (2023)

Şahin, H.: Predictive UAV battery maintenance planning with artificial intelligence. J. Aviat. **9**(2), 260–269 (2025)

Enhancing Multi-Objective Eco-Efficient Flight Route Optimization Under Dynamic Constraint

Joyce Chua Xin Jie[1], Maizatul Akmar Ismail[2](✉), and Artika Arista[2,3]

[1] Master in Data Science, Faculty of Computer Science and Information Technology, Universiti Malaya, 50603 Kuala Lumpur, Malaysia
23093746@siswa.um.edu.my

[2] Department of Information Systems, Faculty of Computer Science and Information Technology, Universiti Malaya, 50603 Kuala Lumpur, Malaysia
{maizatul,artika.arista}@um.edu.my

[3] Information Systems, Faculty of Computer Science, Universitas Pembangunan Nasional Veteran Jakarta, Jakarta 12450, Indonesia

Abstract. This study proposes an integrated AI framework to enhance eco-efficient flight trajectory optimization within the Malaysian domestic aviation sector. To ensure practical feasibility and immediate industrial applicability, this study focuses on velocity and temporal dimensions (4D trajectory) optimization rather than complex horizontal path rerouting. The framework employs a sophisticated multi-stage architecture: utilizing Conditional Generative Adversarial Networks (CGAN) for data augmentation to resolve data imbalances, Multi-Objective Genetic Algorithms (MOGA) for initial speed scheduling and Dynamic Programming Algorithms (DPA) for tactical refinement by adjusting speeds across six segments under dynamic weather constraints. Experimental results have successfully proven that the framework can effectively reduce both fuel consumption and flight time by eliminating historical operational waste. By navigating the Pareto front between fuel efficiency and schedule reliability, the model provides a resilient Decision Support Tool that adheres to the structural limits of aircraft performance. Beyond technical achievements, this study has also achieved sustainability targets through intelligent speed management, which aligns with SDG 13 (Climate Action).

Keywords: Multi-Objective Optimization · Eco-Efficient Flight · Generative Adversarial Networks · 4D Trajectory Profile

1 Introduction

The existing paradigm in flight route optimization is facing significant challenges in developing a framework that can address multiple objectives in dynamic environment. Traditional flight route planning methods are less effective in adapting to unexpected events, such as air traffic congestion or sudden weather changes (Sydorenko et al., 2025). Air traffic congestion, caused by limited control point capacities and unforeseen

N. Mohd Sharef et al. (Eds.): CAMP 2026, CCIS 3093, pp. 111–123, 2027.
https://doi.org/10.1007/978-981-92-4302-0_9

events, often necessitates rerouting or delays, leading to increased costs and inefficiencies (Spyrou et al., 2025). These limitations can significantly hinder efforts to achieve both eco-efficiency and operational resilience. Multi-objective optimization algorithms like MOPSO, MOGA, NSGA-II, and MOETLBO have been used to optimize robotic manipulator trajectories for industrial applications (Rout et al., 2026). Other studies have employed algorithms like MOPO and INSGA-II to optimize robotic arm trajectories based on criteria such as time, energy, and smoothness (Zhang et al., 2025). Others have explored generative AI, such as Tabular Variational Autoencoders (TVAE) and Gaussian Copula (GC), which have been used to create synthetic flight datasets (Aly & Sharpanskykh, 2025).

In previous studies, the methods have typically been used in isolation or to specific static aspects of flight planning (Odat et al., 2024; Petit & Desbiens, 2024). For example, Generative AI can generate realistic data but has not been fully integrated with adaptive optimization techniques to form a unified system. Similarly, MOGAs are effective in initial route generation but struggle to accommodate dynamic constraints. The main challenge is the lack of integrated framework which combines Generative AI for data generation, MOGAs for multi-objective strategic speed planning and adaptive capabilities of DPAs.

In this research, the gap will be filled by developing an integrated AI framework that leverages all three methods to generate flight trajectories that are both eco-efficient and robust to dynamic aviation environment. This study is built upon two key assumptions: the aircraft follows fixed flight path and maintains a constant cruise altitude. By keeping these variables fixed, the study focuses specifically on optimizing the velocity and temporal dimensions (4D trajectory). The key challenge lies in the absence of a unified and integrated approach that combines robust data generation with adaptive optimization techniques to generate speed-optimized trajectories that are both eco-efficient and resilient to the unpredictable and dynamic aviation environment. Addressing this gap allows transformative advancements in aviation sector and decision-making. The main purpose of this research is to develop an integrated AI framework that leverages Generative AI to enhance multi-objective eco-efficient flight trajectory optimization under dynamic constraints. The aim is to create a system that can generate and adapt flight routes to improve sustainability and flight operational efficiency.

The remainder of this paper is organized as follows. Section 2 presents the research methodology, including data acquisition, data integration, feature engineering, preprocessing, CGAN-based data augmentation, and optimization procedures using MOGA and DPA. Section 3 discusses the results of the proposed framework, including comparative performance and model validity. Section 4 provides a broader discussion of the findings and their implications, while Sect. 5 concludes the paper with the main contributions, limitations, and suggestions for future research.

2 Research Methodology

Figure 1 shows the sequential research methodology of this study. The process initiates with the data integration of historical flight and meteorological datasets, followed by feature engineering to derive aviation metrics. The enhanced dataset was cleaned and

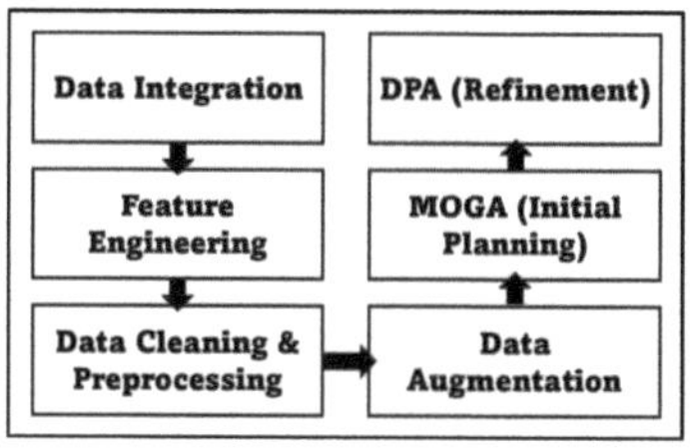

Fig. 1 Research Methodology Flowchart

preprocessed to prepare it for data augmentation using CGAN, initial speed-schedule planning by MOGA, and further refinement by DPA.

2.1 Data Acquisition and Description

This research utilized two primary datasets to build a comprehensive and dynamic data environment for flight route optimization. The first dataset contains historical flight data, while the second one provides meteorological information.

2.1.1 Data Sources

The primary source of flight data is taken from the OpenSky Network, an open-access platform that provides historical flight tracking information. This dataset allows us to analyze actual flight paths and operational behaviors. Besides, the meteorological data is acquired from Open-Meteo Historical Weather API. This source offers historical weather data with specific date range, latitude and longitude. For this study, the datasets only focus in Malaysia area, for the month of September 2022. The high-granularity hourly meteorological data makes it highly suitable for this study.

2.1.2 Data Description

The integrated dataset comprises variables from both sources (Table 1).

Table 1 Dataset Features

No.	Feature/Column name	Description
1	icao24	Aircraft identifier
2	call sign	Flight number
3	estdepartureairport	Origin airport codes
4	estarrivalairport	Destination airport codes
5	firstseen	Flight start time
6	lastseen	Flight end times
7	model	Aircraft model

(*continued*)

Table 1 (*continued*)

No.	Feature/Column name	Description
8	typecode	Aircraft type
9	date	Date of the flight
10	registration	Aircraft registration number
11	temperature_2m	Air temperature at 2 m above ground
12	wind_speed_10m	Wind speed at 10 m above ground
13	wind_direction_10m	Wind direction at 10 m above ground
14	pressure_msl	Atmospheric air pressure reduced to mean sea level (msl)
15	surface_pressure	Pressure at surface
16	cloud_cover	Total cloud cover as an area fraction
17	cloud_cover_low	Low level clouds and fog up to 2 km altitude
18	cloud_cover_mid	Mid level clouds from 2 to 6 km altitude
19	cloud_cover_high	High level clouds from 6 km altitude
20	precipitation	Total precipitation (rain, showers, snow) sum of the preceding hour
21	rain	Only liquid precipitation of the preceding hour including local showers and rain from large scale systems
22	relative_humidity_2m	Relative humidity at 2 m above ground
23	weather_code	Weather condition as a numeric code. Weather code is calculated from cloud cover analysis, precipitation and snowfall
24	wind_gusts_10m	Gusts at 10 m above ground of the indicated hour
25	apparent_temperature	Apparent temperature is the perceived feels-like temperature combining wind chill factor, relative humidity and solar radiation
26	dew_point_2m	Dew point temperature at 2 m above ground
27	vapour_pressure_deficit	Vapor Pressure Deificit (VPD) in kilopascal (kPa)

2.2 Data Integration

The main purpose of data integration is to combine flight data with weather information to understand how environmental factors affect flight performance. Since both datasets come from different sources, they are joined based on date, time and location. To keep the research scope focused only on Malaysia airports, the flight data is filtered with airports starting with 'WM' prefix. The joining time and location are based on the departure airport because the initial weather conditions at the departure airport mostly affect the route planning. The output of this step is a single table with flight details and specific weather conditions like wind speed and temperature.

2.3 Feature Engineering

The integrated raw data only has basic information like time and weather. However, specific aviation metrics like flight distance, fuel consumption and carbon emissions are necessary for flight route optimization. Mathematical physics and aviation standards are applied in this step to derive raw data into useful features.

First, the Haversine formula is used to calculate the Great-Circle Distance (GCD) of the flight. GCD is the shortest distance between two points on the surface of a sphere. This distance can be calculated by using the geographical coordinates of the two points concerned. Haversine Formula:

$$a = \sin^2\left(\frac{\Delta\phi}{2}\right) + \cos(\phi_1)\cdot\cos(\phi_2)\cdot\sin^2\left(\frac{\Delta\lambda}{2}\right) \tag{1}$$

$$c = 2\cdot \operatorname{atan2}\left(\sqrt{a}, \sqrt{1-a}\right)$$

$$\mathrm{d} = \mathrm{R}\cdot \mathrm{c}$$

Then, the aircraft types in the dataset are matched to the equivalent ICAO aircraft types. This step is essential because different types of aircraft consume different amounts of fuel for the same distance. Matching them correctly ensures accurate application of fuel consumption calculations. The fuel consumption is estimated based on the great-circle distance and aircraft type. This approach is consistent with practices who applied ICAO fuel burn tables to estimate aircraft-specific fuel usage.

The formula used is:

$$Fuel\ Burn\ (kg) = f\,(Aircraft\ Type, Distance) \tag{2}$$

Fuel consumption is determined by matching the aircraft types and flight distance with the ICAO Fuel Burn Tables. The ICAO Fuel Burn Tables only list fuel for fixed distances, hence the linear interpolation function is applied to calculate more accurate fuel consumption.

Linear Interpolation Function:

$$\mathrm{F} = \mathrm{F}_1 + (\mathrm{d} - \mathrm{d}_1)\cdot \frac{F2 - F1}{d2 - d1} \tag{3}$$

Carbon emissions are derived directly from the estimated fuel burn using a fixed multiplier of 3.16 kg of CO_2 per 1 kg of jet fuel. This conversion factor is also adopted by ensuring alignment with international environmental reporting standards.

The formula is:

$$CO_2\ Emissions\ (kg) = Fuel\ Burn\ (kg) \times 3.16 \tag{4}$$

Lastly, some additional features are derived for the optimization models. The actual flight duration is derived by taking the difference between the departure and arrival time. The baseline time, which is the minimum possible flight time is set by taking the minimum flight time for each aircraft type. The time deviation ratio, which serves as a performance score, is created by dividing the actual flight time by baseline time.

2.4 Data Cleaning and Preprocessing

Raw data is often inconsistent and contains missing information, weird symbols, or duplicated rows that affect the result. The step serves to transform the raw data into a clean, structured and numerical format suitable for modelling phase. Before performing data cleaning, some functions like *df.info()* and *df.describe()* are used to scan the dataset for anomalies.

The initial dataset was streamlined from 972 to 505 entries. This significant reduction is due to the incomplete records and invalid flight data that might skew the optimization results. After the data is cleaned, some preprocessing techniques have been done to prepare the data for optimization models. The categorical features are encoded using One-Hot Encoding, where these features are converted into numbers 0 and 1. Then, Min-Max scaler is applied to squash all numerical values into a standard range between 0 and 1. This is to ensure that features with larger magnitudes do not disproportionately influence the model's learning process. This phase produced a fully numerical matrix which contains 44 columns of clean and scaled data that is ready for modelling.

2.5 Data Augmentation Using Conditional Generative Adversarial Network (CGAN)

The cleaned and preprocessed historical dataset is limited in size and variety. In real world, there might be rare situations like extreme weather or heavy air traffic that might not be included in the raw dataset. To address the imbalanced nature of historical flight data, the dataset was expanded to 5000 samples. The expanding sample size via Conditional Generative Adversarial Network (CGAN) is essential to generate high-quality synthetic data that reflects the original dataset's distribution. Each flight is divided into segments for DPA preparation.

2.6 Initial Route Optimization Using Multi-Objective Genetic Algorithm (MOGA)

With thousands of possible flight scenarios, Multi-Objective Genetic Algorithm (MOGA) can be applied to find the best possible initial routes for each flight. In this phase, Non-Dominated Sorting Genetic Algorithm (NSGA-II) was implemented due to its efficiency in handling conflicting objectives like fuel consumption and flight time. This algorithm used the survival of the fittest to ensure only routes with the best trade-off survive. Routes that are not beaten by any other route in both fuel and time should be kept.

The main optimization in the proposed framework is done by a structured parallel loop configuration. Specifically, the Multi-Objective Genetic Algorithm (MOGA) is initiated with a population size of 500 candidate solutions for each of the top 100 selected flight routes. This algorithm is exposed to a termination criterion of exactly 100 generations of evolution to iteratively refine these solutions to an optimal Pareto Front with an even balance between fuel and time efficiency using standard genetic operators for selection, crossover and mutation.

2.7 Adaptive In-Flight Optimization Using Dynamic Programming Algorithm (DPA)

In the final phase, the Dynamic Programming Algorithm (DPA) acted as a refinement tool to transform a strategic plan into an operationally perfect flight route. The core methodology is to solve the big problem, the total flight efficiency by making small and optimal adjustments to each of the segments. To initiate this process, the best solution (most fuel-efficient route) for each route selected from MOGA result.

3 Result

3.1 Result and Comparison

This section evaluates the performance of the two-stage optimization framework. By comparing the output from MOGA and DPA against historical baseline, the effectiveness of the system in producing eco-efficient flight routes can be measured.

3.1.1 MOGA Optimization Result

The first stage of the optimization applied MOGA to the top 100 most frequent routes to find the trade-off between fuel and time. This algorithm successfully converged into a series of Pareto-optimal solutions for each route. These solutions represent the best possible trade-offs where fuel cannot be reduced further without increasing flight time. By using a fitness weighting of -5.0 for fuel and -1.0 for time, the algorithm prioritizes the reduction in fuel consumption that can significantly reduce carbon footprint.

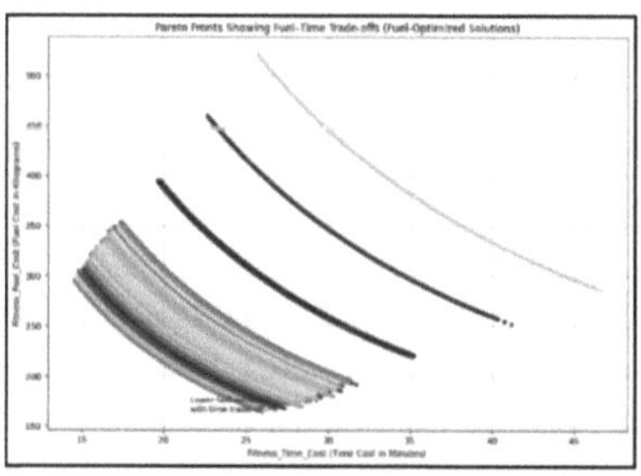

Fig. 2 Pareto Fronts Showing Fuel-Time Trade-offs

Figure 2 shows a collective of results for all 100 optimized routes. Each color represents a different route profile, and each point on the curve represents an optimal speed profile. The downward slope of the lines proves a successful trade-off: for an optimized flight to achieve lower fuel costs requires a higher flight time.

--- Top 10 Solutions ---

Solution_ID	Route_Profile_ID	Fitness_Fuel_Cost	Fitness_Time_Cost	Baseline_Fuel_Cost	Baseline_Time_Cost
R61_SOL_1	R61	164.23	25.98	247.84	32.99
R61_SOL_2	R61	165.17	25.82	247.84	32.99
R61_SOL_3	R61	165.83	25.73	247.84	32.99
R61_SOL_4	R61	165.92	25.72	247.84	32.99
R61_SOL_5	R61	165.95	25.71	247.84	32.99
R61_SOL_6	R61	166.22	25.67	247.84	32.99
R61_SOL_7	R61	166.33	25.66	247.84	32.99
R91_SOL_1	R91	166.41	26.36	247.84	32.99
R91_SOL_2	R91	166.41	26.36	247.84	32.99
R91_SOL_3	R91	166.42	26.36	247.84	32.99

Fig. 3 Comparative Fuel and Time Efficiency: MOGA VS Baseline

However, based on Fig. 3, the key finding in this study is that the optimized solutions successfully reduced both fuel consumption and flight time. This is due to the significant inefficiency of the historical operations, and MOGA has effectively removed this operational waste. In this analysis, the historical baseline was based on the unified mean performance of all historical flights in the dataset.

3.1.2 DPA Refinement Result

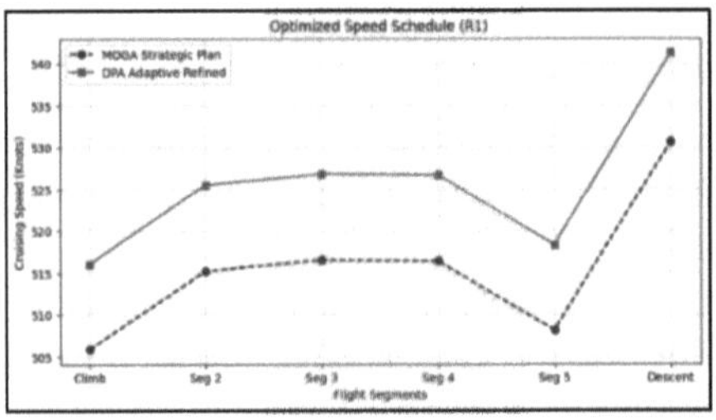

Fig. 4 Optimized Speed Schedule: MOGA VS DPA (Route 1)

The second stage utilized DPA to refine the MOGA results by accounting for wind and temperature efficiency. The DPA selected the best fuel-efficient solution from MOGA results and adjusted the speeds in each segment based on the weather volatility.

Figure 4 illustrates how the aircraft's speed changes during the six segments of a flight. The blue dotted line represents the best calculated speed profile planned by MOGA, which provides a solid foundation for fuel savings. The DPA makes small adjustments to this plan in each of the six flight segments. If the plane encounters a headwind or a change in temperature, the DPA automatically shifts the speed to stay efficient. These small adjustments provided a consistent improvement of 1.5% to 2.8% in additional fuel savings. This demonstrates the tactical refinement by DPA is crucial to achieving eco-efficiency under dynamic weather conditions.

3.1.3 Comparison and Discussion

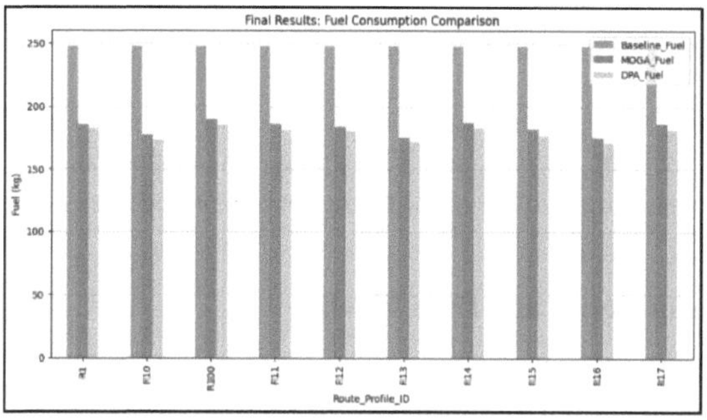

Fig. 5 Fuel Consumption Comparison (Baseline VS MOGA VS DPA)

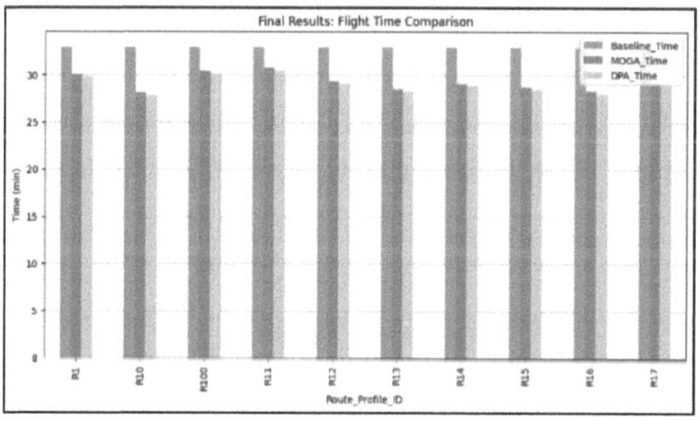

Fig. 6 Flight Time Comparison (Baseline VS MOGA VS DPA)

To evaluate the overall success of the study, a comparative analysis was performed. Figures 5 and 6 display the changes in both fuel consumption and flight time. Both graphs clearly show that these two algorithms successfully reduced fuel consumption, resulting in greener flights while simultaneously making the flight time shorter to achieve operational efficiency.

In many traditional models, there is a penalty for saving fuel, which means saving fuel requires longer flight time. However, due to the inefficiency of the historical baseline, MOGA successfully removed operational waste, resulting in reduction in both fuel consumption and flight time. By selecting smarter speed schedules, this study managed to achieve total fuel reduction up to 35% and time reduction up to 21%.

This double benefit proves that eco-efficiency does not necessarily have to come at the cost of longer flight time. By optimizing suboptimal cruising speeds, the flights are now both environmentally sustainable and operationally faster.

In summary, the results prove that Malaysian domestic flights can be made greener and faster at the same time. This provides a strong justification for airlines to implement

AI-driven route optimization using speed scheduling to maintain eco-efficient flight while ensuring high operational efficiency.

3.2 Model Validity and Stability Results

This section evaluates the performance of the two-stage optimization framework. By comparing the output from MOGA and DPA against historical baseline, the effectiveness of the system in producing eco-efficient flight routes can be measured.

3.2.1 CGAN Performance and Data Fidelity

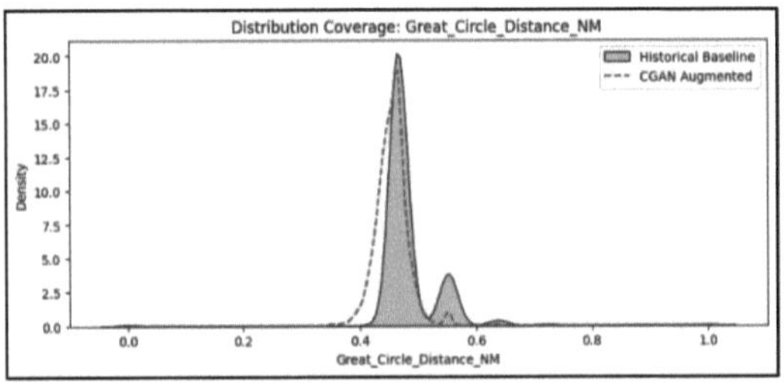

Fig. 7 Distribution Coverage: Great Circle Distance

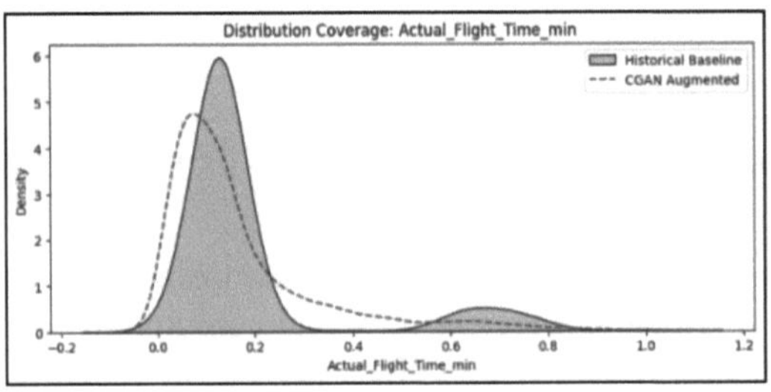

Fig. 8 Distribution Coverage: Actual Flight Time

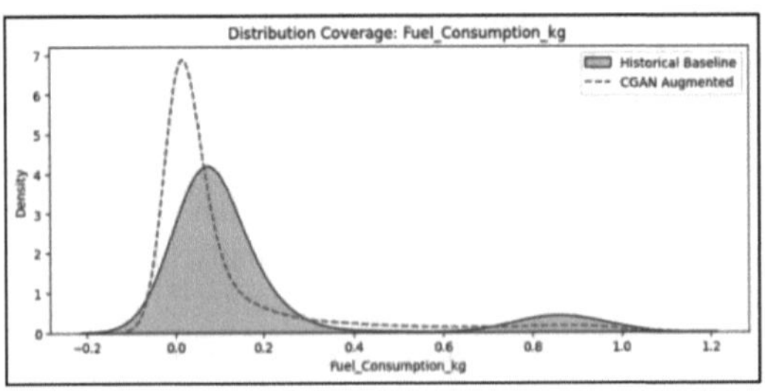

Fig. 9 Distribution Coverage: Fuel Consumption

	Feature	Historical Mean	Synthetic Mean	Mean Similarity (%)	KS Statistic
0	Great_Circle_Distance_NM	0.48	0.46	94.43%	0.6111
1	Actual_Flight_Time_min	0.19	0.17	87.42%	0.2925
2	Fuel_Consumption_kg	0.17	0.11	64.74%	0.4850

Fig. 10 Metric Proximity

Figures 7, 8, and 9 illustrate that the distribution of the augmented dataset closely mirrors the historical baseline, especially for the high-density in flight time and flight distance. Furthermore, the high mean similarity is based on Fig. 10 indicates that the generative model successfully synthesized a robust flight population that remains the statistical characteristics of the original data.

3.2.2 MOGA Stability and Convergence Verification

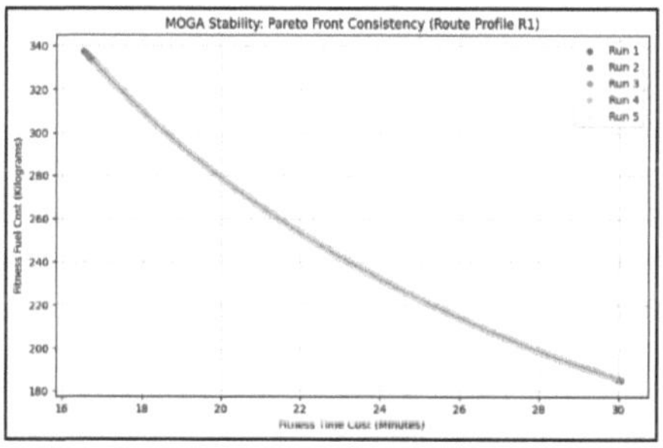

Fig. 11 Pareto Front Consistency Plot (Route 1)

As shown in Fig. 11, all five independent runs converged to a nearly identical optimal boundary. The perfect overlap proves that MOGA is not sensitive to initial random seeding and consistently identifies the most eco-efficient trade-offs between fuel and time costs. This stability ensures that the solutions generated are globally optimal within the defined search space.

The convergence of the MOGA and the data-fidelity of the Conditional Generative Adversarial Network (CGAN) are mainly validated through descriptive trends and rigorous visual inspection of the distribution graphs (a standard qualitative baseline in generative modeling literature), but the initial alignment is highly promising. The visual overlap demonstrates that the synthetic trajectory data captures the core underlying patterns of the operational data. To complement these descriptive findings, future extended studies will include formal statistical validation metrics, such as structural similarity indexing or distribution divergence tests, to mathematically quantify the performance of the framework.

4 Discussion

The integration of Generative AI in aviation and aerospace is revolutionizing various aspects of flight operations, from routing to safety and efficiency. Generative AI models, such as Generative Adversarial Networks (GANs) are being employed in this industry for flight trajectory profile optimization and resource management enhancement in dynamic and uncertain environments. While prior studies have applied GANs for trajectory generation (Zhu et al., 2025) or Multi-Objective Genetic Algorithms (MOGAs) for route optimization, this research uniquely integrates GANs for robust data synthesis, MOGAs for initial eco-efficient speed schedule creation, and Dynamic Programming Algorithms (DPAs) for adaptive in-flight velocity optimization. This offers a more comprehensive approach to dynamic environmental constraints.

This research is important because it addresses critical challenges in the aviation industry, including environmental concerns and the need for adaptability in dynamic environments. It offers a novel approach to overcome the limitations of traditional, static route planning by integrating a multi-stage AI framework. The significance of this study is that the optimization of speed schedules can lead to considerable fuel savings by identifying and eliminating historical operational waste. As a direct result, airline operational costs can be reduced, and global carbon emissions targets can be met. The potential impact includes substantial reductions in fuel consumption and carbon emissions, improved operational efficiency, and enhanced safety.

It is important to mention that the reported fuel savings up to 35% and flight time reductions up to 21% are the baseline performance of the integrated framework in standardized environmental and aircraft configurations. A comprehensive sensitivity analysis over highly dynamic weather conditions, aircraft operating weights and different objective weights is essential for real-world deployment but is beyond the scope of this initial proof-of-concept paper. Future versions of this work will include systematic modelling of these environmental variables and robustness evaluations under different objective constraints.

5 Conclusion

This study has successfully developed a two-stage AI-driven framework to enhance eco-efficient flight route optimization within the Malaysian domestic aviation sector. By combining MOGA for initial route planning with a DPA for further refinement, the study addressed the conflict between fuel consumption and flight time. A CGAN model has been successfully developed to expand the dataset to 5000 high-fidelity flights, maintaining up to a 94.4% similarity to historical routes. MOGA is well utilized to identify Pareto-optimal speed schedules for top 100 route profiles, resulting in a theoretical fuel reduction of up to 35%. DPA is successfully integrated to adjust segments based on environmental variables.

By integrating MOGA and DPA, the system can handle long-term strategic goals while adapting to short-term weather fluctuations. Unlike traditional models that propose a strict trade-off between fuel and time, this research proved that AI could eliminate historical operational inefficiencies, resulting in improvements in both metrics. By performing data augmentation using CGAN, this study demonstrated robustness and reliability of different route profiles in Malaysia. In this study, the focus is on changing the speed of the plane instead of the paths. Hence, it is highly compatible with current Air Traffic Control regulations and easier to be applied in real-world.

As stated in the study strengths, this study primarily focuses on optimizing speed profile, instead of changing paths. This limits the study as changing paths could potentially offer better results in fuel and time efficiency. The historical data was obtained from a specific timeframe (September 2022). Variations in seasonal traffic or extreme weather conditions might not be captured in the dataset, which causes slight changes in the optimization result. This study focused on optimizing speed across fixed flights. Including dynamic altitude changes might provide a more comprehensive optimization outcome.

Integrating OpenSky and weather APIs into a live dashboard could allow pilots to access real-time rerouting suggestions. Integrating horizontal path adjustments with speed scheduling to achieve a full 4D trajectory optimization. By including long-haul international flights, the scalability of the two-stage framework can be tested across different aircraft types and more complex weather systems.

References

Aly, K., Sharpanskykh, A.: Synthetic flight data generation using generative models. In: Integrated Communications, Navigation and Surveillance Conference, ICNS (2025). https://doi.org/10.1109/ICNS65417.2025.10976960

Odat, E., Ghazzai, H., Alsharoa, A.: A WaveGAN approach for mmWave-based FANET topology optimization. Sensors. **24**(1) (2024). https://doi.org/10.3390/s24010006

Petit, L., Desbiens, A.L.: MOAR planner: multi-objective and adaptive risk-aware path planning for infrastructure inspection with a UAV. In: Proceedings - IEEE International Conference on Robotics and Automation, pp. 8422–8428 (2024). https://doi.org/10.1109/ICRA57147.2024.10610907

Rout, A., Kavalli, K., Ramegowda, R.: Path Planning of KR6 R900 Vision Sensor Assisted KUKA Industrial Robot for Pick and Place Application Lecture Notes in Networks and Systems, 1358 LNNS, pp. 277–293 (2026). https://doi.org/10.1007/978-981-96-7614-9_21

Spyrou, E. D., Moustaka, A. M., & Kappatos, V. (2025). Lyapunov drift-plus-penalty method for Atco system making optimal decisions. Integrated Communications, Navigation and Surveillance Conference, ICNS. doi:https://doi.org/10.1109/ICNS65417.2025.10976749

Sydorenko, O., Lysa, N., Sikora, L., Martsyshyn, R., Miyushkovych, Y.: Optimizing aircraft routes in dynamic conditions utilizing multi-criteria parameters. Appl. Sci. (Switz.). **15**(11) (2025). https://doi.org/10.3390/app15116044

Zhang, M., Liu, J., Wu, Y., Hou, T., Li, T.: Multi-objective trajectory planning for robotic arms based on MOPO algorithm. Electronics (Switzerland). **14**(12) (2025). https://doi.org/10.3390/electronics14122371

Zhu, W., Jiang, R., Li, D., Kong, X., & Song, X. (2025). Trajectory generative models: a survey from unconditional and conditional perspectives. GeoInformatica 29(4), pp. 1033–1065. Springer Science and Business Media B.V. doi:https://doi.org/10.1007/s10707-025-00558-8

Feature Selection Using Grey Wolf Optimization for Hypertension Classification with Machine Learning

Muhammad Sam'an[1(✉)], Farikhin[2], and Beta Noranita[3]

[1] Department of Informatics, Universitas Muhammadiyah Semarang, Kedungmundu, Semarang, Central Java, Indonesia
muhammad92sam@unimus.ac.id

[2] Department of Mathematics, Universitas Diponegoro, Prof. Soedarto, Semarang, Central Java, Indonesia

[3] Department of Informatics, Universitas Diponegoro, Prof. Soedarto, Semarang, Central Java, Indonesia

Abstract. Hypertension is one of the major health problems related to cardiovascular disease and requires early detection to support appropriate medical intervention. However, hypertension prediction often involves high-dimensional clinical data containing redundant and irrelevant features that may reduce classification performance. Therefore, an effective feature selection approach is required to identify informative risk factors and improve prediction accuracy. This study proposes hypertension classification using machine learning combined with Grey Wolf Optimization (GWO)-based feature selection. The dataset applied included secondary clinical data containing demographic, lifestyle, anthropometric, and clinical examination variables related to hypertension risk. Five classification algorithms, namely Decision Tree (DT), Naive Bayes (NB), Support Vector Machine (SVM), K-Nearest Neighbor (KNN), and Random Forest (RF), were evaluated before and after feature selection. The experimental results indicated that RF achieved the best classification performance before feature selection, with an accuracy of 79.52%, precision of 69%, recall of 67%, F1-score of 68%, and ROC-AUC of 84.32%. After applying GWO-based feature selection, RF achieved improved performance with an accuracy of 84.34%, precision of 73%, recall of 81%, F1-score of 77%, and ROC-AUC of 90.54%. The selected feature subset successfully reduced irrelevant and redundant variables while preserving important information related to hypertension classification. Feature importance analysis showed that age, body weight, waist circumference, and body height were the most influential features for predicting hypertension. The results demonstrate that GWO produced a more compact feature subset and improved the classification performance of the Random Forest model on the hypertension dataset.

Keywords: Hypertension Classification · Grey Wolf Optimization · Feature Selection · Machine Learning

N. Mohd Sharef et al. (Eds.): CAMP 2026, CCIS 3093, pp. 124–138, 2027.
https://doi.org/10.1007/978-981-92-4302-0_10

1 Introduction

Hypertension remains one of the major health problems globally making contribution to high premature mortality and disability rates across many countries [1,2]. NCD (Non-Communicable Diseases) is the main global death factor recently, including in Indonesia [1,2]. Hypertension is among the five major diseases with a substantial impact on public health [3]. Hypertension is also widely known as a silent killer because many individuals are unaware of their condition until serious like coronary heart disease, kid-ney failure, stroke, and other cardiovascular disorders occur [4].

Many analyses have used ML (machine learning) approaches to enhance the hypertension accuracy classification and prediction applying clinical, lifestyle, and demographic data. The XGBoost (Extreme Gradient Boosting) algorithm was employed to classify hypertension risk using medical records, health examina-tion results, and lifestyle information, achieving a micro-average AUC (Area Under the Curve) of 0.822 for hypertension risk classification [5]. DT (Decision Tree), RF (Random Forest), GBM (Gradient Boosting Machine), XGBoost, LR (Logistic Regression), and LDA (Linear Discriminant Analysis) algorithms were also applied for hypertension prediction in South Asian populations, where XGBoost, LR, LDA and GBM obtained an 90% accuracy, 100%recall, and 95% F1-score [6]. The combination of XGBoost and the SHAP (SHapley Additive Expla-nations) method demonstrated that age, Body Mass Index (BMI), socioeconomic status, and residential region were the main factors influencing hypertension prediction among adults in Bangladesh [7]. XGBoost, Light Gradient Boosting Machine (LightGBM), CatBoost, LR, and RF were further applied to predict hy-pertension risk using data from the SATUSEHAT IndonesiaKu (ASIK) system, where the model incorporating patients' hypertension history achieved an AUC of 0.85, while the model without hypertension history obtained an AUC of 0.78 [8]. A machine learning approach based on 1-day blood pressure data obtained from the ABPM (Ambulatory Blood Pressure Monitor) produced a root mean square value of 0.0512 and demonstrated good hypertension prediction perfor-mance based on confusion matrix evaluation [9]. Several algorithms, including Support Vector Classifier (SVC), DT, LR, RF, GBM, Voting, and Stacking, were also utilized for hypertension prediction using Bioelectrical Impedance Analysis (BIA) data, where AutoMLP, Stacking, and Voting achieved accuracy values of 90%, 84%, and 83% [10]. RFECV (Recursive Feature Elimination with Cross-Validation) was employed to conduct feature selection on physi-cal examination data of hypertension patients before classification using SVM (Support Vector Machine), C4.5 DT, RF, and XGBoost, where XGBoost obtained the most optimal treatment of 94.36% accuracy, 0.875 F1-score, and 0.927 AUC based on 10-fold cross-validation evaluation [11].

Feature selection is an important stage in developing hypertension classification models because not all features in a dataset contribute meaningfully to prediction. A hybrid feature selection approach based on Lasso Regression was applied to obtain factors related to hypertension before the classification process using XGBoost, and the feature optimization process enhanced the model performance with 77.2% accuracy and 84.6% AUC which was approximately 7% greater than the model with no feature optimization [12]. HFSR (Hybrid Feature Selection and Reduction) was utilized to choose and reduce important features extracted from pulse plethysmograph signals for hypertension detection, enabling weighted KNN to achieve 99.4% accuracy, 99.6% sensitivity, and 99.2% specificity based on 10-fold cross-validation evaluation [13]. Factor Analysis (FA), Feature Importance (FI) and Multicollinearity Analysis (MC) were applied to the KNHANES dataset to find the most appropriate features for the prediction of hypertension risk, resulting in the MC-FA-RF combination achieving an accuracy of 80.12%, 83.49% F-score, 85.96% AUC and mean square error of 0.106 [14]. Feature prediction based on RF, LR, DT, XGBoost, and Gradient Boosting (GB) was combined with LSTM (Long Short-Term Memory) to enhance hypertension prediction, and the combination of GB-based Feature Prediction with LSTM achieved an accuracy and F1-score of 98.48% for early hypertension detection [15]. Feature selection based on uni-variate Cox p-value, C-index, RSF (Random Survival Forest), Lasso (Least Absolute Shrinkage and Selection Operator) and Statistically Equivalent Signature (SES) was also used to identify important features in the prediction of hypertension incidence, enabling the feature selection combination and Ridge, Lasso, EN (Elastic Net), RSF, and GB algorithms to achieve good predictive performance with an average C-index ranging 0.76–0.78 in hypertension survival data [16].

Although these studies demonstrate that feature selection can improve hypertension prediction performance, most existing approaches rely on statistical significance measures, correlation-based analysis, or model specific feature importance scores. Such approaches not adequately capture complex nonlinear relationships and interactions among demographic, lifestyle, anthropometric, and clinical risk factors associated with hypertension. Consequently, potentially informative feature combinations may be overlooked, while irrelevant and redundant variables may remain in the dataset, increasing model complexity and limiting classification performance. This limitation is particularly important in hypertension prediction because hypertension is influenced by multiple interrelated risk factors that collectively contribute to disease occurrence. Therefore, a feature selection approach capable of exploring a broader search space and identifying optimal feature combinations is required.

GWO (Grey Wolf Optimization) is a metaheuristic proposed by the social behavior and hunting patterns of grey wolf packs [17]. The algorithm imitates the social hierarchy of wolves, which include α, β, δ, and ω, where the α wolf indicates the most optimal solution in the population, while β and δ act as the 2nd and 3rd optimal solutions that help in guiding the hunt process to the most appropriate solution [18]. The optimization process in GWO is appropriate to main mechanisms: encircling, hunting, and attacking prey. When in encircling stage, the population positions are updated according to the estimated position of the optimal solution [19]. The hunting stage utilizes information from the α, β, and δ solutions to direct the population toward search regions with better solution opportunities [20]. The attacking prey stage is performed by gradually reducing the control parameter so that the search process shifts from exploring the search space to exploiting the best solution obtained [21]. GWO has a simple algorithmic structure, re-quires relatively few parameters, and can handle high-dimensional search spaces [22]. These characteristics make GWO particularly suitable for wrapper-based feature selection, where the search process evaluates numerous feature combinations simultaneously. Unlike filter-based and embedded approaches, wrapper-based feature selection directly assesses candidate feature subsets according to classification performance, enabling the discovery of feature interactions that may not be identified through statistical analysis alone.

Previous studies on hypertension classification have primarily employed filter-based and embedded feature selection techniques, such as Lasso Regression, ANOVA, Feature Importance, ReliefF, mRMR, and RFECV. While these methods have demonstrated promising predictive performance, they generally evaluate features individually or according to model-specific importance measures, which may limit their ability to identify nonlinear interactions and synergistic effects among hypertension risk factors. Consequently, the exploration of optimal feature subsets remains constrained by the characteristics of the applied method. Despite the increasing use of metaheuristic algorithms for feature selection in various healthcare applications, the application of wrapper-based feature selection using GWO in hypertension classification remains limited. Therefore, further investigation is required to evaluate whether GWO can identify more informative feature subsets and improve classification performance in hypertension prediction. In the GWO-based feature selection process, each wolf is represented as a binary vector whose length equals the total number of features in the dataset. A value of "1" indicates that a feature is selected, whereas a value of "0" indicates that the feature is excluded [23,25]. Each feature subset is evaluated using a fitness function that typically considers classification performance metrics, such as accuracy, F1-score, and AUC, together with the number of selected features. The wolf positions are iteratively updated according to the positions of the α, β, and δ wolves until an optimal feature subset is obtained that provides high classification performance with lower model complexity [24].

To address the limitations of existing feature selection methods in capturing complex interactions among hypertension risk factors and the limited application of GWO-based feature selection in hypertension classification, this study proposes a hypertension risk classification framework integrating machine learning and GWO-based feature selection. The framework utilizes demographic, lifestyle, anthropometric, and clinical examination variables to identify an optimal subset of risk factors while reducing feature redundancy. Five machine learning algorithms, namely KNN, NB, DT, SVM, and RF, are evaluated as baseline classifiers using accuracy, precision, recall, F1-score, and AUC metrics. The best-performing classifier is subsequently combined with GWO to optimize feature selection and improve hypertension classification performance.

2 Method

The framework of the proposed study for hypertension risk classification applying ML and GWO-based feature selection is described in 1.

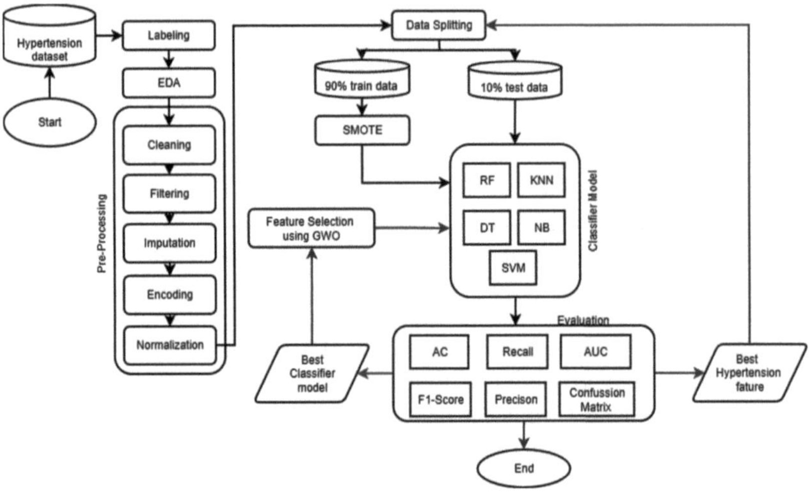

Fig. 1. Framework of proposed study.

2.1 Data Collection

The dataset of the analysis was a secondary clinical dataset obtained from Puskesmas Sumber, Rembang, Indonesia. The dataset consisted of 1,112 records and 32 variables related to demographic characteristics, lifestyle factors, anthropometric measurements, blood pressure indicators, laboratory examination results, and clinical referral attributes associated with hypertension risk assessment. Several variables included age, smoking habits, gender, systolic and diastolic blood pressure, body height, body weight, physical activity, waist circumference, and blood glucose examination results. The dataset was used to de-velop and evaluate hypertension classification models using ML algorithms and GWO. Before classification, the dataset underwent preprocessing stages, including data cleaning, missing value handling, categorical variable transformation, normalization, and data splitting. Table 1 shows the variables applied in this analysis.

2.2 Labeling

The dataset used in this study was obtained from a secondary government health survey database containing demographic, lifestyle, anthropometric, and clinical

Table 1. Description of Hypertension Dataset Variables

Variables 1–15			Variables 16–31		
No	Variable	Description	No	Variable	Description
1	umur	Patient age	16	rujukrs	Hospital referral status
2	jenis_kelamin	Patient gender	17	ktrk_matakanan	Right eye cataract condition
3	merokok	Smoking habit	18	ktrk_matakiri	Left eye cataract condition
4	aktifitas	Physical activity status	19	ktrk_rujukrs	Cataract referral status
5	gula_plus	Excessive sugar consumption	20	klrf_matakanan	Right eye refractive disorder
6	garam_plus	Excessive salt consumption	21	klrf_matakiri	Left eye refractive disorder
7	lemak_plus	Excessive fat consumption	22	klrf_rujukrs	Refractive disorder referral status
8	buah_minus	Low fruit consumption	23	klrf_telingakanan	Right ear hearing disorder
9	alkohol	Alcohol consumption habit	24	klrf_telingakiri	Left ear hearing disorder
10	sistol	Systolic blood pressure	25	klrf_rujukrs.1	Hearing disorder referral status
11	diastol	Diastolic blood pressure	26	cngk_telingakanan	Right ear abnormality condition
12	tinggi__badan	Body height (cm)	27	cngk_telingakiri	Left ear abnormality condition
13	berat_badan	Body weight (kg)	28	cngk_rujukrs	Ear abnormality referral status
14	lingkar_perut	Waist circumference	29	srmn_telingakanan	Right ear cerumen condition
15	cek_gula	Blood glucose examination result	30	srmn_telingakiri	Left ear cerumen condition
			31	srmn_rujukrs	Cerumen referral status

examination records related to hypertension risk factors. The dataset consisted of 1,112 participants who underwent standardized health examinations conducted by authorized healthcare professionals. Hypertension labels were assigned based on recorded systolic blood pressure and diastolic blood pressure measurements according to the adult blood pressure classification guidelines specified in the Decree of the Minister of Health of the Republic of Indonesia Number HK.01.07/MENKES/4634/2021. Participants were categorized as normal, prehypertension, or hypertension, and for binary classification purposes, hypertension cases were assigned a label of '1", while normal and prehypertension cases were assigned a label of '0" [26].

2.3 Exploratory Data Analysis

The EDA was conducted to understand the dataset's characteristics and distribution before the hypertension classification process. The analysis results showed that the dataset contained no missing values across all variables, allowing all attributes to be used in the analysis and classification. Data inspection also identified 3 duplicate records in the dataset, which were handled during preprocessing to prevent them from affecting classification results. The distribution of the target variable showed that the non-hypertension class (label 0) comprised 556 records, while the hypertension class (label 1) comprised 266 records.

2.4 Data Preprocessing

Data preprocessing was performed before training and assessing the hypertension classification model. The process included data cleaning by removing non-informative and leakage-related features, handling missing values using median and mode imputation, transforming binary categorical variables into numerical form, and normalizing numerical features using StandardScaler. After preprocessing, the dataset was grouped into training and testing sets for classification and model assessment.

2.5 Data Splitting and SMOTE

After data preprocessing, the dataset was divided into training and testing sets using a 90:10 train–test split ratio with stratified sampling to preserve the distribution of hypertension classes. A random state of 42 was used to ensure reproducibility. To mitigate class imbalance, SMOTE was applied only to the training data prior to model training, while the testing data remained unchanged to ensure unbiased performance evaluation.

2.6 Classification Models

This study employed 6 machine learning algorithms for hypertension classification, namely NB, KNN, DT, SVM, and RF. NB performs classification based on Bayes' Theorem under statement that each feature is independent. The data instance probability of a particular class is calculated applying Eq. (1).

$$P(C_k \mid X) = \frac{P(X \mid C_k)P(C_k)}{P(X)} \tag{1}$$

where $P(C_k \mid X)$ represents the probability of data X belonging to class C_k, $P(X \mid C_k)$ denotes the probability of features given class C_k, and $P(C_k)$ represents the prior probability of class C_k.

KNN performs classification based on the distance between data points using the Euclidean distance, as presented in Eq. (2).

$$d(x_i, x_j) = \sqrt{\sum_{k=1}^{n}(x_{ik} - x_{jk})^2} \tag{2}$$

where $d(x_i, x_j)$ represents the distance between the data points x_i and x_j, while n indicates the number of features.

DT constructs a decision tree by selecting the best attribute using the *Gini Index*, as shown in Eq. (3).

$$Gini(D) = 1 - \sum_{i=1}^{m} p_i^2 \tag{3}$$

where p_i denotes the proportion of data belonging to class i, while m represents the total number of classes.

SVM works by finding the optimal *hyperplane* to separate two classes, as expressed in Eq. (4).

$$f(x) = w^T x + b \tag{4}$$

where w represents the weight vector, x denotes the input data, and b is the bias term.

RF is an extension of DT that uses an *ensemble learning* approach, where multiple decision trees are constructed using different combinations of data samples and features. Each decision tree independently generates a class prediction, and the final classification result is determined based on the class with the highest number of predictions using the mode function, as shown in Eq. (5).

$$\hat{y} = \text{mode}\{h_1(x), h_2(x), \ldots, h_T(x)\} \tag{5}$$

where $\hat{y}$ represents the final prediction result, $h_t(x)$ denotes the prediction generated by the t-th DT, and T indicates the total number of decision trees in the Random Forest model. The mode function is a statistical function used to select the class that appears most frequently among all DT predictions. Mathematically, the mode function can be expressed as shown in Eq. (6).

$$\text{mode}(Y) = \arg\max_{c \in C} \sum_{t=1}^{T} I(h_t(x) = c) \tag{6}$$

where C represents the set of classes and $I(\cdot)$ is an indicator function that returns 1 if the prediction of a decision tree matches class c, and 0 otherwise.

2.7 Evaluation Metrics

The confusion matrix has some components, namely TP, TN, FP, and FN. TP is total hypertension cases correctly classified as hypertension, TN is total non-hypertension cases correctly classified as non-hypertension, FP is total non-hypertension cases incorrectly classified as hypertension, and FN is total hypertension cases incorrectly classified as non-hypertension. Accuracy measures the overall proportion of correctly classified data and is calculated using Eq. (7).

$$Accuracy = \frac{TP + TN}{TP + TN + FP + FN} \tag{7}$$

Precision measures the proportion of correctly predicted hypertension cases among all data predicted as hypertension, as shown in Eq. (8).

$$Precision = \frac{TP}{TP + FP} \tag{8}$$

Recall measures the ability of the model to correctly identify hypertension cases and is calculated using Eq. (9).

$$Recall = \frac{TP}{TP + FN} \tag{9}$$

F1-score is the harmonic mean of Precision and Recall, as presented in Eq. (10).

$$F1\text{-}Score = \frac{2 \times Precision \times Recall}{Precision + Recall} \tag{10}$$

AUC was used to evaluate the ability of the classification model to distinguish between hypertension and non-hypertension classes. A higher AUC value indicates better classification performance.

2.8 Feature Selection Using Grey Wolf Optimization

Feature selection in this study was performed using GWO to obtain the optimal feature subset for hypertension classification. The feature selection process was conducted after all classification models were evaluated on the original dataset, and the model with the highest accuracy was selected as the base classifier for evaluating feature subsets during optimization. Each solution in GWO represents a combination of selected and unselected features, where 1 indicates a selected feature and 0 indicates an excluded feature. GWO is a metaheuristic algorithm inspired by the hunting behavior and leadership hierarchy of grey wolves, consisting of α, β, δ, and ω wolves. The α wolf represents the best solution, β and δ wolves represent the second and third best solutions, while ω wolves represent the remaining candidate solutions. During optimization, wolf positions are updated based on the positions of α, β, and δ wolves to search for better feature subset solutions. The encircling mechanism is formulated using Eq. (11)–(12).

$$D = |C \cdot X_p(t) - X(t)| \tag{11}$$

$$X(t+1) = X_p(t) - A \cdot D \tag{12}$$

where $X(t)$ represents the current wolf position at iteration t, $X_p(t)$ represents the prey position or the best solution, and D denotes the distance between the wolf and the prey. The coefficient vectors A and C are calculated using Eqs. (13)–(14).

$$A = 2a \cdot r_1 - a \tag{13}$$

$$C = 2 \cdot r_2 \tag{14}$$

where r_1 and r_2 are random values in the range $[0, 1]$, while a is a control parameter that decreases linearly from 2 to 0 throughout the iteration process. The position update process based on alpha, beta, and delta wolves is calculated using Eqs. (15)–(17).

$$D_\alpha = |C_1 \cdot X_\alpha - X|, \quad D_\beta = |C_2 \cdot X_\beta - X|, \quad D_\delta = |C_3 \cdot X_\delta - X| \tag{15}$$

$$X_1 = X_\alpha - A_1 \cdot D_\alpha, \quad X_2 = X_\beta - A_2 \cdot D_\beta, \quad X_3 = X_\delta - A_3 \cdot D_\delta \tag{16}$$

$$X(t+1) = \frac{X_1 + X_2 + X_3}{3} \tag{17}$$

The continuous position output from GWO was converted to binary form using a sigmoid transfer function to determine whether a feature was selected or excluded. The sigmoid function is defined in Eq. (18).

$$S(X_i) = \frac{1}{1 + e^{-10(X_i - 0.5)}} \tag{18}$$

The probability generated by the sigmoid function was then converted into binary representation using Eq. (19).

$$B_i = \begin{cases} 1, \text{ if } rand < S(X_i) \\ 0, \text{ otherwise} \end{cases} \tag{19}$$

where B_i represents the binary status of the i-th feature, while $rand$ is a random value in the range $[0, 1]$. The quality of the selected feature subset was evaluated using the fitness function presented in Eq. (20).

$$Fitness = \alpha(1 - Accuracy) + (1 - \alpha)\left(\frac{N_s}{N}\right) \tag{20}$$

where $Accuracy$ represents the classification accuracy obtained from the base classifier, N_s denotes the number of selected features, N represents the total number of features, and α is a weighting parameter used to balance classification performance and feature reduction. The feature subset with the smallest fitness value was selected as the best feature subset. Algorithm 1 presents the pseudocode of the GWO-based feature selection process used in this study.

Algorithm 1. Feature Selection Using Grey Wolf Optimization

Input: Dataset X, target label y
Set population size
Set maximum iteration
Set fitness weight parameter
Initialize wolf positions randomly
Convert continuous positions into binary representation using Eqs. (18)–(19)
Evaluate each wolf using the fitness function in Eq. (20)
Determine alpha, beta, and delta wolves
for each iteration $t = 1$ to T **do**
  Update parameter a using Eq. (13)
  for each wolf **do**
    Calculate D_α, D_β, and D_δ using Eq. (15)
    Calculate X_1, X_2, and X_3 using Eq. (16)
    Update wolf position using Eq. (17)
  end for
  Convert updated positions into binary representation
  Evaluate fitness values
  Update alpha, beta, and delta wolves
end for
Return the best feature subset

3 Results and Discussion

Hypertension classification was performed using five *machine learning* algorithms, namely NB, KNN, DT, SVM, and RF, using all features from the preprocessing stage without feature selection. Model performance was evaluated using Accuracy, Precision, Recall, and F1-Score to assess the classification capability for hypertension and non-hypertension cases. The classification results before feature selection are presented in Table 2. RF achieved the best classification performance among all evaluated algorithms. This performance is related to the characteristics of the hypertension dataset, which contains numerical and categorical variables with non-linear relationships among features. RF can handle these characteristics by constructing multiple decision trees from different combinations of data samples and features, allowing the model to learn more diverse patterns while reducing the risk of *overfitting*. In addition, the *ensemble learning* mechanism in RF produces more stable classification results than single decision tree and linear-based models. Based on these results, RF was selected as the *base classifier* for the GWO-based feature selection process.

Table 2. Performance Comparison of Classification Models Before Feature Selection

Classifier	Accuracy	Precision		Recall		F1-Score	
		Class 0	Class 1	Class 0	Class 1	Class 0	Class 1
Naive Bayes	0.3253	0.00	0.33	0.00	1.00	0.00	0.49
KNN	0.6024	0.73	0.41	0.64	0.52	0.69	0.46
Decision Tree	0.6627	0.76	0.48	0.73	0.52	0.75	0.50
SVM	0.6747	0.78	0.50	0.71	0.59	0.75	0.54
Random Forest	0.7952	0.84	0.69	0.86	0.67	0.85	0.68

The parameter settings used in the GWO process consisted of 30 wolves ($n\_wolves = 30$), 30 iterations ($max\_iter = 30$), and a fitness weighting parameter of 0.99 ($\alpha = 0.99$). The parameter setting was designed to prioritize classification performance during feature selection while also considering the reduction in the number of selected features. The optimization process used RF as the *base classifier* to evaluate the quality of each feature subset generated by the wolf population. Figure 2c shows the diversity of the initial candidate feature subsets. Figure 2b demonstrates the convergence of GWO toward improved solutions, while Fig. 2a indicates that the number of selected features stabilized at eight after the second iteration. These results suggest that GWO successfully identified a compact and stable feature subset for RF-based hypertension classification. Based on the results in Table 3, RF achieved the best classification performance after GWO-based feature selection. The selected feature subset effectively supported RF in identifying hypertension-related patterns while reducing irrelevant features. KNN and DT maintained stable performance, whereas SVM and NB showed lower performance on the reduced feature set.

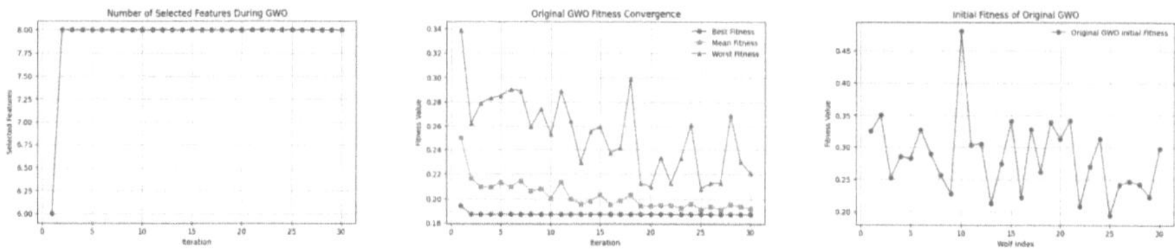

(a) Number of selected. (b) Fitness convergence (c) Initial fitness.

Fig. 2. Feature selection process using Grey Wolf Optimization. (Color figure online)

Table 3. Performance Comparison of Classification Models After GWO Feature Selection

Classifier	Accuracy	Precision		Recall		F1-Score	
		Class 0	Class 1	Class 0	Class 1	Class 0	Class 1
Naive Bayes	0.3253	0.00	0.33	0.00	1.00	0.00	0.49
KNN	0.6265	0.77	0.44	0.64	0.59	0.70	0.51
Decision Tree	0.6867	0.83	0.51	0.68	0.70	0.75	0.59
SVM	0.5904	0.73	0.40	0.62	0.52	0.67	0.45
Random Forest	0.8434	0.91	0.73	0.86	0.81	0.88	0.77

Figure 3 compares the ROC curves before and after GWO-based feature selection. RF achieved the highest AUC and showed improved class discrimination after feature selection. DT and KNN also exhibited slight improvements, whereas SVM showed reduced performance. NB obtained the lowest AUC, likely due to violations of the feature independence assumption.

Figure 4 presents the confusion matrix comparison before and after feature selection. RF achieved the highest number of correctly classified cases and showed

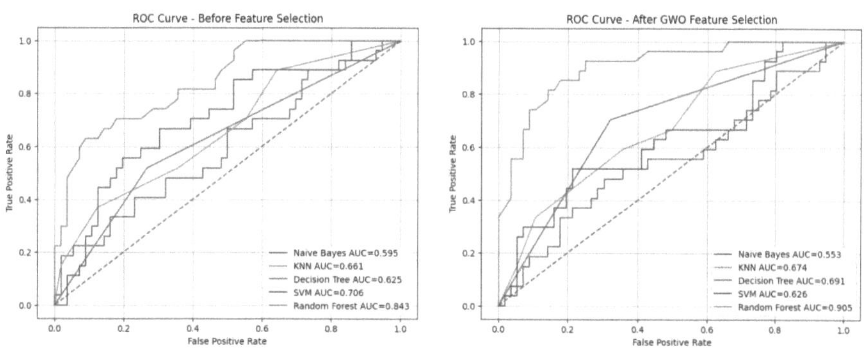

(a) Before. (b) After.

Fig. 3. Comparison of ROC curves GWO-based feature selection.

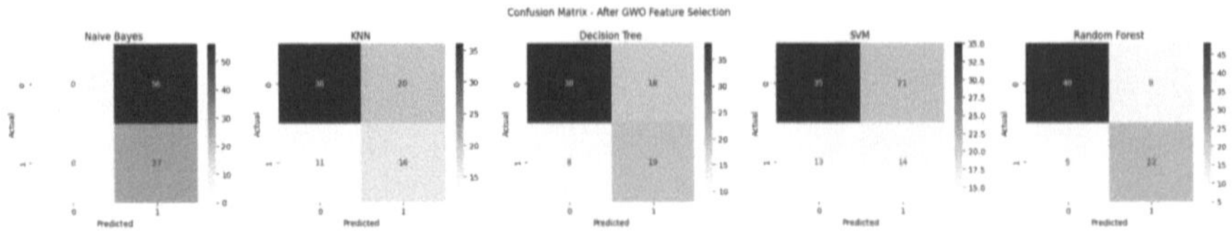

(a) Before.

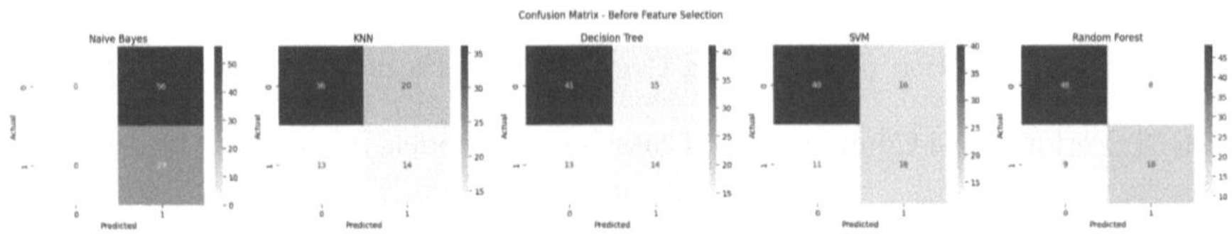

(b) After.

Fig. 4. Comparison of confusion matrices GWO-based feature selection.

a reduction in false negatives after applying GWO, indicating improved hypertension detection. DT and KNN also exhibited improved classification performance, whereas SVM produced more misclassifications, consistent with its lower evaluation metrics and ROC results. The feature importance analysis showed that both RF and the GWO-based feature selection model consistently identified age (*umur*), waist circumference (*lingkar_perut*), body height (*tinggi__badan*), and body weight (*berat_badan*) as the most influential features in hypertension classification. After applying GWO-based feature selection, the importance values of dominant features, particularly age and waist circumference, increased, while features with smaller contributions, such as alcohol consumption and physical activity, obtained lower importance values after the selection process.

4 Conclusion and Future Work

This study demonstrated that GWO-based feature selection improved RF performance for hypertension classification by reducing redundant features while preserving important predictive information. Age, waist circumference, body weight, and body height were identified as the most influential predictors. Future studies may investigate alternative metaheuristic methods, hybrid optimization strategies, and deep learning models on larger datasets.

Acknowledgments. This work was supported by a research grant from the Ministry of higher education, science and technology under the Regular Fundamental Research scheme 2026 [FPR 2026]

References

1. Djasri, H., Alfajri, N.Z., Asdary, R.N., Wardoyo, M.H.P.: Strategies for increasing effective coverage of hypertension services in health care facilities: a scoping review. BKM Public Health Commun. Med. (2024)
2. Health, W., O.: Cardiovascular diseases (CVDs). World Health Organization (2025)
3. World Health Organization.: Global Status Report on Noncommunicable Diseases 2014. WHO Press, Geneva (2014)
4. World Health Organization.: A global brief on hypertension: Silent killer, global public health crisis, WHO Press (2015)
5. Chen, N., Zhang, P., Fan, F., et al.: Development of a risk assessment model for multimorbidity of diabetes, hypertension, and coronary heart disease with XGBoost in primary care in Shanghai, China. BMC Cardiovasc. Disord. (2026)
6. Islam, S.M.S., Talukder, A., Awal, M.A., et al.: Machine learning approaches for predicting hypertension and its associated factors using population-level data from three South Asian countries. Front. Cardiovasc. Med. **9**, 1–15 (2022)
7. Ridoy, R.R., Mothashin, M., Hossain, M.A., et al.: Machine learning based identification of key predictors and socioeconomic inequalities in the co-existence of diabetes and hypertension among Bangladeshi adults. Sci. Rep. **16**(1), 1–15 (2026)
8. Septian, E., Khaefi, M.R., Athoillah, A., et al.: Prediction of personalised hypertension using machine learning in Indonesian population. J. Med. Syst. **49**(1), 1–15 (2025)
9. D'Silva, V., Desai, G., D'Silva, V. A., et al.: Prediction of hypertension using machine learning. In: IEEE Bombay Section Signature Conference, pp. 1–6. IEEE, Bombay (2021)
10. Nematollahi, M.A., Jahangiri, S., Asadollahi, A., et al.: Body composition predicts hypertension using machine learning methods: a cohort study. Sci. Rep. **13**(1), 1–14 (2023)
11. Chang, W., Liu, Y., Xiao, Y., et al.: A machine-learning-based prediction method for hypertension outcomes based on medical data. Diagnostics **9**(4), 1–15 (2019)
12. Peng, Y., Xu, J., Ma, L., Wang, J.: Prediction of hypertension risks with feature selection and XGBoost. J. Mech. Med. Biol. **21**(5), 1–15 (2021)
13. Khan, M.U., Aziz, S., Akram, T., et al.: Expert hypertension detection system featuring pulse plethysmograph signals and hybrid feature selection and reduction scheme. Sensors **21**(1), 1–32 (2021)
14. Khongorzul, D., Kim, M.: Comparison of feature selection methods applied on risk prediction for hypertension. KIPS Trans. Softw. Data Eng. **11**(3), 107–114 (2022)
15. Abbas, S., Sampedro, G.A., Krichen, M., et al.: Effective hypertension detection using predictive feature engineering and deep learning. IEEE Access **12**, 89055–89068 (2024)
16. Chowdhury, M.Z.I., Leung, A.A., Walker, R.L., et al.: A comparison of machine learning algorithms and traditional regression-based statistical modeling for predicting hypertension incidence in a Canadian population. Sci. Rep. **13**(1), 1–14 (2023)
17. Panda, M., Das, B.: Grey Wolf Optimizer and its applications: a survey. Lect. Notes Electr. Eng. **556**, 179–194 (2019)
18. Mostafaei, K., Yousefi, M., Kreuzer, O., Kianpour, M.N.: Simulation-based mineral prospectivity modeling and Gray Wolf optimization algorithm for delimiting exploration targets. Ore Geol. Rev. **177**, 1–14 (2025)

19. Ye, Z., Huang, R., Zhou, W., et al.: Hybrid rice optimization algorithm inspired grey wolf optimizer for high-dimensional feature selection. Sci. Rep. **14**(1), 1–17 (2024)
20. Dagal, I., Demirci, A., Harrison, A., et al.: Prioritized multi-step decision-making Gray Wolf Optimization algorithm for engineering applications. Eng. Rep. **7**(5), 1–15 (2025)
21. Zhang, Y., Li, T., Yang, H., Chen, W.: An improved Grey Wolf Optimizer based on mutation operator, evolutionary population dynamics, and nonlinear population size reduction strategy. Sci. Rep. **15**, 1–18 (2025)
22. Wang, J., Lin, D., Zhang, Y., Huang, S.: An adaptively balanced grey wolf optimization algorithm for feature selection on high-dimensional classification. Eng. Appl. Artif. Intell. **114**, 1–13 (2022)
23. Sam'an, M., Farikhin, M.M.: An improved decision tree model through hyperparameter optimization using a modified gray wolf optimization for diabetes classification. Comput. Methods Biomech. Biomed. Eng. (2025)
24. Sam'an, M., Mat Deris, M., Farikhin: An improved Grey Wolf Optimization based heuristic initialization algorithm for feature selection in P2P lending default prediction. Int. J. Comput. Appl. (2025)
25. Sam'an, M., Mat Deris, M., Farikhin, N.B.: Hyperparameter optimization of Decision Tree using modified Grey Wolf Optimizer for diabetes classification. Lect. Notes Data Eng. Commun. Technol. **257**, 251–264 (2026)
26. Kementerian Kesehatan Republik Indonesia.: Keputusan Menteri Kesehatan Republik Indonesia Nomor HK.01.07/MENKES/4634/2021 tentang Pedoman Nasional Pelayanan Kedokteran Tata Laksana Hipertensi Dewasa. Kementerian Kesehatan, Jakarta (2021). Accessed 19 Nov 2024. https://kemkes.go.id/id/pnpk-2021---tata-laksana-hipertensi-dewasa

Firefly Species Classification Using Deep Learning and Image Augmentation

Zafirah Binti Basri and D. N. F. Awang Iskandar(✉)

Faculty of Computer Science and Information Technology, Universiti Malaysia Sarawak, 94300 Kota Samarahan, Malaysia
dnfaiz@unimas.my

Abstract. Fireflies are charismatic insects that belong to the Lampyridae family which important to ecosystem where they act as an indicator to photopollution and as a biological control agent to trematode parasites infestation. However, it is quite difficult to distinguish the species for fireflies by using standard methods due to high similarity in morphology. This study proposed a method by using deep learning techniques to classify firefly species by using firefly images. Dataset in this study consists of 300 images divided into training, validation and test sets containing 200, 60, and 40 images respectively. In this study, the transfer learning model called InceptionV3 was used. Due to the limited size of the dataset, data augmentation was applied to increase the datasets size. Augmentation technique was applied on train and validation images. Two experiments were carried out which at first experiment was aimed to find the best classification model in classifying the following single Pteroptyx species: Pteroptyx bearni, Pteroptyx malaccae, Pteroptyx tener, Pteroptyx testacea and Pteroptyx valida, and to evaluate the effectiveness of data augmentation techniques. Second experiment was aimed to evaluate the effectiveness of the classification model in experiment 1 for the classification of Pteroptyx against non-Pteroptyx species. Based on the results obtained from the first experiment, it showed that after applied the augmentation technique, the test accuracy shows an improvement. The best augmentation technique (Ori + Flip + Shear + Zoom) achieved the highest average test accuracy of 85%. For the second experiment, no augmentation technique outperformed other methods with an average test accuracy of 49.75%. These research outcomes can be used by other researcher to carry out their research that related with fireflies and deep learning in the future.

Keywords: Fireflies · Deep Learning · Data Augmentation · Image Classification · InceptionV3

1 Introduction

Fireflies are beetles that classified under Coleoptera insect order and there are many species in this order are still unidentified (Seri & Rahman, 2022). Fireflies are one of earth's most charismatic insects and very attractive. They have bioluminescent display specialty which made them have a potential to become an ecotourism attraction. The growth of firefly tourism has positively impacted the local economic development (Lewis et al., 2020).

N. Mohd Sharef et al. (Eds.): CAMP 2026, CCIS 3093, pp. 139–151, 2027.
https://doi.org/10.1007/978-981-92-4302-0_11

However, the population of fireflies have shown a significant decline globally due to the most serious threats and one of it is habitat loss. Thus, a group of firefly experts suggested to preserve this iconic insect and preserve suitable habitat by developing a suitable sanctuary to protect firefly habitats (Lewis et al., 2020). A biodiversity inventory and an expert taxonomist are crucial for the effective conservation and preservation of this iconic insect. The limited number of expert taxonomists has caused gaps in taxonomic knowledge which has affected the biodiversity inventory (Valan, 2021).

Seri & Rahman (2022) suggested to conduct further research that related with fireflies especially on their species richness, their distribution, seasonal variation and habitat in Southeast Asia after they found out the two species of fireflies called *Pteroptyx balingiana* and *Pteroptyx malaccae* have similarities in morphological and genetic. This issue has led to a debate on the validity of the former. Due to the lack of basic information on biodiversity and its evolutionary history, it is important to know more about the biodiversity of the species, so they also suggest to future researcher to broaden study on geographical, genetic sampling and taxonomic of Southeast Asian fireflies.

Quick and accurate morphological classification of organisms is an important step before developing databases for conservation, ecological and pest control studies. Taxonomic identification requires trained expertise and time, but these limitations could be reduced through automated image processing systems trained with machine learning techniques that automatically extracting and recognizing meaningful patterns from image data. Identifying firefly species is quite difficult because closely related species share high similarity of external morphological characteristics. Automatic identification systems developed using machine learning, an application of artificial intelligence is a promising tool for plant and animal classification. Image-based machine learning is widely used for species classification (Tuda & Luna-Maldonado, 2020). However, the effectiveness of computer vision in classification of smaller and taxonomically challenging species is still not well studied (Spiesman et al., 2024).

An integrative approach has potential to improve the efficiency and accuracy of firefly taxonomic identification. Hence, this study aims to propose a deep learning approach using convolutional neural networks (CNNs) to automatically classify firefly species and to discover the effectiveness of data augmentation techniques on model performance.

2 Related Works

Deep learning (DL) is an important area of machine learning and is widely used for image classification. Recent developments show that deep learning has been widely used in many fields and has significantly contributed to artificial intelligence. Deep learning enables computational models with multiple processing layers to learn complex data representations. These approaches have significantly improved performance in speech recognition, visual recognition, object detection and other fields such as drug discovery and genomics (Moshayedi et al., 2022).

In the field of DL, Convolutional neural networks (CNNs) have become one of the most widely used for image classification and show an outstanding performance in the ImageNet competition (Chai et al., 2021). Alexnet, ZefNet, VGG, GoogleNet, ResNet, Inception: ResNet and Inception-V3/4, DenseNet are several lists of CNN architecture.

Spiesman et al. (2021) carried out research to identify bumble bee species and compared four convolutional neural network classification models; ResNet, Wide ResNet, InceptionV3 and MnasNet models. Among these models, InceptionV3 has outperformed others. This model could be used for future research to identify other insect species such as firefly. Species classification is a widely used application of image-based machine learning. However, as to our knowledge, currently there is not much available approach to automatically classify the firefly species based on image.

There are few research on firefly were carried out by using different kind of method. Some researchers classify firefly by using flash pattern. Some of research study the flash communication pattern by using computer vision. A study by Tathawee et al. (2020) was aimed to reduce the time of data collection and analysis by developing a program that could help in analyzing firefly flashing behavior using computer vision approach. They stated that different species exhibit different patterns of light.

Another method called geometric morphometrics (GM) was used to identify firefly species. This method was used by Sumruayphol & Chaiphongpachara (2019) in their research. GM method consists of two methods, landmark-based and outline-based GM. GM is a cheap method to study on shape and size variation. Their dataset consists of 200 and 157 fireflies from 3 genera and four species, including Luciola aquatilis Thancharoen, Pteroptyx valida Olivier, Pteroptyx malaccae Gorham and Pyrocoelia praetexta Olivier. Based on their results, the outline-based method has outperformed the landmark-based method. However, outline-based method was time-consuming and requires many samples.

Martin et al. (2024) applied recurrent neural network (RNN) to develop an automated firefly species classification based on their flash pattern. In their research, they able to extract flash trajectories of individuals within swarm through recordings from commodity cameras and able to classify the firefly species with an accuracy of approximately 70%. Their study has several limitations such as the effects of temperature variation, sympatric firefly species, data acquisition methodology and model methodology. For example, data acquisition methodology, a given trajectory might not fully capture the entire flash pattern of an individual due to overlapping bioluminescent signals from nearby fireflies.

Besides deep learning-based methods, other researchers used other method called molecular analysis to identify firefly species. In research by Urtgam et al. (2024), they used molecular analysis combined with morphological studies to identify firefly species. They employed the nucleotide sequencing of two mitochondrial genes; Cytochrome c oxidase subunit I (COI) and 16 s rRNA and the phylogenetic trees were constructed and compared with the firefly nucleotide sequences in the NCBI database. Based on their results, COI nucleotides are one of the most effective methods in classification of firefly species. Their findings indicated that an addition of the 16 s rRNA gene analysis enhanced the accuracy of firefly identification. Although this molecular method can be used to classify firefly species and has high level of taxonomic accuracy, this method involves high cost and requires trained molecular biology expertise (Sumruayphol & Chaiphongpachara, 2019).

3 Materials and Methods

3.1 Datasets

The dataset used in this study consists of digital images of fireflies categorized into five species classes. The five firefly species included in this study are: Pteroptyx bearni, Pteroptyx malaccae, Pteroptyx tener, Pteroptyx testacea and Pteroptyx valida. The total number of images used in this study were 300. Some species has 62 images whereas the other species has more than 200 images. To ensure a balanced dataset, only 60 images were used for each firefly species. This approach to prevent classification bias and enhance the model's capability to distinguish among all species effectively. The images consist of two types: the upper and underside views of fireflies. The images of firefly were obtained from the Faculty of Resource Science and Technology UNIMAS.

In this study, the dataset was divided into Train (66.67%), Validation (20%) and Test (13.33%) sets. Table 1 show the dataset distribution for Pteroptyx fireflies species and Table 2 show the dataset distribution for single Pteroptyx species with augmentation. Two experiments were designed in this study. The first experiment focused on classifying a single Pteroptyx species and evaluating the effectiveness of data augmentation techniques. In the second experiment, we will evaluate the effectiveness of models developed in Experiment 1 in classification of combined Pteroptyx against non-Pteroptyx species.

In this study, binary classification approach was employed. To classify Target species against Remaining species, in positive class which labeled as Target species, all images are from one target species, while all images in negative class that labeled as Remaining species are images randomly selected from the remaining four species. In Train set consists of 40 images of the target species and 40 images in Remaining class formed by combining 10 images from each of the remaining four species. For example, when we want to classify single PV species, all 40 images in Target species folder were PV images, and in Remaining species folder contain 10 images of PB, 10 images of PM, 10 images of Pten and 10 images of Ptes. In Validation set, consists of 12 images of the target species and 12 images in Remaining class formed by combining three images from each of the remaining four species. In Test set, consists of eight images of the target species and eight images in Remaining class formed by combining two images from each of the remaining four species. In the augmentation experiment, image augmentation was performed on the training and validation sets. All images were augmented using geometric transformation methods, including flip, zoom, and shear, to enhance model generalization.

Dataset distribution shown in Table 3 is a Test set that was constructed by combining five Pteroptyx species into one class and compares them with non-Pteroptyx class. This Test set will be used for evaluation in the second experiment. In this Test set, the positive class labeled as Pteroptyx was created by combining eight images from each of the five target species used in Experiment 1. As for non-Pteroptyx class, all 40 images were collected from five different species: 9 from Lucidota, 22 from Photinus, 1 from Photuris, 4 from Pyractomena angulata, and 4 from Pyropyga. The 40 images of non-Pteroptyx were obtained from iNaturalist (n.d.)

Figure 1 presents microscopic images of both upperside and underside for each species of fireflies in this study. Images were acquired under conditions that accurately

Table 1 Dataset distribution for *Pteroptyx* Fireflies species

Species Name	Train set (66.67%)	Validation set (20%)	Test set (13.33%)	Total no. of images
Pteroptyx bearni	40	12	8	60
Pteroptyx malaccae	40	12	8	60
Pteroptyx tener	40	12	8	60
Pteroptyx testacea	40	12	8	60
Pteroptyx valida	40	12	8	60
Total number of images in datasets	200	60	40	300

Table 2 Datasets distribution for single *Pteroptyx* species with augmentation

	Classification type	Train set	Validation set	Test set
No augmentation	Ori	80	24	16
With augmentation F – Flip S – Shear Z - Zoom	Ori + (F/S/Z)	160	48	16
	F + S + Z	240	72	16
	Ori + F + S + Z	320	96	16

Table 3 Dataset distribution in Test set of combined *Pteroptyx* against non-*Pteroptyx* species

	Test set
Pteroptyx	40
Non-Pteroptyx	40

reflect the fireflies'- under standardized laboratory conditions. Distance, lighting and magnification are carefully controlled to maintain consistency and realism in the images.

Figure 2 presents images of non-Pteroptyx fireflies in this study which were retrieved from iNaturalist, (n.d.).

3.2 Methods

This section outlines the methodological procedures and computational framework applied for firefly species classification using deep learning and image augmentation.

The methodological framework show in Fig. 3 was designed to enables consistent and reliable taxonomic identification of fireflies. The process begins with image acquisition, followed by preprocessing, and then classification techniques. The acquired images underwent preprocessing to enhance image quality and standardize input data. All images

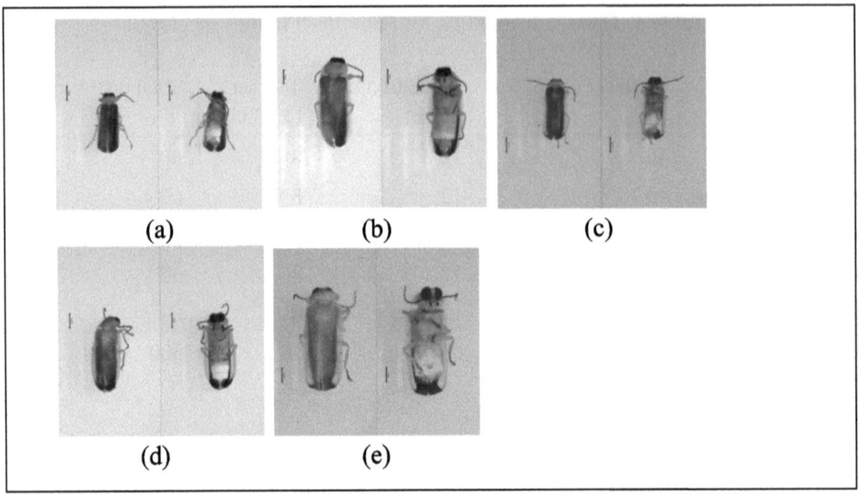

Fig. 1 Images of Pteroptyx species. (a) Pteroptyx bearni, (b) Pteroptyx malaccae, (c) Pteroptyx tener, (d) Pteroptyx Testacea and (e) Pteroptyx valida

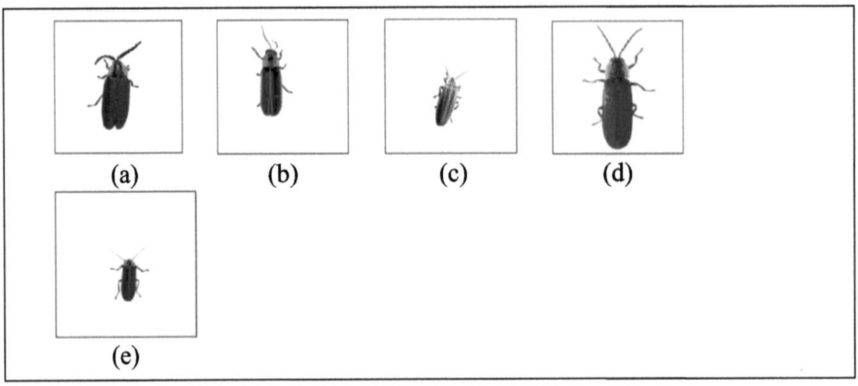

Fig. 2 Images of non-Pteroptyx species. (a) Lucidota, (b) Photinus, (c) Pyractomena angulata (d) Pyropyga and (e) Photuris

were resized to 299x299 pixels to match the input dimensions required by InceptionV3. In this study, two experiments were conducted. The first experiment aimed is to classify single Pteroptyx species and to assess the effectiveness of data augmentation techniques in improving model performance. Second experiment aimed to evaluate the effectiveness of the model developed from the first experiment in classification of combined Pteroptyx against non-Pteroptyx species.

An augmentation experiment was conducted to examine the effectiveness of data augmentation. All images in train and validation sets were augmented. Every image was augmented using three types of geometric transformation techniques: flip, shear and zoom. Geometric Transformations are employed to counteract positional biases in the training set resulting in well-regularized and robust solutions (Alzubaidi et al., 2021).

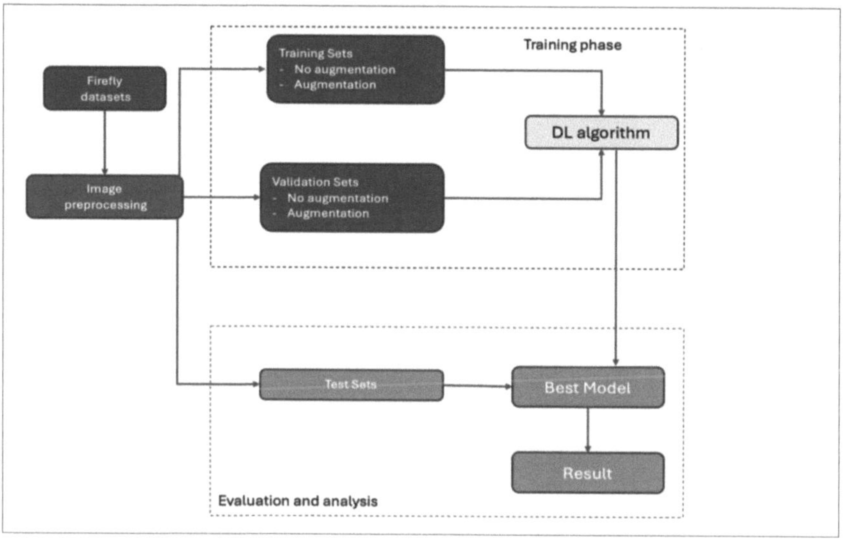

Fig. 3 The proposed framework for Firefly Species Classification using Deep Learning and Image Augmentation

Image augmentation is to expand our datasets and to reduce overfitting: shear range = 0.2, zoom range = 0.2, random horizontal flip. In base model is added with CNN layers with ImageNet weights. Three levels of augmentation were tested: (1) Original and one transformation (flip, shear, or zoom), (2) original and all three transformations and (3) transformations only (no original images).

The classification of firefly species was performed using a deep learning approach based on the InceptionV3 architecture, which was adapted to perform binary classification. InceptionV3 is a family of CNN that achieves Top 5 error rate on the ImageNet dataset when a deep learning performance compared to human (Alzubaidi et al., 2021). InceptionV3 has smaller size, quick training time and has small number of parameters if to be compared with InceptionResNetV2 (Keras, 2025). InceptionV3 was proposed with aim to minimize the computational cost and no effect on the deeper network generalization (Alzubaidi et al., 2021). Thus, InceptionV3 was chosen due to advantages that has been stated. The final dense layer consisted of a single neuron with a sigmoid activation function to calculate classification probabilities. The model was trained using Stochastic Gradient Descent (SGD) with a fixed learning rate of 0.001 to update the network weights to minimize the loss function. Training was conducted for 100 epochs to allow the model sufficient iterations to converge. The binary cross-entropy loss function was employed to measure the difference between predicted probabilities and true class labels. The test set was used to evaluate the effectiveness of the trained model. Table 4 presents a summary of the model's hyperparameters. Table 5 presents the model-parameter details for model used in research. The model was trained on Google Colab using an NVIDIA T4 GPU for 100 epochs. The total training time was approximately 10–15 min.

Table 4 Hyperparameters of the model

Input: Fireflies image dataset (Breakdown to 66.67:20:13.33) with size 299 × 299 × 3
Epoch: 100
Optimizer: SGD
Filter: Conv2D
Loss: binary_crossentropy
Metric: Accuracy
Batch Size: 32
Learning rate: 0.001

Table 5 Model and parameter details

Model	Image size	Parameter detail
InceptionV3	299	Total parameters: 23, 903, 010 (91.18 MB) Trainable parameters: 2, 100, 226 (8.01 MB) Non-trainable parameters: 21, 802, 784 (83.17 MB)

4 Results and Discussion

The performance of the classification model was evaluated using standard metrics: Accuracy, Precision, Recall, F1 score, Loss and Mean Absolute Error. The model.evaluate() function was used to evaluate the fit performance of the built model, while model.predict() function was used for classification. Below is the formal definition of accuracy, precision, recall, and F1 Score (Hossin and Sulaiman, 2015)

$$Accuracy = \frac{(Tp + Tn)}{(Tp + Fp + Fn + Tn)}$$

$$Precision = \frac{Tp}{(Tp + Fp)}$$

$$Recall = \frac{Tp}{Tp + Fn}$$

$$F1\ Score = \frac{2 \times (Recall \times Precision)}{(Recall + Precision)}$$

4.1 Deep Learning Model to Classify Fireflies Species

All model implementation and development were performed using python 3.12.12 on the Google Colab platform with GPU acceleration and TensorFlow version 2.19. Table 6 shows the classification results in terms of accuracy, precision, recall, loss, MAE and F1-Score.

4.2 Results Analysis for Single Pteroptyx Species

The following section presents the average results of model evaluation obtained from Experiment 1 in classifying of single Pteroptyx species. Table 6 show the average results for model evaluation.

In Table 6 presents the average results that obtained from Experiment 1. In Experiment 1, the performance of the proposed model was evaluated under two settings: without data augmentation and with data augmentation. The value for average test accuracy before augmentation was 71.25%.

Table 6 Average results of model evaluation for Single *Pteroptyx* species

Augmentation type	Training Accuracy	Test Accuracy	Test Precision	Test Recall	Test Loss	Test MAE	Test F1 Score
No augmentation	0.7850	0.7125	0.7329	0.6875	0.6172	0.4496	0.6774
Ori + Flip	0.9000	0.8000	0.8138	0.7750	0.5597	0.4187	0.7945
Ori + Shear	**0.9113**	0.8125	0.7438	**0.8375**	0.5587	0.4181	0.8100
Ori + Zoom	0.9037	0.7875	0.7856	0.7250	0.5662	0.4217	0.7840
Flip + Shear + Zoom	0.8825	0.8250	0.8256	0.7750	0.5239	0.3928	0.8236
Ori + Flip + Shear + Zoom	0.8950	**0.8500**	**0.8385**	**0.8375**	**0.4904**	**0.3700**	**0.8490**

After applying data augmentation, the model shows an improvement in performance especially when using a combined technique (Ori + Flip + Shear + Zoom) by achieving the highest average test accuracy of 85%. Among the three types of augmentation technique, zoom yielded the lowest average accuracy of 78.75%. This might be due to the loss of fine morphological details caused by excessive zooming. These results indicate that data augmentation enhanced the model's ability to classify firefly images correctly.

A significant test was carried out to discover whether the performance of the best model was significantly better than the others. To perform significant test, the highest average accuracy was compared against the average accuracy of the remaining models. Based on the significant test, the average accuracy value of model (Ori + Flip + Shear + Zoom) was significantly better than the others with a p-value, 0.0171($p < 0.05$).

K-fold cross validation (CV) is a widely used method for estimating the true performance of machine learning models, allowing model selection and parameter tuning. The dataset is divided into K equally sized subsets, allowing each subset to serve as a validation set while the remaining subsets are used for training (Moss et al., 2018). In this research we applied K-fold method on the best model only from Experiment 1 which was PV (Ori + Flip + Shear + Zoom). This model achieved the highest test accuracy of 93.75% with loss value, 0.3745 without K-fold method. After applied K-fold method, the best model achieved 93.75% test accuracy and loss value, 0.4068. Table 7 presents the comparison value of the best model PV (Ori + Flip + Shear + Zoom) with and without K-fold. Based on this result, it showed that model without K-fold able to achieved a better performance when we compared the loss and MAE value. In the K-fold cross validation experiment, the training and validation images were combined.

Table 7 Comparison value of the best model PV (Ori + Flip + Shear + Zoom) with and without K-fold

Augmentation type	Species	K-fold	Test Accuracy	Test Precision	Test Recall	Test Loss	Test MAE	Test F1 Score
Ori + Flip + Shear + Zoom	PV	No	0.9375	0.9375	0.9375	**0.3745**	**0.3031**	0.9373
		Yes	0.9375	**1.0000**	0.9375	0.4068	0.3258	0.9373

4.3 Results Analysis of Using Combined Pteroptyx against Non-Pteroptyx Species Test Sets

This section discussed on the results of Experiment 2 which aimed to evaluate the effectiveness of the model developed from the first experiment in classification of combined Pteroptyx against non-Pteroptyx species. Table 8 presents the average prediction results obtained using the new test set consisting of Combined Pteroptyx and non-Pteroptyx species. In this second experiment, results indicate that when no data augmentation was applied, the proposed model achieved the highest average test accuracy of 49.75%. This result suggests that augmentation techniques may not always enhance classification accuracy, as the original images already provide sufficient discriminative morphological features to classify Pteroptyx against non-Pteroptyx species. Although augmentation technique increases data diversity, but this may have introduced variations that altered fine structural details.

The highest average test accuracy obtained was below 50% indicating that the proposed model has not yet achieved the best performance as a classification model. This moderate performance maybe caused by limited training dataset, particularly non-Pteroptyx images were not included during training phase and were only introduced during testing. As a result, the model may have learned features that are specific to Pteroptyx species and lacked with sufficient knowledge to identify characteristics of non-Pteroptyx specimens. Model generalization may be enhanced by adding more images of Pteroptyx and non-Pteroptyx in training dataset.

A significant test also being carried out in Experiment 2 to discover whether the performance of the best model was significantly better than the other model. From the significant test result it showed the highest average test accuracy (no augmentation) was significantly better than the others with p-value, 0.0026 ($p < 0.05$).

Table 9 presents the comparison value of the best model in Experiment 2, Ptes (no augmentation). The model without K-fold achieved 67.50% test accuracy. When we employed K-fold method, the model achieved 65.00% test accuracy. Hence, from this result, we decided to proceed our experiment without K-fold method.

Table 8 Average results of model prediction for combined *Pteroptyx* against non-*Pteroptyx* species

Augmentation type	Test Accuracy	Test Precision	Test Recall	Test F1 Score
No augmentation	**0.4975**	0.5222	**0.6400**	**0.5651**
Ori + Flip	0.4775	**0.5732**	0.3950	0.4346
Ori + Shear	0.4650	0.5490	0.3650	0.4121
Ori + Zoom	0.4475	0.5176	0.3700	0.4074
Flip + Shear + Zoom	0.4400	0.5442	0.3650	0.3990
Ori + Flip + Shear + Zoom	0.4675	0.5641	0.3550	0.4046

Table 9 Comparison value of the best model Ptes (no augmentation) with and without K-fold

Augmentation type	Species	K-fold	Test Accuracy	Test Precision	Test Recall	Test F1 Score
No augmentation	Ptes	No	**0.6750**	0.7059	**0.6000**	**0.6486**
		Yes	0.6500	**0.7308**	0.4750	0.5758

5 Conclusion and Future Work

This study proposed a deep learning-based framework for the classification of firefly species using firefly images. The InceptionV3 architecture was employed for feature extraction. Two experiments were carried out to evaluate the effectiveness of the proposed approach. The first experiment was to classify single Pteroptyx species and to assess the effectiveness of data augmentation techniques in improving model performance. The combined technique (Ori + Flip + Shear + Zoom) achieved the highest average accuracy of 85%. Thus, data augmentation enhanced the performance of single-species classification. In second experiment, which aimed to evaluate the effectiveness of the model developed from the first experiment in classification of combined Pteroptyx against non-Pteroptyx species, the model without data augmentation achieved the highest average accuracy of 49.75%, suggesting the importance of maintaining original morphological characteristics.

In this research used one type of CNN model, InceptionV3. Hence our next research paper we decided to carry out an experiment where we will compare InceptionV3 with other CNN models. In this research we only aim to focus on one CNN model and to discover the effectiveness of data augmentation technique.

The proposed framework provides valuable insights into development of automated taxonomic identification system for firefly species. A limited dataset could lead to overfitting thus to mitigate this issue, in this research employed augmentation technique. Even though our experiment has applied augmentation technique to address our limited dataset size, augmented images are generated from the original images and may not fully capture real-world variations. Future research should aim to expand the dataset by adding more species of firefly and images that was captured under varying environmental conditions. In this research we only use microscopic images. Include images under diverse environmental conditions could improve the robustness and generalization capability of the model.

Acknowledgments. The authors acknowledge Universiti Malaysia Sarawak for supporting the project under the VC-High Impact Research Grant, Image-Based Taxonomic Identification of Fireflies from Sarawak Using Deep Learning (UNI/F08/VC-HIRG/85483/P02-03/2022). We express gratitude to the Faculty of Resource Science and Technology, UNIMAS for the firefly samples captured and used in this study.

References

Alzubaidi, L., et al.: Review of deep learning: concepts, CNN architectures, challenges, applications, future directions. J. Big Data. **8**(1) (2021). https://doi.org/10.1186/s40537-021-00444-8

Chai, J., Zeng, H., Li, A., Ngai, E.W.T.: Deep learning in computer vision: a critical review of emerging techniques and application scenarios. Mach. Learn. Appl. **6**, 100134 (2021). https://doi.org/10.24433/CO.0411648.v1

iNaturalist: Observations. Retrieved July 6, 2025, from https://www.inaturalist.org/observations?page=7&taxon_id=47731 (n.d.)

Keras: Keras Applications. https://keras.io/api/applications/ (2025, May 6)

Lewis, S.M., et al.: A global perspective on firefly extinction threats. Bioscience. **70**(2), 157–167 (2020). https://doi.org/10.1093/biosci/biz157

Hossin, M., Sulaiman, M.N.: A review on evaluation metrics for data classification evaluations. Int. J. Data Min. Knowl. Manag. Process. **5**(2), 01–11 (2015). https://doi.org/10.5121/ijdkp.2015.5201

Martin, O., et al.: Embracing firefly flash pattern variability with data-driven species classification. Sci. Rep. **14**(1) (2024). https://doi.org/10.1038/s41598-024-53671-3

Moshayedi, A.J., Roy, A.S., Kolahdooz, A., Shuxin, Y.: Deep Learning Application Pros and Cons Over Algorithm (2022). https://doi.org/10.4108/10.4108/airo.v1i.19

Moss, H. B., Leslie, D. S., Rayson, P.: Using J-K-fold Cross Validation to Reduce Variance When Tuning NLP Models (2018)

Seri, N.A., Rahman, A.A.: The compilation Records of Fireflies (Coleoptera: Lampyridae) diversity and distribution and display trees throughout Malaysia. Pertanika J. Sci. Technol. **30**(3), 1963–1987 (2022). https://doi.org/10.47836/pjst.30.3.11

Spiesman, B.J., Gratton, C., Gratton, E., Hines, H.: Deep learning for identifying bee species from images of wings and pinned specimens. PLoS One. **19**(5), e0303383 (2024). https://doi.org/10.1371/journal.pone.0303383

Spiesman, B.J., et al.: Assessing the potential for deep learning and computer vision to identify bumble bee species from images. Sci. Rep. **11**(1) (2021). https://doi.org/10.1038/s41598-021-87210-1

Sumruayphol, S., Chaiphongpachara, T.: Geometric morphometrics as a tool for three species identification of the firefly (Coleoptera: Lampyridae) in Thailand. Biodiversitas. **20**(8), 2388–2395 (2019). https://doi.org/10.13057/biodiv/d200837

Tathawee, T., Wattanachaiyingcharoen, W., Suwannakom, A., Prasarnpun, S.: Flash communication pattern analysis of fireflies based on computer vision. Int. J. Adv. Intell. Inf. **6**(1), 60–71 (2020). https://doi.org/10.26555/ijain.v6i1.367

Tuda, M., Luna-Maldonado, A.I.: Image-based insect species and gender classification by trained supervised machine learning algorithms. Eco. Inform. **60** (2020). https://doi.org/10.1016/j.ecoinf.2020.101135

Urtgam, S., Thawathotsapakorn, S., Wattanachaiyingcharoen, W.: The partial nucleotide sequences of the mitochondrial genes, COI and 16s rRNA, of fireflies in the genera pygolu-ciola, trisinuata, and medeopteryx (Coleoptera: Lampyridae). J. Curr. Sci. Technol. **14**(2), 22 (2024). https://doi.org/10.59796/jcst.V14N2.2024.22

Valan, M.: Automated image-based taxon identification using deep learning and citizen-science contributions (2021)

Fungal Infection Detection Using Custom Fusion-Based Deep Learning Features

Irum Nehvi[1,2], D. N. F. Awang Iskandar[1], Jaafar Alghazo[3], Ghazanfar Latif[4(✉)], and Insha Nahvi[5]

[1] Faculty of Computer Science and Information Technology, Universiti Malaysia Sarawak, Kota Samarahan 94300, Sarawak, Malaysia

[2] Department of Computer Science, Prince Mohammad Bin Fahd University, Khobar, Saudi Arabia

[3] Math, Science and Technology Department, University of Minnesota Crookston, Crookston, MN 56716, USA

[4] Department of Computing Science, Thompson Rivers University, Kamloops, BC V2C 0C8, Canada

glatif@tru.ca

[5] Department of Biological Sciences, College of Science, King Faisal University, AlAhsa, Saudi Arabia

Abstract. Fungal infections pose a significant global health challenge, ranging from superficial skin conditions to life-threatening systemic diseases, particularly in immunocompromised individuals. With an estimated 1.5 million deaths annually, accurate and timely diagnosis is critical. Traditional diagnostic methods, such as microscopy and culture, are often slow and prone to errors. Recent advances in deep learning (DL) have shown promise in analyzing microscopic fungal images for high-accuracy classification. This study focuses on classifying five clinically important fungal species, Candida albicans, Aspergillus niger, Trichophyton rubrum, Trichophyton mentagrophytes, and Epidermophyton floccosum, using 5,000 high-quality microscopic images from the publicly available DeFungi dataset. The proposed ensemble approach combining EfficientNet-B1, DenseNet201, and ResNet18 achieved the highest overall accuracy and balanced performance across all classes, particularly improving the classification of Aspergillus niger. These findings highlight the effectiveness of ensemble learning in leveraging diverse feature representations to enhance fungal image classification, offering a reliable and automated diagnostic tool to support clinical decision-making and improve patient outcomes.

Keywords: Fungal infection · Deep Learning · Candida albicans · Aspergillus niger · Trichophyton rubrum · Medical Diagnostics

1 Introduction

Fungal infections pose a significant global health threat, ranging from mild skin conditions to severe systemic diseases [1], and contribute to over 3.75 million deaths annually, with candidiasis accounting for about 2.55 million deaths [2]. Accurate identification of

N. Mohd Sharef et al. (Eds.): CAMP 2026, CCIS 3093, pp. 152–167, 2027.
https://doi.org/10.1007/978-981-92-4302-0_12

fungal species is essential for effective diagnosis and treatment [3]. However, traditional microscopy and culture-based methods are often slow and limited in precision. Deep learning and machine learning approaches offer improved accuracy by analyzing large-scale fungal image data and identifying subtle patterns for reliable classification [4, 5]. Ensemble deep learning models further enhance diagnostic performance and support faster, more effective clinical decision-making.

This study focuses on five clinically important fungal species: *Candida albicans*, *Aspergillus niger*, *Trichophyton rubrum*, *Trichophyton mentagrophytes*, and *Epidermophyton floccosum*. These pathogens cause infections ranging from mild skin and nail conditions to severe systemic diseases, particularly in immunocompromised patients [6–8]. Given their global prevalence and clinical impact, accurate and timely diagnosis is essential. However, traditional microscopic and culture-based methods are slow and less precise, whereas deep learning enables faster and more accurate automated fungal classification using image analysis [9].

This paper investigates DL-based classification of five clinically significant fungal species using state-of-the-art models, including GoogleNet, ResNet18, DenseNet201, EfficientNet-B0/B1/B2, and ResNet50. An ensemble of EfficientNet-B1, ResNet18, and DenseNet201 is proposed to improve performance by leveraging complementary strengths: EfficientNet-B1 for balanced scaling, DenseNet201 for dense feature reuse and gradient flow, and ResNet18 for stable residual learning. The study aims to identify effective DL strategies to enhance classification accuracy and support more reliable automated fungal infection diagnosis.

The remainder of this paper is structured as follows: Sect. 2 reviews recent literature on fungal image classification, focusing on models, datasets, accuracies, and limitations. Section 3 describes the proposed methodology, including the DeFungi dataset, data preprocessing (scaling, normalization, and augmentation), and input requirements for different models. Section 4 describes the results and analysis.

2 Literature Review

Fungal detection has shifted from traditional microscopy to CNN-based deep learning and machine learning approaches for automated classification of fungal pathogens. Table 1 compares recent studies using different models [5–26]. Some works use only train/test splits without validation, which may lead to optimistic results; for example, [20] does not include a separate validation set. Despite strong performance, challenges such as limited data, overfitting, and high computational cost remain [27]. Transfer learning improves efficiency on small datasets compared to custom CNNs [28], while reducing classes may improve accuracy but limits discrimination. Overall, deep learning remains a promising approach for fungal detection in medical applications.

Table 1 Classification performance comparison of various research papers based on the models used

Ref.	Proposed method	Data set	Accuracy	Critical issue
[6]	CNN-based Transfer Learning	DeFungi	91%	Challenges in data quality, image variability, and model performance.
[7]	Inception V3, VGG16	DeFungi	85.04%	Collection and processing of labeled data.
[8]	Modified Deep Neural Network (DNN) and DSURF	PlantDoc	90%+	Overfitting deep models, computational complexity, and dataset imbalances are impacting generalization.
[9]	Vision and Swin Transformer	microscopic images	98.36%	Model performance need for extensive fine-tuning strategies.
[10]	CNN, Fisher Vector approach	Fungus Species Images	95.8%	Dataset imbalance, challenges in distinguishing similar fungal species
[11]	Histopathological examination	clinical samples	NA	Misidentification, overlapping features with non-fungal conditions.
[12]	Microscopic examination	clinical data	NA	Difficulty in diagnoses
[13]	FungiVision, ViT	Danish Fungi 2020 (DF20)	93%	Data consistency and reliance on citizen participation
[5]	Regional CNN	Microscopic images	95%	Detection of diverse infection patterns and generalization.
[14]	convolutional neural network (CNN)	Microscopic fungal images	97%	Image quality variations, limited dataset size, and differentiating between morphologically similar fungal species.

(*continued*)

Table 1 (*continued*)

Ref.	Proposed method	Data set	Accuracy	Critical issue
[15]	HLNNC combining CNN and SVM	Eyes images of mucormycosis	NA	Dataset quality, integration of preprocessing techniques for enhanced diagnostic accuracy.
[16]	traditional and DNA-based/ molecular methods	No specific dataset	NA	Incomplete species identification, limited culturing methods, and high reliance on skilled personnel.
[17]	CNN	*Aspergillus* fungi dataset	95%	High computational cost, potential overfitting, and data variability
[18]	CNN	toenail specimen images	92.5%	Limited dataset size, difficulty in accurately segmenting the infected.
[19]	VGG16, ResNet50, Inceptionv3	PlantVillage dataset	99.07%	limited dataset sizes, model performance, and variability in symptoms.
[20]	CNN	DeFungi (UCI Medical Repository)	97.19%	model training and data quality
[21]	AI with visualization tools such as dermoscopy and microscopy	KOH microscopic images	NA	variable sensitivity, reliance on human interpretation, high costs, and long turnaround times for culture results.
[22]	SVM, k-NN, Trees, RF, CNN, DBNs, RNNs, VGG16, ResNet	Public and proprietary microscopic images	98%	Limited datasets, imbalanced data, computational complexity, and model generalizability across diverse types.

(*continued*)

Table 1 (*continued*)

Ref.	Proposed method	Data set	Accuracy	Critical issue
[23]	back-propagation neural network (BPNN)	Walnut Leaf images with fungi infections	95.3%	Computational complexity due to pre-processing and segmentation, sensitivity to image quality, and variability in leaf images.
[24]	MeFunX	DeFungi	96%	Limited generalizability, small dataset size, overfitting, and image quality.
[25]	CNNs-based Transfer Learning	Microscopic images	95%	Limited dataset, challenges in detecting rare fungal species
[26]	CNNs, VGG16 and ResNet-50	DeFungi	92.7%	Limited dataset diversity and potential overfitting

3 Methodology

This paper presents a hybrid transfer learning approach for fungal image classification using GoogleNet [30] and ResNet18 [28], along with ResNet50, EfficientNet (B0–B2), and DenseNet-201. To improve performance, several ensemble strategies were applied, including combinations of GoogleNet, ResNet18, DenseNet201, and a TTA-based weighted ensemble of EfficientNet-B4, ResNet18, and DenseNet201.

The proposed ensemble combines EfficientNet-B1, ResNet18, and DenseNet-201, initialized with ImageNet weights and fine-tuned for five-class classification. Models were trained using Adam (lr = 0.0001), batch size 42, for 90 epochs with cross-entropy loss. Class imbalance was handled using weighted random sampling, and data augmentation (cropping, flipping, rotation) was applied for better generalization. Final predictions were obtained by averaging softmax outputs from all three models under identical training settings.

All experiments were conducted on a workstation equipped with an NVIDIA RTX A4500 GPU (20 GB VRAM), an Intel Xeon Silver 4410 T processor (2.70 GHz), 64 GB RAM, and approximately 1.86 TB storage, running a 64-bit Windows operating system. The models were implemented using the PyTorch deep learning framework with the torchvision library. Training and evaluation were performed on the same system to ensure consistency across all experiments, and all reported results correspond to this computational setup.

3.1 Experimental Dataset

This study utilizes the DeFungi dataset, which is available on Kaggle [35]. There are 5000 high-quality images in this dataset for training. The number of images and the sample of training images are shown below in Table 2. Images of superficial fungal infections resulting from dermatophyte fungus, yeasts, or molds are shown. With the help of a subject matter expert, the images have been painstakingly categorized into five classes: H1 (Candida albicans), H2 (Aspergillus Niger), H3 (Trichophyton rubrum), H5 (Trichophyton mentagrophytes), H6 (Epidermophyton floccosum), and carefully selected [7].

Table 2 Summary of Images in DeFungi dataset.

Class	H1	H2	H3	H5	H6
Training Images	1000	1000	1000	1000	1000
Test Images	437	232	81	80	69
Validation Images	437	233	82	80	70
Example Image	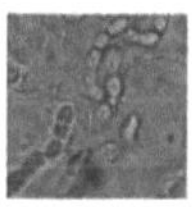	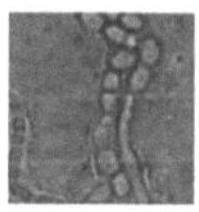		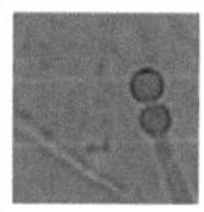	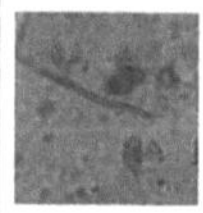

The dataset was organized into predefined training, validation, and testing subsets using separate directory structures. This ensured a consistent data split across all experiments and models. To address class imbalance in the training set, a weighted random sampling strategy was employed, allowing underrepresented classes to be sampled more frequently during training. No additional resplitting was performed, and the same dataset partitions were used for all models to ensure a fair comparison. All results were obtained from a single training run under these fixed conditions.

3.2 Data Preprocessing

Image preprocessing standardizes inputs and improves classification performance through rescaling and normalization to a 0–1 range, reducing computational complexity [29]. Data augmentation (rotation, flipping, zooming, and shearing) is applied to enhance generalization and robustness, as summarized in Table 3. To further reduce overfitting, techniques such as dropout, batch normalization, and transfer learning using ImageNet pre-trained models are widely used [30]. In data-scarce scenarios, approaches such as one-shot learning (Siamese networks) and zero-shot learning using semantic embeddings (Word2Vec, GloVe) can also be employed.

To address class imbalance, weighted random sampling was used to give higher probability to underrepresented classes, improving minority class learning. Data augmentation was also applied to enhance generalization. Performance was evaluated using per-class precision, recall, and F1-score to assess class-wise effectiveness.

Table 3 Input Image Requirement Comparison

Aspect	EfficientNet-B1	DenseNet-201	ResNet18
Image Size	240x240	224x224	224x224
Channels	3(RGB)	3(RGB)	3(RGB)
Aspect Ratio	1:1 (Square)	1:1 (Square)	1:1 (Square)
Normalization	Mean: [0.485, 0.456, 0.406]	Mean: [0.485, 0.456, 0.406]	Mean: [0.485, 0.456, 0.406]
	STD: [0.229, 0.224, 0.225]	STD: [0.229, 0.224, 0.225]	STD: [0.229, 0.224, 0.225]
Preprocessing	Resize, crop	Resize, crop	Resize, crop

3.3 CNN Based Transfer Learning

Transfer learning adapts pre-trained models to new tasks, improving performance on limited data [28]. In this study, EfficientNet (Fig. 1) uses MBConv blocks with depth-wise separable convolutions and SE optimization, while ResNet (Fig. 2) employs residual blocks with shortcut connections and identity mapping using zero-padding when dimensions differ.

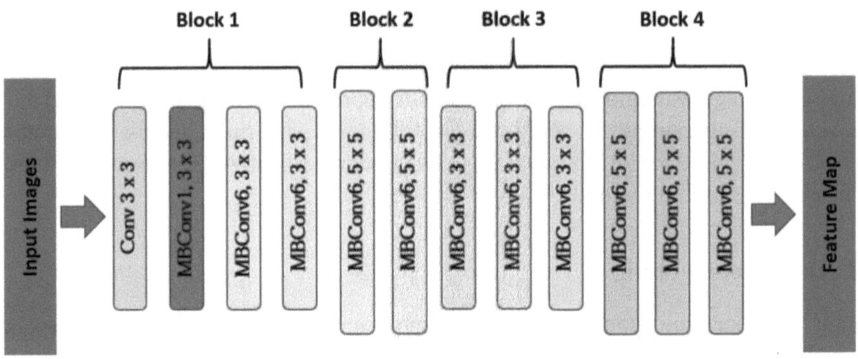

Fig. 1 EfficientNet b1 model

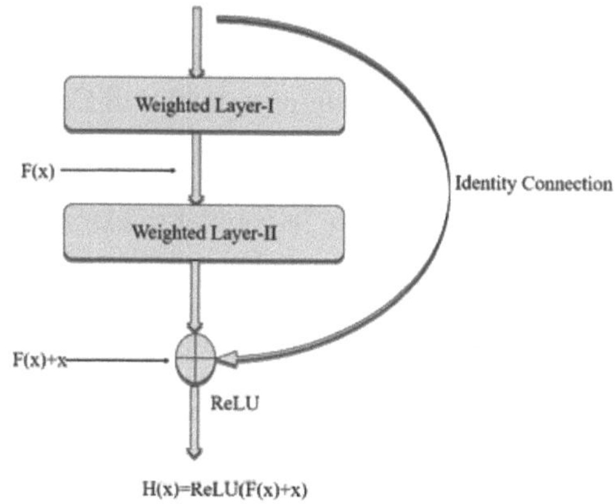

Fig. 2 ResNet's Basic Building block

As shown in Fig. 3, DenseNet-201 maximizes feature reuse by connecting each layer to every subsequent layer within its dense blocks via feature map concatenation. Following an initial convolution and pooling stage, the architecture processes information through four core dense blocks. Between these blocks, transition layers apply convolution and pooling to reduce both the number of feature maps and spatial dimensions.

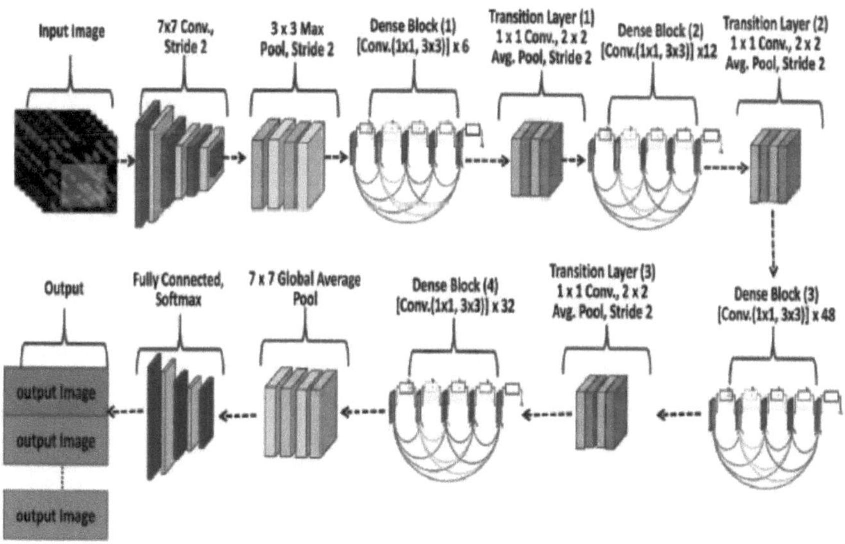

Fig. 3 DenseNet architecture

Ensemble CNN feature extraction [34] aims to create a more powerful feature representation by combining the strengths of multiple convolutional networks, leading to improved performance in tasks such as classification, detection, and recognition. The convolution operation is a core component of CNNs [36]. The selected CNN utilizes standard convolution operations, Rectified Linear Unit (ReLU) activations for non-linearity, and max pooling to reduce spatial dimensions, culminating in fully connected layers for final class prediction. In this study, an ensemble of Convolutional Neural Networks (CNNs) was utilized to improve the automated classification of fungal infections. Rather than relying on a single architecture, pre-trained EfficientNet-B1, ResNet18, and DenseNet201 models (Fig. 4) operate in parallel to extract diverse features from microscopic images. These complementary architectures maximize hierarchical learning, feature reuse, and dimensional scaling to capture both fine-grained local textures and high-level global patterns.

Final classification is achieved via a soft voting strategy that averages the probability outputs. While requiring three independent forward passes increases computational inference costs, this approach leverages complementary feature extraction to significantly enhance overall robustness and generalization.

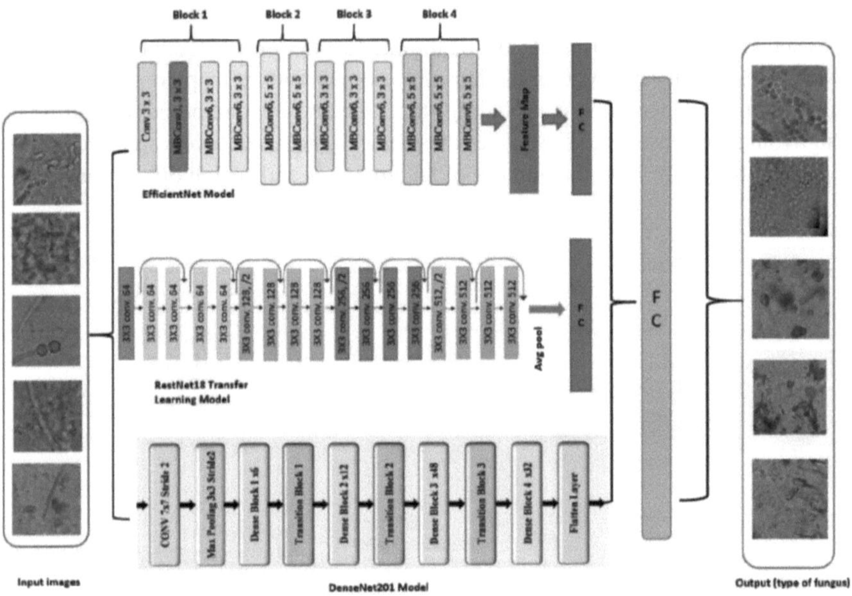

Fig. 4 The proposed model for extracting features from fungus images.

4 Results and Analysis

The evaluation was done on various individual deep learning models (GoogleNet, ResNet18/50, DenseNet201, EfficientNet B0-B2) and ensemble strategies to classify the dataset. Tables 4, 5, and 6 detail the precision, recall, and F1-score comparisons. Overall, ensembles consistently outperformed standalone architectures. Among individual models, DenseNet201 achieved the highest accuracy (86.59%) and specificity (95.65%), whereas GoogleNet performed the lowest (73.06% accuracy). By combining the strengths of EfficientNet-B1, ResNet18, and DenseNet201, the proposed ensemble achieved the peak overall performance of 91.35% accuracy and 96.92% specificity (Table 7).

Table 4 Precision Comparison of Individual and Hybrid Deep Learning Models

Class	H1	H2	H3	H5	H6
GoogleNet	0.796	0.509	0.820	0.920	0.901
ResNet 18	0.846	0.766	0.854	0.949	1.000
ResNet 50	0.882	0.605	0.845	0.973	0.958
EfficientNet b1	0.855	0.703	0.886	0.914	0.971
DensetNet201	0.880	0.760	0.960	0.950	1.000
GoogleNet + ResNet18	0.821	0.657	0.716	0.881	0.984
EfficientNet B2 + ResNet18	0.833	0.771	0.958	0.949	0.944
EfficientNet B4 + ResNet18 + DenseNet201	0.888	0.770	0.952	0.938	0.958
EfficientNet B1 + ResNet18 + DenseNet201	0.886	0.882	0.975	0.951	0.945

Table 5 Comparative Recall Analysis of the models

Class	H1	H2	H3	H5	H6
GoogleNet	0.776	0.588	0.610	0.863	0.914
ResNet 18	0.895	0.704	0.854	0.938	0.929
ResNet 50	0.895	0.704	0.854	0.938	0.929
EfficientNet b1	0.863	0.700	0.854	0.925	0.957
DensetNet201	0.900	0.780	0.780	0.960	0.930
GoogleNet + ResNet18	0.849	0.584	0.829	0.925	0.900
EfficientNet B2 + ResNet18	0.911	0.665	0.842	0.938	0.971
EfficientNet B4 + ResNet18 + DenseNet201	0.886	0.760	0.963	0.950	0.986
EfficientNet B1 + ResNet18 + DenseNet201	0.941	0.773	0.951	0.963	0.986

Table 6 Comparative F1 score analysis

Class	H1	H2	H3	H5	H6
GoogleNet	0.786	0.546	0.699	0.890	0.908
ResNet 18	0.870	0.734	0.854	0.943	0.963
ResNet 50	0.800	0.690	0.855	0.942	0.965
EfficientNet b1	0.859	0.701	0.870	0.919	0.964

(continued)

Table 6 (*continued*)

Class	H1	H2	H3	H5	H6
DensetNet201	0.890	0.770	0.860	0.960	0.960
GoogleNet + ResNet18	0.835	0.618	0.768	0.902	0.940
EfficientNet B2 + ResNet18	0.870	0.714	0.896	0.943	0.958
EfficientNet B4 + ResNet18 + DenseNet201	0.887	0.765	0.958	0.944	0.972
EfficientNet B1 + ResNet18 + DenseNet201	0.912	0.824	0.963	0.957	0.965

Table 7 Performance comparison of models on Accuracy and Average Specificity

Models	Accuracy	Specificity (avg)	Result
GoogleNet	73.06%	91.73%	lowest
ResNet18	84.81%	95.06%	strong
ResNet 50	79.71%	94.07%	weaker H1/H2
EfficientNet B1	83.26%	94.75%	strong
DenseNet	**86.59%**	**95.65%**	**strong**
GoogleNet + ResNet18 Ensemble	78.94%	93.47%	weaker H1
EfficientNet-B2 + ResNet18 Ensemble	84.81%	94.92%	better
EfficientNet-B4 + ResNet18 + DenseNet201	87.36%	94.92%	very close
EfficientNet-B1 + ResNet18 + DenseNet-201	**91.35%**	**96.92%**	**best**

The confusion matrix (Fig. 5) confirms high classification accuracy with strong diagonal alignment. Despite an inherent class imbalance (ranging from 437 samples in H1 to 70 in H6), the model maintains robust true positive rates across all categories. Minor misclassifications occur primarily in class H2, but the model successfully avoids significant cross-class leakage into the minority categories.

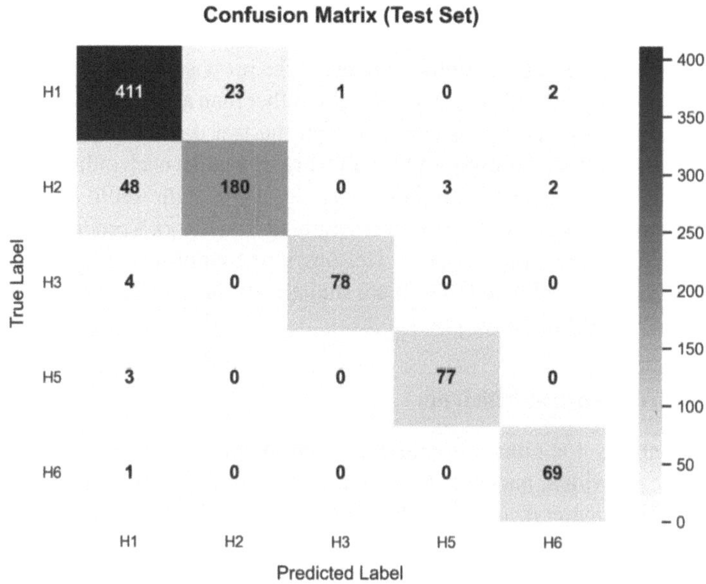

Fig. 5 Confusion matrix of 5 Fungal classes

The One-vs-Rest (OvR) ROC curves (Fig. 6) demonstrate the strong discriminative ability of the proposed ensemble, achieving AUC scores between 0.9729 and 0.9989. The minority classes (H3, H5, and H6) attained particularly high performance (AUC = 0.9931), indicating robust sensitivity and specificity despite dataset imbalance.

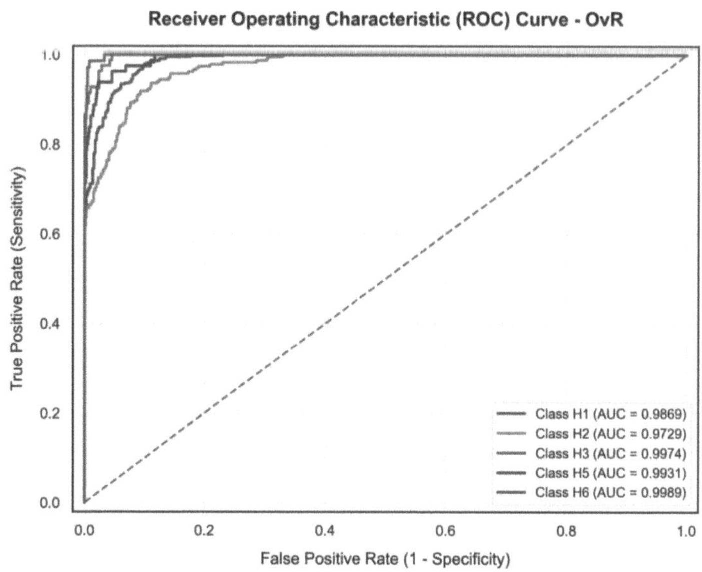

Fig. 6 Depiction of 5 Fungal classes using ROC

4.1 Statistical Significance Analysis

To validate that the performance improvements of the proposed ensemble (EfficientNet-B1 + ResNet18 + DenseNet201) are meaningful rather than a result of random variation, a formal statistical evaluation was conducted on the test dataset (N = 902). Pairwise McNemar's tests comparing the ensemble's error rates against each individual backbone architecture yielded p-values strictly below the significance threshold ($p < 0.05$). This rejects the null hypothesis, proving that the integrated framework's predictive superiority is statistically significant. Furthermore, DeLong's test confirmed that the ensemble's elevated AUC values (0.9729 to 0.9989) are statistically superior to the baseline curves, confirming robust discriminative power.

4.2 Model Architectural Efficiency

To assess suitability for clinical deployment, the model's structural complexity was calculated. The individual backbones EfficientNet-B1 (6,519,589 parameters), ResNet-18 (11,179,077 parameters), and DenseNet-201 (18,102,533 parameters) combine for an aggregate ensemble footprint of 35,801,199 parameters. The framework achieves a high 91.35 accuracy while maintaining a total parameter volume that is highly competitive; for context, the entire three model ensemble requires only about 10 million more parameters than a single standard ResNet-101 architecture (44.5 parameters). Crucially, because these baseline models operate independently during the initial feature extraction phase, their execution can be parallelized across modern GPU multi-threading architectures. This structural independence ensures that the overall inference latency is bounded only by the heaviest single network (DenseNet-201) rather than the sequential sum of all three models. These findings highlight the effectiveness of combining diverse architectures to leverage their complementary strengths for superior classification results while ensuring computational feasibility for real-time diagnostic workflows.

5 Conclusion

This study demonstrates the effectiveness of a hybrid transfer learning approach for classifying microscopic fungal infections using the DeFungi dataset. The proposed framework achieved high classification performance across five clinically significant fungal species. Several deep learning architectures and their ensemble combinations were evaluated, with ensemble models consistently improving prediction robustness and class balance. The best results were obtained using an EfficientNet-B1, ResNet18, and DenseNet201 ensemble, achieving the highest accuracy, F1-score, and specificity. These findings highlight the potential of ensemble learning for enhancing multi-class fungal infection classification and supporting early, accurate clinical diagnosis. Additionally, the study provides a foundation for future investigations into AI-powered mycological diagnostic instruments. Future research may focus on expanding the dataset, adding more fungus species, and creating real-time diagnostic tools that can be used in medical settings.

Although the proposed ensemble improves classification performance, its deployment cost is higher than that of a single backbone because it combines three deep CNNs.

Therefore, the framework should be viewed as a high-accuracy diagnostic system rather than a fully optimized real-time solution. Future work will focus on reducing computational cost through model compression techniques such as pruning, quantization, and knowledge distillation to enable efficient clinical deployment. We may also plan to extend our evaluation by incorporating state-of-the-art transformer-based architectures to provide a more comprehensive comparative analysis in medical image classification tasks.

References

1. Nahvi, I., Rehman, S., Kaushik, A., Jermy, B.R., Akhtar, S., Farhan, M., et al.: Candidacidal effect of Moringa stabilized silver nanomaterials reveal disruption of cell wall integrity, efflux pump, vacuole homeostasis and virulence traits in Candida auris. PLoS One. **20**(11), e0336309 (2025)
2. The Conversation: Academic rigour, journalistic flair. Accessed on: June 2025, Global deaths from fungal disease have doubled in a decade – new study
3. Mendonca, A., Santos, H., Franco-Duarte, R., Sampaio, P.: Fungal infections diagnosis–past, present and future. Res. Microbiol. **173**(3), 103915 (2022)
4. Fang, W., Wu, J., Cheng, M., Zhu, X., Du, M., Chen, C., et al.: Diagnosis of invasive fungal infections: challenges and recent developments. J. Biomed. Sci. **30**(1), 42 (2023)
5. Koo, T., Kim, M.H., Jue, M.S.: Automated detection of superficial fungal infections from microscopic images through a regional convolutional neural network. PLoS One. **16**(8), e0256290 (2021)
6. Mohamed, H., Nagib, A.E., Hany, M.: Comparative study of microscopic fungii classification using transfer learning models. In: 2024 Intelligent Methods, Systems, and Applications (IMSA), Giza, Egypt, pp. 87–92 (2024). https://doi.org/10.1109/IMSA61967.2024.10652826
7. Bashar, A., Latif, G., Ben Brahim, G., Mohammad, N., Alghazo, J.: COVID-19 pneumonia detection using optimized deep learning techniques. Diagnostics. **11**(11), 1972 (2021)
8. Saraswat, S., Singh, P., Kumar, M., Agarwal, J.: Advanced detection of fungi-bacterial diseases in plants using modified deep neural network and DSURF. Multimed. Tools Appl. **83**(6), 16711–16733 (2024)
9. Gümüş, A.: Classification of microscopic fungi images using vision transformers for enhanced detection of fungal infections. Türk Doğa ve Fen Dergisi. **13**(1), 152–160 (2024)
10. Zieliński, B., Sroka-Oleksiak, A., Rymarczyk, D., Piekarczyk, A., Brzychczy-Włoch, M.: Deep learning approach to describe and classify fungi microscopic images. PLoS One. **15**(6), e0234806 (2020)
11. Howell, S.A.: Dermatopathology and the diagnosis of fungal infections. Br. J. Biomed. Sci. **80**, 11314 (2023)
12. Nokdhes, Y.N., Leeyaphan, C., Jirawattanadon, P., Pongkittilar, B., Sereeaphinan, C., Bunyaratavej, S.: Prevalence and characteristics of Epidermophyton floccosum skin infections: a 12-year retrospective study. Mycoses. **67**(2), e13702 (2024)
13. Picek, L., Šulc, M., Matas, J., Heilmann-Clausen, J., Jeppesen, T.S., Lind, E.: Automatic fungi recognition: deep learning meets mycology. Sensors. **22**(2), 633 (2022)
14. Rahman, M.A., Clinch, M., Reynolds, J., Dangott, B., Villegas, D.M.M., Nassar, A., et al.: Classification of fungal genera from microscopic images using artificial intelligence. J. Pathol. Inf. **14**, 100314 (2023)
15. Karthikeyan, S., Ramkumar, G., Aravindkumar, S., Tamilselvi, M., Ramesh, S., Ranjith, A.: A novel deep learning-based black fungus disease identification using modified hybrid learning methodology. Contrast. Media Mol. Imaging. **2022**(1), 4352730 (2022)

16. Gautam, A.K., Verma, R.K., Avasthi, S., Sushma, Bohra, Y., Devadatha, B., et al.: Current insight into traditional and modern methods in fungal diversity estimates. J. Fungi. **8**(3), 226 (2022)
17. Ma, H., et al.: Deep convolutional neural network: a novel approach for the detection of Aspergillus fungi via stereomicroscopy. J. Microbiol. **59**, 563–572 (2021)
18. Jansen, P., Creosteanu, A., Matyas, V., Dilling, A., Pina, A., Saggini, A., et al.: Deep learning assisted diagnosis of onychomycosis on whole-slide images. J. Fungi. **8**(9), 912 (2022)
19. Kumar, D., Kukreja, V.: Deep learning in wheat diseases classification: a systematic review. Multimed. Tools Appl. **81**(7), 10143–10187 (2022)
20. Cinar, I., Taspinar, Y.S.: Detection of fungal infections from microscopic fungal images using deep learning techniques. Proc Int Conf Adv Technol. (2023)
21. Lim, S.S., Ohn, J., Mun, J.H.: Diagnosis of onychomycosis: from conventional techniques and dermoscopy to artificial intelligence. Front. Med. **8**, 637216 (2021)
22. Rani, P., Kotwal, S., Manhas, J., Sharma, V., Sharma, S.: Machine learning and deep learning based computational approaches in automatic microorganisms image recognition: methodologies, challenges, and developments. Arch. Computat. Methods Eng. **29**(3), 1801–1837 (2022)
23. Khan, M.A., Ali, M., Shah, M., Mahmood, T., Ahmad, M., Jhanjhi, N.Z., et al.: Machine learning-based detection and classification of walnut fungi diseases. Intell. Autom. Soft Comput. **30**(3), 771–785 (2021)
24. Rawat, S., Bisht, B., Bisht, V., Rawat, N., Rawat, A.: Mefunx: a novel meta-learning-based deep learning architecture to detect fungal infection directly from microscopic images. Franklin Open. **6**, 100069 (2024)
25. Gao, W., Li, M., Wu, R., Du, W., Zhang, S., Yin, S., et al.: The design and application of an automated microscope developed based on deep learning for fungal detection in dermatology. Mycoses. **64**(3), 245–251 (2021)
26. Álvarez, M.A.V., Sopó, L., Sopo, C.J.P., Hajati, F., Gheisari, S.: P456 Defungi: direct mycological examination of microscopic fungi images. Med. Mycol. **60**(Supplement_1), myac072P456 (2022)
27. Zhou, Y., Feng, Y., Zhang, H.: Human fungal infection image classification based on convolutional neural network. In: Wang, Y., Huang, Q., Peng, Y. (eds.) Image and graphics technologies and applications: IGTA 2019, vol. 1043, pp. 3–13 (2019)
28. Zhuang, F., Qi, Z., Duan, K., Xi, D., Zhu, Y., Zhu, H., et al.: A comprehensive survey on transfer learning. Proc. IEEE. **109**(1), 43–76 (2020)
29. Yang, S., Xiao, W., Zhang, M., Guo, S., Zhao, J., Shen, F.: Image data augmentation for deep learning: a survey. arXiv preprint https://arxiv.org/abs/2204.08610. (2022)
30. Shorten, C., Khoshgoftaar, T.M.: A survey on image data augmentation for deep learning. J. Big Data. **6**(1), 1–48 (2019)
31. Das, N.N., Kumar, N., Kaur, M., Kumar, V., Singh, D.: Automated deep transfer learning-based approach for detection of COVID-19 infection in chest X-rays. Irbm. **43**(2), 114–119 (2022)
32. Ma, Y., Chen, T., Zheng, J., Wang, W., Gao, S.: Research on identification method and improvement of various crop diseases based on ResNet18. Int. Core J. Eng. **8**(6), 115–124 (2022)
33. Sengupta, A., Ye, Y., Wang, R., Liu, C., Roy, K.: Going deeper in spiking neural networks: VGG and residual architectures. Front. Neurosci. **13**, 95 (2019)
34. Goumiri, S., Benboudjema, D., Pieczynski, W.: A new hybrid model of convolutional neural networks and hidden Markov chains for image classification. Neural Comput. & Applic. **35**(24), 17987–18002 (2023)
35. Microscopic Fungi Image - DeFungi Dataset. https://www.kaggle.com/datasets/anshtanwar/microscopic-fungi-images. Accessed 1 Oct 2024

36. Khan, A., Sohail, A., Zahoora, U., Qureshi, A.S.: A survey of the recent architectures of deep convolutional neural networks. Artif. Intell. Rev. **53**, 5455–5516 (2020)
37. Nair, V., Hinton, G.E.: Rectified linear units improve restricted Boltzmann machines. In: Icml, pp. 807–814 (2010)
38. Chandra, M.A., Bedi, S.S.: Survey on SVM and their application in image classification. Int. J. Inf. Technol. **13**(5), 1–11 (2021)
39. Latif, G., Al Anezi, F.Y., Zikria, M., Alghazo, J.: EEG-ECG signals classification for arrhythmia detection using decision trees. In: Proceedings of the 2020 Fourth International Conference on Inventive Systems and Control (ICISC), Coimbatore, India (8–10 January 2020)
40. Ren, Z. et al.: EEG-based driving fatigue detection using a two-level learning hierarchy radial basis function. Front. Neurorobot. **15**, 618408 (2021)
41. Wickramasinghe, I., Kalutarage, H.: Naive Bayes: applications, variations and vulnerabilities: a review of literature with code snippets for implementation. Soft. Comput. **25**, 2277–2293 (2021)

Fuzzy Computational Models for Public Health Forecasting to Support SDG 3 Implementation

Tri Wijayanti Septiarini[1(✉)], Tofan Agung Eka Prasetya[2], and Sahidan Abdulmana[3]

[1] Mathematics Study Program, Faculty of Science and Technology, Universitas Terbuka, 15437 South Tangerang, Indonesia
tri.wijayanti@ecampus.ut.ac.id

[2] Health Department, Universitas Airlangga, Surabaya 60115, Indonesia

[3] Department of Data Science and Analytics, Fatoni University, Pattani 94160, Thailand

Abstract. Achieving SDG 3 requires accurate forecasting tools capable of handling uncertainty and nonlinear dynamics in public health data. This study aims to develop a multivariate fuzzy time series model to forecast key health indicators, including Maternal Mortality Ratio (MMR), Under-5 Mortality Rate (U5MR), tuberculosis incidence, non-communicable disease mortality, and immunization coverage. The study utilizes time series data from 2000 to 2021 obtained from the World Bank and WHO, processed through data cleaning, normalization, and fuzzification. The modeling framework involves defining the universe of discourse, interval partitioning, constructing multivariate fuzzy logical relationships, and defuzzification, followed by evaluation using Mean Absolute Percentage Error (MAPE) and Root Mean Squared Error (RMSE). The results demonstrate high predictive accuracy, with MAPE of 2.98% and RMSE of 7.95, outperforming multivariate regression and showing strong capability in capturing nonlinear interactions among health indicators. Forecast projections toward 2030 indicate a declining trend in MMR; however, the projected values remain above the SDG target threshold, suggesting that current progress may be insufficient without intensified interventions. This study contributes a robust and policy-relevant computational framework for public health forecasting and highlights the importance of integrating multivariate and uncertainty-aware models, with future work directed toward hybrid and machine learning-based approaches.

Keywords: Computational modeling · Fuzzy time series · Health indicators · Public health forecasting · SDG 3

1 Introduction

Achieving United Nations Sustainable Development Goal 3 (SDG 3: Good Health and Well-being) remains a central global challenge, particularly in reducing preventable mortality and improving population health outcomes (Singh et al. 2025). Key indicators such as Maternal Mortality Ratio (MMR), Under-5 Mortality Rate (U5MR), tuberculosis (TB) incidence, non-communicable disease (NCD) mortality, and immunization coverage are widely used to monitor progress toward SDG targets (Bharuka et al. 2023;

N. Mohd Sharef et al. (Eds.): CAMP 2026, CCIS 3093, pp. 168–178, 2027.
https://doi.org/10.1007/978-981-92-4302-0_13

Singh et al. 2025). However, these indicators exhibit complex and nonlinear dynamics influenced by multiple socio-economic and healthcare-related factors, making accurate forecasting a challenging task (Adama et al. 2025).

Reliable forecasting models are essential to support evidence-based policy and long-term planning. Recent studies highlight the importance of time series forecasting in public health, particularly for predicting disease trends and assessing progress toward SDG targets (Chhabra et al. 2024). Classical statistical models such as Autoregressive Integrated Moving Average (ARIMA) and Exponential Smoothing (ETS) have been widely applied due to their strong predictive capabilities and interpretability (Arumugam and Natarajan 2023; Canama and Talirongan 2025). These models have demonstrated effectiveness in various health-related forecasting problems, including epidemic trends and mortality rates (Kalahasti et al. 2025). However, their reliance on assumptions of linearity and stationarity limits their ability to capture nonlinear patterns and uncertainty inherent in public health data.

To overcome these limitations, fuzzy computational models have emerged as an alternative approach capable of handling uncertainty and imprecision (Alyousifi et al. 2021). Based on the theory of fuzzy sets introduced by Lotfi A. Zadeh, fuzzy time series (FTS) models transform numerical data into linguistic variables, allowing more flexible representation of complex systems. Recent studies show that fuzzy models can effectively capture irregular patterns and noisy data without requiring strict statistical assumptions (Orang et al. 2022; Marín Díaz 2025). Moreover, hybrid fuzzy approaches have been shown to improve forecasting performance by incorporating adaptive mechanisms and handling dynamic variability in time series data (Enke and Mehdiyev 2014; Gupta and Kumar 2019).

Despite these advantages, existing literature reveals several critical gaps. First, many studies focus on univariate fuzzy time series models, which consider only a single variable and ignore interactions among related indicators (Bitencourt et al. 2023). This is a significant limitation in public health systems, where indicators such as mortality, disease incidence, and immunization coverage are strongly interdependent (Papanicolas and Ledesma 2025). Second, comparative studies often show that classical models can outperform basic fuzzy models in certain contexts, particularly when dealing with well-structured time series data (Septiarini et al. 2026). This suggests that conventional fuzzy approaches may lack robustness when applied to complex real-world datasets. Third, there remains limited research integrating fuzzy models with multivariate frameworks and policy-oriented forecasting, particularly in the context of SDG monitoring.

The novelty of this study is proposing a multivariate fuzzy computational framework for forecasting key SDG 3 indicators which is different with a standard fuzzy time series. Unlike traditional univariate approaches, the proposed model incorporates multiple health variables into fuzzy logical relationships, enabling the capture of both temporal dependencies and cross-variable interactions. Furthermore, the advantage of fuzzy models is in capturing non-linear patterns and uncertainties that cannot be handled by classical models such as ARIMA or regression (Hopfe et al. 2024). And the importance of this model as a dynamic decision support tool (Zyoud et al. 2025). Moreover, this study evaluates the performance of the proposed model against classical approaches to provide a comprehensive comparison across different modeling paradigms. In addition, this study extends the analysis by generating long-term projections of health indicators toward the year 2030. These projections are used to assess whether current trends are sufficient to meet SDG targets and to identify potential gaps requiring policy intervention. By integrating computational modeling with policy-relevant analysis, this research contributes to the development of a robust and practical framework for public health forecasting.

The contributions of this study are threefold. First, it introduces a multivariate fuzzy time series model that captures interactions among multiple health indicators. Second, it demonstrates the comparative advantage of fuzzy modeling in handling nonlinear and uncertain data relative to classical statistical approaches. Third, it provides forward-looking projections that support strategic planning and resource allocation toward achieving SDG 3 targets.

2 Methods

2.1 Data Description

This study utilizes time series data of key public health indicators associated with SDG 3. The dataset covers the period from 2000 to 2021 and is compiled from internationally recognized sources, including the World Bank and the World Health Organization. Details of the dataset are compiled in Table 1. Furthermore, these indicators were selected due to their direct relevance to SDG 3 targets and their representation of both mortality and healthcare system performance.

Table 1 Variable indicators

Variable	Code	Measured	Sources
Maternal Mortality Ratio	MMR	Per 100,000 live births	https://data.worldbank.org/
Under-5 Mortality Rate	U5MR	Per 1,000 live births	https://data.worldbank.org/
Tuberculosis Incidence	TB	Per 100,000 population	https://data.who.int/
NCD Mortality	NCD	Percentage (%)	https://data.who.int/
Immunization Coverage	IMM	Percentage (%)	https://data.worldbank.org/

The descriptive analysis of the dataset reveals that the selected health indicators exhibit distinct temporal patterns over the period 2000–2021 (Fig. 1). The Maternal Mortality Ratio (MMR) shows a general declining trend, indicating gradual improvements in maternal health services. Similarly, the Under-5 Mortality Rate (U5MR) demonstrates a consistent decrease, reflecting enhanced child healthcare and immunization programs. In contrast, tuberculosis (TB) incidence and non-communicable disease (NCD) mortality display more fluctuating patterns, suggesting the influence of external factors such as healthcare accessibility, socio-economic conditions, and policy interventions. Immunization coverage (IMM) appears relatively stable with slight improvements over time.

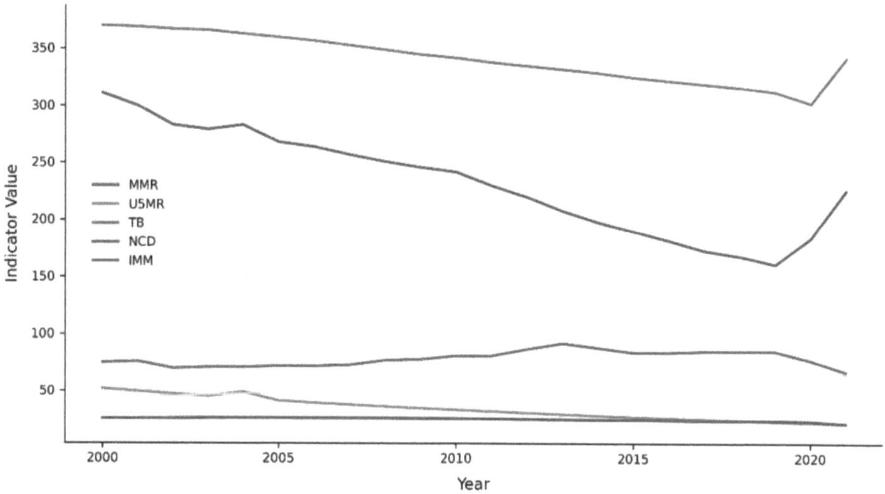

Fig. 1 Plot of trend SDG 3 indicators during 2000–2021

2.2 Data Preprocessing

To ensure data quality and consistency, several preprocessing steps were performed:

(a) Data cleaning and alignment. All variables were aligned temporally to ensure a consistent time range (2000–2021).
(b) Handling missing values. Missing observations were addressed using linear interpolation, which is appropriate for time series data with gradual changes.
(c) Normalization. For multivariate modeling, min-max normalization was applied to scale all variables into a common range:

$$X' = \frac{X - X_{min}}{X_{max} - X_{min}}$$

(d) Splitting data. To evaluate the generalization capability of the proposed model, the dataset was divided into training and testing subsets using an 80:20 ratio. Observations from 2000 to 2017 were used for model development and parameter estimation, while data from 2018 to 2021 were reserved for out-of-sample evaluation. This approach enables an objective assessment of forecasting performance on unseen data and reduces the risk of overestimating model accuracy.

2.3 Fuzzy Time Series Model

This study employs a multivariate fuzzy time series (FTS) model to capture uncertainty and nonlinear relationships among health indicators.

(a) Universe of Discourse

The universe of discourse U is defined as:

$$U = [D_{\min}, D_{\max}]$$

where $D_{\min}$ and $D_{\max}$ represent the minimum and maximum values of the dataset.

(b) Interval

The universe of discourse is divided into a finite number of intervals. In this study, seven intervals are used to achieve an optimal balance between model granularity and computational efficiency. While fewer intervals can give chance to oversimplify the complex interactions between health indicators, a higher number of intervals could lead to sparse fuzzy logical relationship groups, especially given the refined dataset of 22 observation points. This partitioning level ensures that the model remains sensitive to the nonlinear dynamics of SDG 3 indicators, such as the fluctuations in TB incidence and NCD mortality without compromising the interpretability of the fuzzy linguistic variables. This approach aligns with the principle of streamlining the evaluation process while retaining predictive accuracy in uncertain public health environments (Zyoud et al. 2025).

(c) Fuzzification

Each numerical observation is transformed into a fuzzy set based on the defined intervals. This process converts crisp values into linguistic representations.

(d) Multivariate Fuzzy Logical Relationships (FLR)

Multivariate Fuzzy Logical Relationships (FLR) is constructed to capture the holistic dynamics of public health indicators by defining a unified Multivariate Fuzzy State Vector (Züge and Coelho 2024). For every time point, the status of the public health system is represented as a vector $S_t = \{f_{MMR,\,t}, f_{U5MR,\,t}, f_{TB,\,t}, f_{NCD,\,t}, f_{IMM,\,t}\}$, where each element corresponds to the linguistic fuzzy set of the respective indicator. Unlike univariate models that treat indicators in isolation, this approach represents interactions by treating the entire vector S_t as a single antecedent. Moreover, the model can capture interactions among health indicators. The relationships are stored using a mapping mechanism where each unique antecedent vector S_t is linked to its observed consequent vector S_{t+1}. During the forecasting phase, the model identifies the current multivariate fuzzy state and retrieves the corresponding transition from the stored FLRG. By predicting the entire vector simultaneously, the model ensures that the temporal dependencies and cross-variable interactions between all five SDG 3 indicators are utilized to produce a coherent system-wide projection toward 2030.

(e) Defuzzification

The predicted fuzzy values are converted back into numerical values using the midpoint (centroid) method which is used to find the fuzzy set's center of mass. It uses the membership degrees of the output values to produce a clear value that is represented by the weighted average of the output values. Since the crisp value is irreversible, the average output value in relation to the membership values of the fuzzy set represents its output. The centroid can be found analytically by integrating the product of the variable and its corresponding membership function over the whole range of possible output values (Dejprayoon et al. 2025). The centroid method was presented using the following equation(Yu et al. 2024a):

$$Z_{centroid} = \frac{\sum_Z z \cdot \mu_C(z)}{\sum_Z \mu_C(z)}$$

where $Z_{centroid}$ is defined the crisp output value gained by calculating the weighted average of z and $\mu_C(z)$ is the membership degree of z in the output fuzzy set C.

2.4 Model Evaluation

The performance of the forecasting model is evaluated using two standard metrics. These metrics provide complementary insights into model accuracy and error magnitude where A is defined as the actual value and F defines as forecasted value.

- Mean Absolute Percentage Error (MAPE)

$$\text{MAPE} = \frac{1}{n}\sum\nolimits_{t=1}^{n} \left| \frac{A_t - F_t}{A_t} \right| \times 100\%$$

- Root Mean Squared Error (RMSE)

$$\text{RMSE} = \sqrt{\frac{1}{n}\sum\nolimits_{t=1}^{n} (A_t - F_t)^2}$$

2.5 Comparative Models

To validate the effectiveness of the proposed model, its performance is compared with several baseline approaches:

- Autoregressive Integrated Moving Average (ARIMA). A classical statistical model for time series forecasting.
- Multivariate Linear Regression. A linear model that incorporates multiple predictors but assumes linear relationships.

These models represent different forecasting paradigms and provide a comprehensive benchmark.

2.6 Forecasting Towards 2030

To assess progress toward SDG targets, long-term forecasting is conducted up to the year 2030 using a recursive forecasting approach. In addition, a scenario-based framework is applied:

1. Optimistic scenario assumes accelerated improvement.
2. Moderate scenario is based on current trends.
3. Pessimistic scenario assumes slower progress which projections enable evaluation of whether current trajectories are sufficient to achieve SDG targets and provide insights for policy intervention and resource allocation.

3 Results and Discussion

3.1 Model Performance

The performance of the proposed multivariate fuzzy time series model is evaluated using MAPE and RMSE metrics. The results show that MAPE is 2.98% and RMSE is 7.95. These results indicate excellent predictive accuracy, as the MAPE value is well below the commonly accepted threshold of 10%. The low RMSE further confirms that the model produces predictions with minimal deviation from actual values. The high accuracy demonstrates the capability of the multivariate fuzzy model to capture both temporal dynamics and interactions among multiple health indicators (Fig. 2).

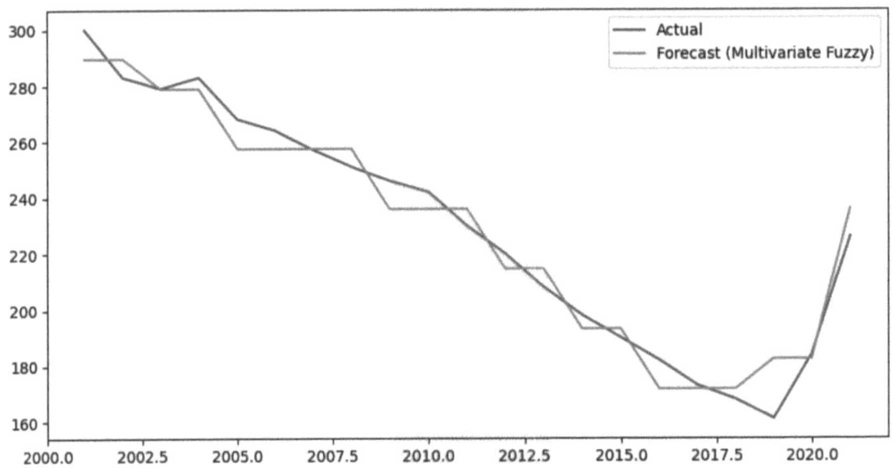

Fig. 2 Plot of forecasting result multivariate fuzzy model

To validate the effectiveness of the proposed model, its performance is compared with classical forecasting approaches, including ARIMA, Exponential Smoothing (ETS), and multivariate linear regression as represented in Table 2. Although Exponential Smoothing achieved the lowest MAPE (2.71%), the proposed multivariate fuzzy model produced the lowest RMSE (7.95), indicating superior stability in capturing the underlying dynamics of public health indicators. Unlike ETS and ARIMA, which rely primarily on historical patterns of a single variable, the multivariate fuzzy approach incorporates interactions among multiple SDG 3 indicators, providing a more comprehensive representation of complex health systems. Therefore, the proposed model offers a promising framework for uncertainty-aware public health forecasting and SDG monitoring.

Table 2 Evaluation value

Model	MAPE (%)	RMSE
Multivariate Fuzzy	2.98	7.95
Regression	3.51	8.61
ARIMA	9.78	78.85
Exponential Smoothing	2.71	9.83

3.2 Forecast Towards SDG

The forecasting results indicate a continued downward trend in MMR toward 2030. However, the projected values remain above the SDG target threshold of MMR < 70 per 100,000 births. This suggests that although progress has been made, the current rate of improvement may not be sufficient to fully achieve SDG 3 targets by 2030. Scenario-based projections further indicate under optimistic conditions which the reduction rate improves but may still fall short of the target, and under pessimistic conditions which progress slows significantly as shown in Fig. 3.

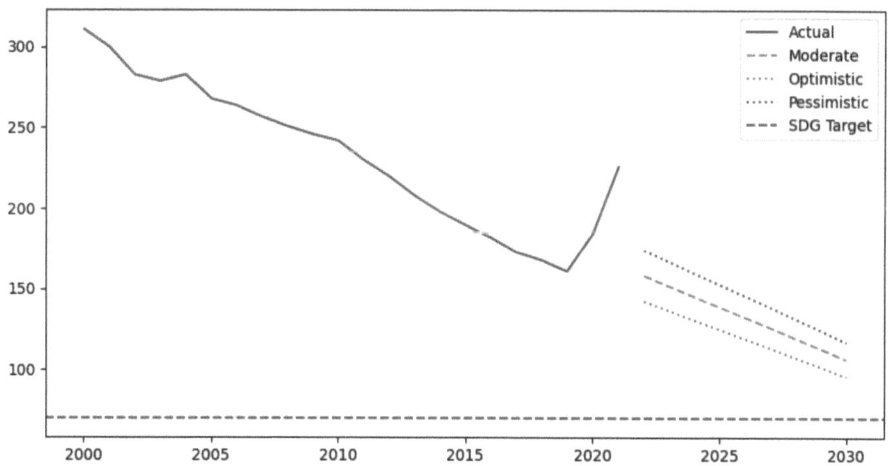

Fig. 3 Plot of forecasting projection towards SDG

3.3 Policy Implication

The forecasting results provide several important policy insights which are acceleration of intervention programs is required to meet SDG targets, strengthening maternal healthcare systems is essential, Integrated strategies across indicators (e.g., immunization and disease control) can enhance outcomes, and resource allocation should be prioritized based on projected gaps. The results demonstrate that data-driven forecasting can support more effective planning and decision-making in public health.

3.4 Discussion

The findings of this study confirm that multivariate fuzzy time series models are highly effective for forecasting public health indicators characterized by uncertainty and non-linear relationships (Bitencourt et al. 2023; Yu et al. 2024b). By incorporating multiple variables into the modeling process, the proposed approach captures interdependencies that are often overlooked in univariate models. Compared to classical approaches, the fuzzy model offers greater flexibility and adaptability, particularly in handling imprecise and fluctuating data (Tang and Ahmad 2024; Vasudevan et al. 2025). The superior performance observed in this study supports previous findings that fuzzy-based methods are well-suited for complex real-world applications (Abdalla et al. 2024; Goyal and Sakale 2024; Sathya et al. 2024).

However, several limitations should be noted. First, the relatively small dataset may affect the generalizability of the results. Second, the recursive forecasting approach may introduce error accumulation in long-term projections. Future research can explore hybrid models, such as integrating fuzzy systems with machine learning or deep learning techniques, to further improve forecasting performance. Overall, this study provides a robust computational framework that not only achieves high predictive accuracy but also delivers meaningful insights for policy and decision-making in the context of SDG 3.

4 Conclusion

This study proposes a multivariate fuzzy time series model for forecasting SDG 3 health indicators under uncertainty and nonlinear conditions. The results demonstrate high predictive accuracy (MAPE 2.98%, RMSE 7.95), outperforming regression and showing strong capability in capturing interdependencies among health variables. Forecast projections toward 2030 indicate that, despite improvements, the current trajectory may not fully achieve SDG targets without enhanced policy interventions. The proposed framework provides a practical and data-driven tool to support public health planning and resource allocation. Future work may focus on integrating hybrid models to improve long-term forecasting performance.

References

Abdalla, A.Y., Abdalla, T.Y., Chyaid, A.M.: Internet of things-based fuzzy systems for medical applications: a review. IEEE Access. **12**, 163883–163902 (2024)

Adama, Z.K., Mettle, F.O., Baiden, B.M., Bii, N.K.: Forecasting progress: analyzing the trajectory of under-five child mortality for Ghana, Niger, Nigeria, and Sierra Leone towards SDG3 using ARIMA time series model. BMC Public Health. **25**, 1607 (2025). https://doi.org/10.1186/s12889-025-22869-z

Alyousifi, Y., Othman, M., Almohammedi, A.A.: A novel stochastic fuzzy time series forecasting model based on a new partition method. IEEE Access. **9**, 80236–80252 (2021). https://doi.org/10.1109/ACCESS.2021.3084048

Arumugam, V., Natarajan, V.: Time series modeling and forecasting using autoregressive integrated moving average and seasonal autoregressive integrated moving average models. Instrum. Mes. Metrol. **22** (2023). https://doi.org/10.18280/i2m.220404

Bharuka, S., Gupta, A., Preethiya, T.: Comparative analysis of time series models for maternal mortality ratio forecasting. In: International Conference on Sustainable Communication Networks and Application, ICSCNA 2023 - Proceedings (2023)

Bitencourt, H.V., Orang, O., de Souza, L.A.F., et al.: An embedding-based non-stationary fuzzy time series method for multiple output high-dimensional multivariate time series forecasting in IoT applications. Neural Comput. Appl. **35**, 9407–9420 (2023)

Canama, N.V., Talirongan, F.J.: Data-driven-forecasting-of-morbidity-trends-using-ARIMA-models-statistical-approach-to-public-health-planning. Int. J. Multidiscip. Res. **7** (2025). https://doi.org/10.36948/ijfmr.2025.v07i04.51813

Chhabra, A., Singh, S.K., Sharma, A., et al.: Sustainable and intelligent time-series models for epidemic disease forecasting and analysis. Sustain. Technol. Entrepr. **3**, 100064 (2024). https://doi.org/10.1016/j.stae.2023.100064

Dejprayoon, K., Lekcharoen, S., Pankham, S.: Determinants of customer intention for sustainable tourism packages in Thailand using rough set-fuzzy theory. J. Fuzzy Ext. Appl. **6**, 522–554 (2025). https://doi.org/10.22105/jfea.2025.513722.1852

Enke, D., Mehdiyev, N.: A hybrid neuro-fuzzy model to forecast inflation. Procedia Comput. Sci. **36**, 254–260 (2014). https://doi.org/10.1016/j.procs.2014.09.088

Goyal, N., Sakale, R.K.: An assessment of inventory models utilizing fuzzy theory: an application and approaches. J. Adv. Schol. Res. Allied Educ. **21** (2024). https://doi.org/10.29070/e2fz5x23

Gupta, K.K., Kumar, S.: Fuzzy time series forecasting method using probabilistic fuzzy sets. In: Advances in Intelligent Systems and Computing (2019)

Hopfe, D.H., Lee, K., Yu, C.: Short-term forecasting airport passenger flow during periods of volatility: comparative investigation of time series vs. neural network models. J. Air Transp. Manag. **115**, 102525 (2024). https://doi.org/10.1016/j.jairtraman.2023.102525

Kalahasti, S., Faucher, B., Wang, B., et al.: Foundation time series models for forecasting and policy evaluation in infectious disease epidemics. medRxiv. (2025)

Marín Díaz, G.: Supporting reflective AI use in education: a fuzzy-explainable model for identifying cognitive risk profiles. Educ. Sci. (Basel). **15**, 923 (2025). https://doi.org/10.3390/educsci15070923

Orang, O., de Lima e Silva, P.C., Guimarães, F.G.: Time series forecasting using fuzzy cognitive maps: a survey. Artif. Intell. Rev. **56**, 7733–7794 (2022). https://doi.org/10.1007/s10462-022-10319-w

Papanicolas, I., Ledesma, J.: Measuring health system resilience: understanding the relationship between excess mortality and health system performance. Health Policy (New York), 105430 (2025)

Sathya, D., Saravanan, G., Thangamani, R.: Fuzzy logic and its applications in mechatronic control systems. In: Computational Intelligent Techniques in Mechatronics. Wiley (2024)

Septiarini, T.W., Kharis, S.A.A., Jayanegara, A., Abdulmana, S.: Comparison of Arima, exponential smoothing, and Chen-Singh fuzzy models for inflation forecasting in Asean countries. BAREKENG: J. Math. Appl. **20**, 619–636 (2026)

Singh, R., Singh, H., Singh, A., et al.: Advancing maternal and child health: a comprehensive analysis of United Nations' sustainable development goals. Majmaah J. Health Sci. **13**, 88 (2025). https://doi.org/10.5455/mjhs.2025.01.009

Tang, H.H., Ahmad, N.S.: Fuzzy logic approach for controlling uncertain and nonlinear systems: a comprehensive review of applications and advances. Syst. Sci. Control Eng. **12**, 2394429 (2024)

Vasudevan, A., Yogeesh, N., Mohammad, S.I., et al.: Unlocking insights of fuzzy mathematics for enhanced predictive modelling. Appl. Math. Inf. Sci. **19**, 49–60 (2025). https://doi.org/10.18576/amis/190105

Yu, H.-C., Wang, Q.-A., Li, S.-J.: Fuzzy logic control with long short-term memory neural network for hydrogen production thermal control system. Appl. Sci. **14**, 8899 (2024a). https://doi.org/10.3390/app14198899

Yu, J., Wang, H., Chen, M., et al.: A novel method to select time-varying multivariate time series models for the surveillance of infectious diseases. BMC Infect. Dis. **24**, 832 (2024b). https://doi.org/10.1186/s12879-024-09718-x

Züge, C.V., Coelho, L.d.S.: Granular weighted fuzzy approach applied to short-term load demand forecasting. Technologies (Basel). **12**, 182 (2024). https://doi.org/10.3390/technologies12100182

Zyoud, S., Omair, S.M., Jarrad, S.A.: A two-stage model of the fuzzy analytic hierarchy process and the fuzzy synthetic evaluation technique to prioritize sustainable sanitation services under uncertainty. Sci. Rep. **15**, 3736 (2025). https://doi.org/10.1038/s41598-025-88236-5

GT-RL: Graph Transformer-Based Reinforcement Learning for Airport Apron Scheduling Optimization

Devia Kartika[1(✉)], Sarjon Defit[2], and Syafri Arlis[3]

[1] Department of Informatics Engineering, Universitas Putra Indonesia YPTK, Padang, Indonesia
deviakartika@upiyptk.ac.id

[2] Doctoral Program of Computer Science, Universitas Putra Indonesia YPTK, Padang, Indonesia
Sarjon_defit@upiyptk.ac.id

[3] Department of Information Systems, Universitas Putra Indonesia YPTK, Padang, Indonesia
syafri_arlis@upiyptk.ac.id

Abstract. Airport apron scheduling is a complex combinatorial optimization problem involving aircraft stand allocation under dynamic operational constraints, including stand compatibility, temporal conflicts, limited apron capacity, and flight delays. Conventional heuristic and reinforcement learning approaches often rely on simple numerical state representations that are insufficient to model complex spatial and temporal relationships among operational entities. To address this limitation, this study proposes GT-RL, a Graph Transformer-Based Deep Reinforcement Learning framework for intelligent airport apron scheduling optimization. The proposed approach models the airport apron environment as a graph, where nodes represent aircraft and stands, while edges represent operational relationships such as compatibility and scheduling conflicts. A Graph Transformer encoder with multi-head self-attention is utilized to capture both local and global dependencies within the scheduling environment, enabling richer spatial–temporal state representation for reinforcement learning-based decision making. The proposed model is evaluated using real-world airport operational data under two scheduling scenarios consisting of 1,511 and 4,460 flight movements, respectively. Performance evaluation is conducted using reward, average delay, conflict rate, stand utilization, and On-Time Performance (OTP). Experimental results demonstrate that the proposed GT-RL model consistently outperforms baseline methods, including FCFS, Greedy, and Deep Q-Network (DQN), across different levels of scheduling complexity. In the high-density scenario involving 4,460 flights, GT-RL achieves the highest reward of 226.43, reduces average delay to 0.09, and decreases scheduling conflicts to 847 while maintaining an OTP of 99.87%. Furthermore, scheduling simulation results confirm the capability of the proposed framework to generate adaptive and conflict-free stand allocation decisions under realistic airport operational conditions. Overall, the proposed framework demonstrates strong effectiveness, robustness, and scalability for intelligent airport apron scheduling optimization.

N. Mohd Sharef et al. (Eds.): CAMP 2026, CCIS 3093, pp. 179–195, 2027.
https://doi.org/10.1007/978-981-92-4302-0_14

Keywords: Airport Apron Scheduling · Graph Transformer · Deep Reinforcement Learning · Stand Allocation · Spatial-Temporal Representation · Airport Optimization

1 Introduction

The global aviation industry continues to experience significant growth in line with increasing human mobility, international trade, and economic integration between regions. This growth is driving increased demand for air transportation, which directly impacts the capacity of airport infrastructure in various countri [1]. According to Airports Council International 2025, the number of global passenger movements is projected to continue to increase until 2053, so this condition requires airports to optimize operational resources more efficiently and adaptively to address the imbalance between increasing demand and capacity. A modern airport is a complex system consisting of integrated runways, taxiways, terminals, gates, and aprons. One of the most critical areas in this system is the apron, or stand, which serves as the primary location for aircraft parking and various ground handling activities, such as passenger boarding and deboarding, fueling, baggage handling, catering, cabin cleaning, and departure preparation. The complexity of activities on the apron makes it a bottleneck in airport operations, where resource constraints and scheduling inefficiencies can degrade the quality of flight service and increase the potential for delays. Therefore, efficient apron management has a direct impact on punctuality of departures, aircraft stand utilization, turnaround time efficiency, and airline operational costs [2]. The apron scheduling problem is essentially a complex combinatorial optimization problem and is NP-hard, as it involves various operational constraints such as limited number of stands, aircraft type compatibility, overlapping arrival and departure schedules, flight delays, and safety regulations between stands. Furthermore, each aircraft can only occupy one stand at a time, increasing scheduling complexity in a dynamic operational environment. At high-traffic airports, suboptimal aircraft placement decisions can lead to aircraft queues, parking conflicts, fleet rotation delays, and disruption to the airport's overall operational performance [3].

Various approaches have been used to solve the apron scheduling problem, ranging from rule-based methods to heuristic and learning approaches. Heuristic methods, such as greedy and metaheuristic, are widely used because they can produce solutions quickly, but they tend to be static and less adaptable to complex operational dynamics [4]. To overcome these limitations, learning-based approaches such as Deep Reinforcement Learning, including DQN, have begun to be developed to address decision-making in dynamic environments. Several studies have shown that reinforcement learning-based methods are capable of producing better performance than classical methods such as Genetic Algorithms and DQNs in complex airport operational scenarios [5] However, the apron scheduling problem remains a challenge because it involves various operational constraints and complex relationships between entities, such as interactions between flights, resource limitations, and temporal conflicts, which significantly affect operational efficiency and potential delays [6]. In recent years, Deep Reinforcement Learning (DRL) has emerged as an effective approach for decision-making in complex systems. DRL enables agents to learn through direct interaction with the environment to derive optimal long-term reward-based policies. Various studies have shown that DRL is capable of handling sequential decision-making problems and adapting to real-time environmental dynamics, such as flight scheduling and airport resource allocation. Furthermore, DRL-based approaches have been shown to outperform classical methods such as Genetic Algorithms and Deep Q-Networks (DQN), both in terms of solution quality, stability, and computational efficiency [7]. However, most DRL models for scheduling problems still use simple numerical state representations that only capture statistical correlations, making them less able to explicitly represent structural relationships such as inter-task dependencies, temporal dynamics, and resource constraints. In the context of airport aprons, stand allocation decisions are influenced not only by current conditions but also by operational dependencies, service sequences, resource conflicts, and ongoing time dynamics [8]. Graph-based approaches have proven effective in representing complex relationships in scheduling problems. Previous studies have shown that integrating Graph Neural Networks (GNNs) with DRLs can better capture spatial and temporal dependencies [9]. However, conventional GNNs are still limited in capturing global dependencies due to their local message-passing mechanism. To address this limitation, the Graph Transformer architecture was developed, utilizing a self-attention mechanism that can more effectively model global interactions between nodes [10]. The integration of graph-based representation and deep reinforcement learning (DRL) has shown significant improvements in resource allocation efficiency and cross-domain generalization capabilities across various transportation systems, particularly in dynamic scheduling scenarios involving complex constraints and operational uncertainty [11].

Based on the limitations of previous approaches in capturing complex relationships in apron scheduling, this study proposes a Graph Transformer-Based Reinforcement Learning method for Airport Apron Scheduling. This model combines a Graph Transformer as a state representation and Reinforcement Learning as a decision maker. Compared to conventional Graph Neural Network approaches, this method is able to capture global relationships between entities more effectively through a self-attention mechanism. The proposed approach is expected to improve aircraft placement accuracy, reduce stand usage conflicts, suppress operational delays, and increase apron utilization. Furthermore, integration with Reinforcement Learning allows the system to learn adaptively from data and adjust decisions to dynamic operational conditions. This study contributes to developing a Graph Transformer-based representation to model relationships between apron entities, and integrating it with Reinforcement Learning to produce more adaptive stand allocation. The model performance is then compared with baseline methods such as FCFS, Greedy, and DQN to demonstrate the resulting performance improvements. Overall, this study offers a data-driven framework that can be used to support more efficient and intelligent airport apron management.

2 Related Works

2.1 Airport Apron Scheduling

Airport apron scheduling is the process of optimizing aircraft stand allocation by considering various operational constraints, such as capacity limitations, aircraft type compatibility, operating time conflicts, and apron resource utilization. The primary goal of this process is to minimize stand conflicts and operational delays while increasing the efficiency of apron utilization [12]. This problem falls into the category of NP-hard combinatorial optimization, as it involves a large solution space, numerous operational constraints, and complex interactions between aircraft in modern airport systems. Recent studies have shown that optimization problems in air transportation are complex, dynamic, and difficult to solve with conventional mathematical methods, especially in large-scale scenarios with many constraints [13].

2.2 Conventional Optimization Approaches

Optimization approaches in airport apron scheduling include rule-based methods such as First Come First Served (FCFS) and Greedy, as well as optimization and learning approaches such as DQN and metaheuristics such as Genetic Algorithm (GA) and Particle Swarm Optimization (PSO). Rule-based methods are still used in operational practice due to their simplicity, but they often produce suboptimal solutions because they do not take into account global interactions between resources [14]. Meanwhile, metaheuristic approaches have been widely used to handle the complexity of scheduling problems with various operational constraints, but are generally still limited in dealing with real-time system dynamics [15]. On the other hand, learning-based approaches such as DQN demonstrate capabilities in adaptive decision making, but still have limitations in representing complex relationships between operational entities [8]. In addition, airport scheduling problems are generally complex and involve many constraints and uncertainties, such as flight delays and changes in operational capacity, which cannot yet be fully addressed by conventional approaches [16]. Therefore, a more adaptive approach is needed that is able to represent the relationships between entities more comprehensively.

2.3 Reinforcement Learning for Scheduling

Reinforcement Learning (RL) is a powerful method for solving decision-making problems in dynamic environments. It enables agents to learn optimal strategies by interacting with their environment to maximize long-term rewards. In scheduling problems, RL is often formulated as a Markov Decision Process (MDP), which provides a structured framework to model and solve these challenges effectively [17]. In the context of transportation and airport systems, this learning-based approach has been widely used to improve operational efficiency, including in reducing delays and optimizing resource allocation [18]. In addition, approaches such as DQN have shown the ability to produce more adaptive decisions than traditional methods in complex environments. However, RL performance is highly dependent on the quality of the state representation. Simple representations often fail to capture complex relationships between operational entities, such as inter-plane dependencies, resource conflicts, and temporal relationships. This is because most RL approaches still model the problem as a state-action mapping without explicitly representing the underlying relational structure of the system.

2.4 Graph-Based Representation and Transformers

Graph-based representations are becoming an important approach in modeling complex systems such as airport aprons, where the relationships between aircraft and stands are interdependent and involve various operational constraints. In the transportation context, graphs are used to represent the structure, topology, and dynamics of systems involving multi-entity interactions and spatial-temporal dependencies [19, 20]. In a graph, nodes represent entities such as airplanes or stands, while edges represent relationships such as conflicts, compatibilities, or operational dependencies. On the other hand, the development of the Transformer architecture allows for more effective modeling of complex dependencies through a self-attention mechanism, which is able to capture global relationships between elements in a system [21]. This approach has proven effective in modeling complex patterns in modern transportation systems. The integration between graph representation and Transformer produces Graph Transformer, which combines the ability of graphs to represent relationships between entities with the ability of Transformers to comprehensively capture global dependencies [22, 23]. This approach is considered more suitable for systems with complex interactions such as apron scheduling, which involves heterogeneous relationships, resource conflicts, and interrelated temporal dependencies [24].

2.5 Research Gap and Contribution

Recent studies have demonstrated the effectiveness of RL, Graph Neural Networks (GNN), and Transformer architectures in solving complex scheduling problems [25, 26] showed that graph-based reinforcement learning approaches can improve scheduling efficiency by modeling relational dependencies among operational entities. Meanwhile, [27–29] demonstrated that Transformer-based architectures are effective in capturing long-range spatial–temporal dependencies through self-attention mechanisms. In addition, [30, 31] highlighted the potential of integrating graph representation and reinforcement learning for adaptive scheduling optimization in dynamic environments. However, several limitations remain unresolved. Most existing reinforcement learning approaches rely on vector-based representations or conventional graph neural networks that primarily capture local neighborhood information, limiting their ability to model global operational dependencies. In addition, Transformer-based scheduling models often focus on temporal sequence learning without explicitly representing graph-based relationships among scheduling entities. Furthermore, only limited research has investigated Graph Transformer-based reinforcement learning for airport apron scheduling, where heterogeneous entities, stand compatibility requirements, temporal overlap constraints, and dynamic aircraft movements must be considered simultaneously. Unlike existing Graph Transformer-based scheduling approaches, the proposed GT-RL framework models aircraft and stand entities within a heterogeneous graph structure and employs a Graph Transformer encoder to capture compatibility relationships, temporal conflicts, stand occupancy dynamics, and operational dependencies through multi-head self-attention mechanisms. This enables the framework to learn conflict-aware stand allocation policies under realistic airport operational conditions. Therefore, this study proposes a Graph Transformer-Based Reinforcement Learning (GT-RL) framework for airport apron scheduling optimization. The main contributions of this study are three-fold: (1) the development of a heterogeneous graph representation for modeling aircraft–stand interactions and operational constraints, (2) the utilization of a Graph Transformer encoder to capture both local and global scheduling dependencies, and (3) the integration of graph-based representations with reinforcement learning to generate adaptive and conflict-aware stand allocation policies. Experimental results demonstrate that the proposed framework consistently improves scheduling performance compared with FCFS, Greedy, and DQN baselines.

3 Methodology

The proposed framework for airport apron scheduling is illustrated in Fig. 1. The framework integrates Graph Transformer as a graph-based state representation module with Reinforcement Learning (RL) as a decision-making mechanism within a unified intelligent scheduling pipeline. The proposed approach aims to optimize aircraft stand allocation under dynamic airport operational constraints, including stand compatibility, temporal overlap, operational conflicts, and apron resource utilization.

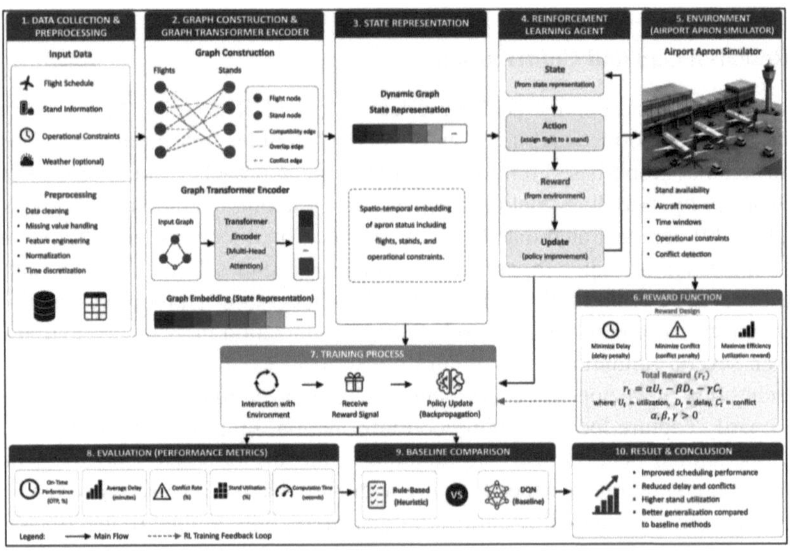

Fig. 1. Framework of Graph Transformer-Based Reinforcement Learning for Airport Apron Scheduling

3.1 Data Collection and Preprocessing

The proposed model utilizes airport operational data consisting of flight schedules, stand information, and operational constraints. The dataset contains flight arrival and departure times, aircraft types, stand allocation information, and operational status required for apron scheduling optimization. Before model training, the raw data is processed through several preprocessing stages, including data cleaning, missing value handling, normalization, feature engineering, and time discretization. Feature engineering is performed to generate relevant operational attributes such as turnaround duration, stand occupancy duration, delay status, and scheduling conflict indicators. This preprocessing stage ensures that the generated dataset is consistent and suitable for graph-based reinforcement learning.

3.2 Graph Construction

The airport apron environment is modeled as a graph G = (V, E), where nodes V represent operational entities including aircraft and apron stands, while edges E represent operational relationships such as stand compatibility, temporal overlap, and scheduling conflicts. Aircraft nodes contain operational features including aircraft type, arrival time, departure time, turnaround duration, and delay status. Meanwhile, stand nodes contain stand-related information such as stand availability and stand capacity. An edge is constructed when operational relationships or potential conflicts exist between entities. The adjacency relationship is defined as follows: A_{ij} is assigned a value of 1 if nodes iii and j are operationally related, and 0 otherwise. This graph representation enables the model to explicitly capture structural dependencies and operational interactions within the airport apron environment.

3.3 Graph Transformer Encoder

After graph construction, the generated graph is processed using a Graph Transformer encoder to learn graph-based spatial–temporal representations. The Graph Transformer combines graph representation learning with multi-head self-attention mechanisms to capture both local and global operational dependencies among aircraft and stands. The self-attention mechanism is formulated as [28]:

$$\mathrm{Attention}(Q, K, V) = \mathrm{softmax}\left(\frac{QK^T}{\sqrt{d_k}}\right)V \tag{1}$$

Where Q, K, and V denote the query, key, and value matrices, respectively, while dk represents the dimensionality of the key vector. The Graph Transformer encoder generates embedding representations that encode scheduling information, operational conflicts, stand utilization conditions, and temporal dependencies. These learned embeddings are then used as state representations for the reinforcement learning agent.

3.4 Reinforcement Learning Formulation

The airport apron scheduling problem is formulated as a Markov Decision Process (MDP), defined as:

$$(\mathrm{S}, \mathrm{A}, \mathrm{R}, \mathrm{P}, \gamma) \tag{2}$$

Where S denotes the state space, A represents the action space, R denotes the reward function, P represents the state transition probability, and γ is the discount factor. In this framework, the state corresponds to the Graph Transformer embedding representation of the airport apron environment, while the action represents assigning aircraft to available stands.The reinforcement learning agent aims to maximize cumulative long-term rewards defined as [31];

$$R_t = \sum\nolimits_{k=t}^{T-1} \gamma^{k-t} r_{k+1} \tag{3}$$

The reward function is designed to balance multiple operational objectives, including maximizing stand utilization while minimizing delay and conflicts;

$$R = \alpha U\text{--}\beta D\text{--}\gamma C \tag{4}$$

Where U denotes utilization, D represents delay, and C indicates conflict, while α, β, γ are weighting parameters.

3.5 Policy Learning and Environment Interaction

The generated graph-based state representation is used as input for the reinforcement learning agent to determine optimal scheduling actions. The environment simulates airport apron operations, including stand availability, aircraft movement, operational constraints, and scheduling conflicts. Based on the selected action, the agent receives a reward signal from the environment and updates its scheduling policy iteratively using a Q-learning mechanism formulated as [32];

$$Q(s, a) \leftarrow Q(s, a) + \alpha\left[r + \gamma \max_{a'} Q(s', a') - Q(s, a)\right] \tag{5}$$

Through iterative interaction between the agent and the environment, the model learns an optimal scheduling policy. The reinforcement learning component of the proposed GT-RL framework is implemented using a Deep Q-Network (DQN). The graph embeddings generated by the Graph Transformer encoder are used as state representations for the DQN agent. The agent learns stand allocation policies through experience replay and target network updates to improve training stability and convergence. The integration of Graph Transformer-based representations with DQN enables the framework to capture complex operational dependencies while maintaining effective sequential decision-making capabilities for airport apron scheduling optimization.

3.6 Performance Evaluation

The performance of the proposed GT-RL framework is evaluated using several operational metrics, including reward, average delay, conflict rate, stand utilization, and On-Time Performance. The proposed model is compared against baseline approaches including FCFS, Greedy, and DQN to evaluate its effectiveness under different scheduling scenarios. Additionally, scheduling simulations are carried out to assess the suggested framework's ability to provide adaptive and conflict-aware scheduling decisions under actual airport operating situations and to visualize aircraft stand allocation results.

4 Experimental Setup

4.1 Dataset and Experimental Scenario

The proposed GT-RL framework is evaluated using real-world airport operational data obtained from the OASYS dataset. The dataset contains operational information related to aircraft movement, flight schedules, stand allocation, and airport apron operations. Several preprocessing stages are performed before model training, including data cleaning, normalization, missing value handling, and feature engineering to generate structured scheduling data suitable for reinforcement learning. The OASYS dataset includes several operational attributes such as flight identifiers, aircraft types, scheduled arrival and departure times, stand allocation information, turnaround duration, and delay status indicators. These features provide essential information for modeling aircraft–stand interactions and operational constraints within airport apron environments. After preprocessing, the data are transformed into graph structures where aircraft and stands are represented as nodes, while compatibility relationships, temporal overlaps, and potential scheduling conflicts are represented as edges. To evaluate the robustness and scalability of the proposed framework, two experimental scenarios with different operational complexities are constructed. The first scenario consists of 1511 flight movements representing a moderate-density scheduling environment, while the second scenario contains 4460 flight movements representing a high-density and more complex operational environment. Each scenario includes aircraft scheduling information, stand allocation constraints, operational conflicts, and temporal dependencies used for graph construction and reinforcement learning training. The scheduling objective is to allocate aircraft to available stands while minimizing operational conflicts and delays under dynamic airport operational constraints.

4.2 Training Configuration

Reinforcement learning is employed to train the proposed GT-RL framework under various airport scheduling conditions. The Graph Transformer encoder consists of two Transformer layers with an embedding dimension of 32 and four attention heads, enabling the model to learn graph-based spatial–temporal representations through multi-head self-attention mechanisms. Structural and positional information is represented through node attributes, including arrival time, departure time, turnaround duration, and stand occupancy information. The reinforcement learning component is implemented using a Deep

Q-Network (DQN) optimized with the Adam optimizer and a learning rate of 0.001. To improve training stability, experience replay and target network updates are employed. A mini-batch size of 64 and a replay buffer containing 5,000 transitions are used during training. The discount factor (γ) is set to 0.95, while the target network is synchronized every 10 episodes to stabilize Q-value estimation. The model is trained for 100 episodes in each experimental scenario until a stable scheduling policy is achieved. The selected hyperparameter values were determined through preliminary experimentation to balance model complexity, learning stability, and computational efficiency. These settings consistently produced stable scheduling performance across different operational scenarios while maintaining reasonable training time.

4.3 Baseline Methods

To evaluate the effectiveness of the proposed GT-RL framework, several baseline scheduling approaches are used for comparison, including FCFS, Greedy scheduling, and DQN. FCFS is a rule-based scheduling approach that allocates stands sequentially according to aircraft arrival order. Greedy scheduling assigns aircraft to stands based on immediate local optimization without considering long-term operational impact. Meanwhile, DQN represents a reinforcement learning-based baseline that utilizes conventional vector-based state representation without graph-based dependency modeling. The comparison with these baseline methods aims to evaluate the capability of the proposed GT-RL framework in improving scheduling efficiency, conflict minimization, and operational performance under different airport operational conditions.

4.4 Evaluation Metrics

The performance of the proposed framework is evaluated using several operational metrics commonly used in airport scheduling optimization, including reward, average delay, conflict rate, stand utilization, and On-Time Performance. The reward metric evaluates the overall scheduling performance generated by the reinforcement learning agent. Average delay measures the average operational delay experienced during scheduling. Conflict rate indicates the number of stand allocation conflicts occurring during airport operations. Stand utilization measures the effectiveness of apron resource usage, while OTP evaluates scheduling punctuality based on operational time constraints. These evaluation metrics provide a comprehensive assessment of the effectiveness, robustness, and scalability of the proposed GT-RL framework under different scheduling scenarios.

5 Results and Discussion

5.1 Performance Evaluation

The performance of the proposed Graph Transformer-Based Reinforcement Learning (GT-RL) framework is evaluated under two scheduling scenarios with different levels of operational complexity. The proposed model is compared with several baseline methods, including First Come First Served (FCFS), Greedy scheduling, and Deep Q-Network (DQN), using evaluation metrics such as reward, average delay, On-Time Performance (OTP), stand utilization, and conflict rate (Table 1).

Table 1. Performance Comparison In Scenario 1 (1511 Flights)

Method	Reward	Delay	OTP (%)	Utilization	Conflict
FCFS	−1400.89	9.30	87.95	100	28094
Greedy	99.60	0.94	98.94	100	2847
DQN	103.16	0.91	98.48	100	2747
GT-RL	**251.28**	**0.15**	**99.74**	**100**	**459**

In Scenario 1, which represents a moderate operational environment with 1,511 flight movements, the proposed GT-RL framework achieves the best performance across all evaluation metrics.

The model obtains the highest reward value of 251.28, outperforming Greedy (99.60) and DQN (103.16), while reducing the average delay to 0.15. In addition, GT-RL significantly decreases scheduling conflicts to only 459, compared with 28,094 in FCFS, 2,847 in Greedy, and 2,747 in DQN. These results demonstrate the effectiveness of the Graph Transformer-based state representation in capturing complex operational dependencies among aircraft and apron stands, enabling the reinforcement learning agent to generate more adaptive and conflict-aware stand allocation decisions.

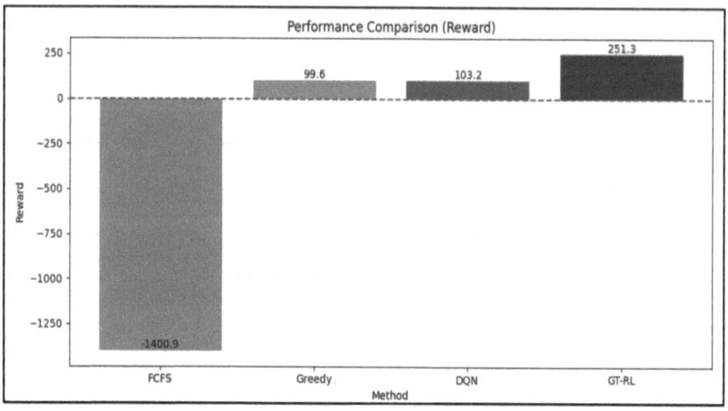

Fig. 2. Reward comparison of scheduling methods in Scenario 1

Figure 2 illustrates the reward comparison among different scheduling methods in Scenario 1. The proposed GT-RL framework achieves the highest reward value, significantly outperforming FCFS, Greedy, and DQN. This result confirms the effectiveness of integrating Graph Transformer representation with reinforcement learning for airport apron scheduling optimization under moderate operational conditions (Table 2).

Scenario 2 represents a high-density operational environment consisting of 4,460 flight movements. In this scenario, the proposed GT-RL framework consistently outperforms all baseline methods, achieving the highest reward value of 226.43 and the lowest average delay of 0.09. The model also significantly reduces scheduling conflicts from

Table 2. Performance Comparison in Scenario 2 (4460 Flights)

Method	Reward	Delay	OTP(%)	Utilization	Conflict
FCFS	−3988.68	7.99	88.16	100	71257
Greedy	138.37	0.25	99.66	100	2226
DQN	−242.81	0.95	98.18	100	8486
GT-RL	**226.43**	**0.09**	**99.87**	**100**	**847**

71,257 in FCFS and 8,486 in DQN to only 847, demonstrating strong robustness and scalability under highly constrained airport operational conditions. Furthermore, GT-RL achieves the highest On-Time Performance (OTP) of 99.87%. These results indicate that the proposed Graph Transformer-based representation effectively captures complex spatial–temporal dependencies, enabling more adaptive and conflict-aware scheduling decisions.

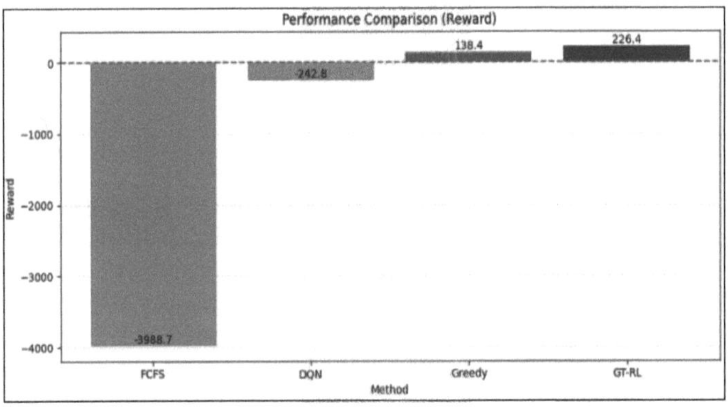

Fig. 3. Reward comparison of scheduling methods in Scenario 2

Figure 3 presents the reward comparison among scheduling methods in Scenario 2. The proposed GT-RL framework consistently achieves the highest reward while maintaining stable performance despite increased scheduling complexity and flight density. Compared with the DQN baseline, GT-RL improves the reward value by approximately 143.6% in Scenario 1 and 193.2% in Scenario 2. These results demonstrate the scalability and effectiveness of the proposed framework in high-density airport operational environments. The substantial performance improvement suggests that the Graph Transformer-based state representation enables the reinforcement learning agent to better capture complex spatial–temporal dependencies and generate more effective conflict-aware stand allocation decisions (Table 3).

To further evaluate the robustness of the proposed framework, the GT-RL model was executed five times using different random seeds. Table III presents the mean and standard deviation of the obtained performance metrics. The results show that the proposed

Table 3. Statistical Validation Results of GT-RL

Metric	Mean ± Std
Reward	243.78 ± 22.29
Delay	0.06 ± 0.04
OTP (%)	99.88 ± 0.06
Conflict	565.20 ± 371.96

framework achieved an average reward of 243.78 ± 22.29 while maintaining a low average delay of 0.06 ± 0.04 and a high OTP of 99.88 ± 0.06. Although some variability is observed due to the stochastic exploration process inherent in reinforcement learning, the model consistently demonstrates stable scheduling performance across multiple executions. These findings further confirm the robustness and reliability of the proposed GT-RL framework for airport apron scheduling optimization.

5.2 Scheduling Simulation Result

To further validate the effectiveness of the proposed GT-RL model, a scheduling simulation is conducted using real airport operational data. The simulation aims to visualize the stand allocation results generated by the proposed approach under realistic operational conditions.

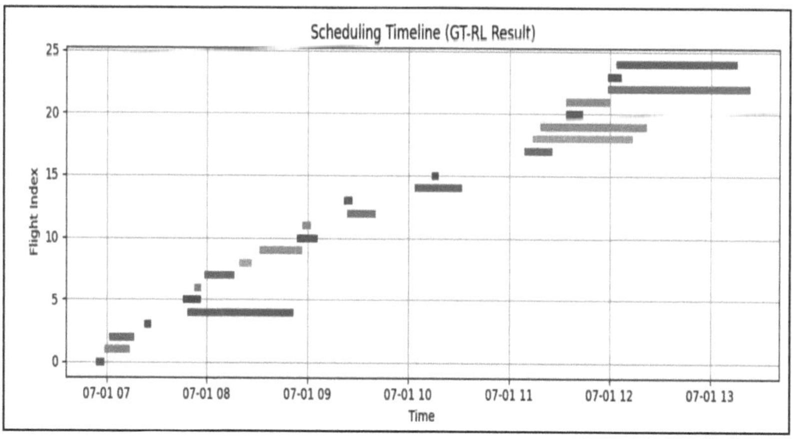

Fig. 4. Scheduling timeline generated by the proposed GT-RL model

Figure 4 presents the scheduling timeline generated by the proposed GT-RL model. The visualization shows that aircraft are distributed efficiently across the scheduling horizon without significant overlap or conflict. The generated schedule demonstrates the capability of the proposed approach to allocate stands adaptively while maintaining minimal delay and operational feasibility. This result further confirms the effectiveness of the Graph Transformer-based representation in capturing temporal and operational dependencies within airport apron scheduling.

5.3 Embedding Analysis

Since the proposed approach relies on Graph Transformer-based state representation, embedding analysis is conducted to evaluate the capability of the encoder in learning meaningful latent representations from operational scheduling data. To further analyze the effectiveness of the proposed Graph Transformer encoder, an embedding visualization is performed to observe how the model represents operational scheduling data in the latent feature space. The visualization aims to evaluate whether the learned representations are able to capture structural relationships and hidden patterns among scheduling entities.

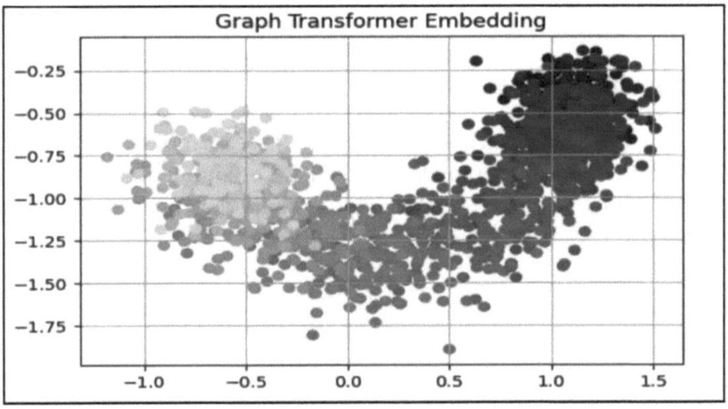

Fig. 5. Graph Transformer embedding visualization generated from airport scheduling data.

Figure 5 illustrates the embedding distribution generated by the Graph Transformer encoder. Several well-defined clusters can be observed, indicating that the encoder successfully learns meaningful latent representations from airport operational data. The embedding structure suggests that the Graph Transformer effectively captures spatial–temporal relationships and operational dependencies among aircraft and apron stands. These graph-based representations provide informative state features for the reinforcement learning agent, contributing to more adaptive and conflict-aware scheduling decisions under dynamic airport operational conditions.

6 Conclusion

This study proposed a Graph Transformer-Based Reinforcement Learning (GT-RL) framework for airport apron scheduling optimization. The proposed framework integrates Graph Transformer-based state representation learning with a Deep Q-Network

(DQN) to generate adaptive and conflict-aware aircraft stand allocation decisions under dynamic airport operational conditions. By modeling the airport apron environment as a graph structure, the framework effectively captures complex spatial–temporal relationships among aircraft and apron stands. Experimental results demonstrated that the proposed GT-RL framework consistently outperformed FCFS, Greedy, and DQN baseline methods across different operational scenarios. In the most complex scenario involving 4,460 flight movements, the proposed framework achieved a reward value of 226.43, reduced scheduling conflicts to 847, and maintained an On-Time Performance (OTP) of 99.87%. Statistical validation using multiple experimental runs further confirmed the robustness and stability of the proposed approach. Despite these promising results, several limitations remain. The current framework focuses primarily on stand allocation optimization and does not explicitly consider real-time operational uncertainties such as weather disruptions, taxiway congestion, and dynamic flight rescheduling. In addition, a detailed ablation study was not conducted to quantify the individual contribution of the Graph Transformer encoder and reinforcement learning components. Future work will focus on real-time adaptive scheduling, multi-agent coordination, and integrated airport operational optimization involving gate assignment, taxiway management, and ground handling operations. Further studies will also incorporate comprehensive ablation analyses and broader statistical evaluations to provide deeper insights into the contribution and reliability of the proposed GT-RL framework.

Acknowledgments. The author would like to express sincere gratitude to the Putra Indonesia University YPTK Padang Foundation for the scholarship support, academic opportunities, facilities, and encouragement provided throughout the doctoral study, which greatly contributed to the successful completion of this research. The author also gratefully acknowledges the supervisors for their invaluable guidance, constructive feedback, and continuous support during the research process. Appreciation is further extended to all lecturers, academic staff, colleagues, and individuals who directly or indirectly contributed to the completion of this study.

References

1. Jacquillat, A., Odoni, A.R.: A roadmap toward airport demand and capacity management. Transp. Res. Part A Policy Pract. **114**, 168–185 (2018). https://doi.org/10.1016/j.tra.2017.09.027
2. Yao, M., Hu, M., Yin, J., Su, J., Yin, M.: A two-stage optimization model for airport stand allocation and ground support vehicle scheduling. Appl. Sci. (Switz.). **14**(23), 11407 (2024). https://doi.org/10.3390/app142311407
3. Bouras, A., Ghaleb, M.A., Suryahatmaja, U.S., Salem, A.M.: The Airport Gate Assignment Problem: A Survey. Hindawi Publishing Corporation. (2014). https://doi.org/10.1155/2014/923859
4. Ding, C., Bi, J., Wang, Y.: A hybrid genetic algorithm based on imitation learning for the airport gate assignment problem. Entropy. **25**(4), 565 (2023). https://doi.org/10.3390/e25040565
5. Zhu, C., Wei, Z., Lyu, Z., Yuan, X., Hang, D., Feng, L.: Gate assignment algorithm for airport peak time based on reinforcement learning. Transp. Res. Rec. **2678**(11), 750–760 (2024). https://doi.org/10.1177/03611981241242352

6. Jiang, Y., Wang, Y., Xiao, Y., Xue, Q., Shan, W., Zhang, H.: Joint runway–gate assignment based on the branch-and-price algorithm. Transp. Res. Part C Emerg. Technol. **162**, 104605 (2024). https://doi.org/10.1016/j.trc.2024.104605
7. Li, H., Wu, X., Ribeiro, M., Santos, B., Zheng, P.: Deep reinforcement learning approach for real-time airport gate assignment. Oper. Res. Perspect. **14**, 100338 (2025). https://doi.org/10.1016/j.orp.2025.100338
8. Wang, S., Chi, T.: Semantic-constrained planning for airport vehicle scheduling. Appl. Sci. (Switz.). **16**(5) (2026). https://doi.org/10.3390/app16052536
9. Maru, V.: A Graph-Enhanced Deep-Reinforcement Learning Framework for the Aircraft Landing Problem. [Online]. http://arxiv.org/abs/2502.12617 (Mar. 2025)
10. Soykan, B.: HGT-Scheduler: Deep Reinforcement Learning for the Job Shop Scheduling Problem via Heterogeneous Graph Transformers. [Online]. http://arxiv.org/abs/2603.06777 (Mar. 2026)
11. Zhang, Z., Zhong, Q., Pan, W., Ai, Y., Wang, Q.: Cross-domain deep reinforcement learning for real-time resource allocation in transportation hubs: from airport gates to seaport berths. Aerospace. **13**(1), 108 (2026). https://doi.org/10.3390/aerospace13010108
12. Zhang, J., Chong, X., Wei, Y., Bi, Z., Yu, Q.: Optimization of apron support vehicle operation scheduling based on multi-layer coding genetic algorithm. Appl. Sci. (Switz.). **12**(10) (2022). https://doi.org/10.3390/app12105279
13. Huang, R., Chen, Z.G.: Evolutionary Computation for Air Transportation: A Survey. Multidisciplinary Digital Publishing Institute (MDPI) (Sep. 01, 2025). https://doi.org/10.3390/math13172867
14. Thenarasu, M., Rameshkumar, K., Di Mascolo, M., Anbuudayasankar, S.P.: Multi-criteria scheduling of realistic flexible job shop: a novel approach for integrating simulation modelling and multi-criteria decision making. Int. J. Prod. Res. **62**(1–2), 336–358 (2024). https://doi.org/10.1080/00207543.2023.2238084
15. Ding, K., Lan, H., Zhang, J., Zhang, S., Shi, H., Cao, Z.: Sustainable optimization in air transport: hybrid particle swarm and tabu search algorithm for the multi-objective airport gate assignment problem. Sustainability (Switzerland). **18**(7), 3331 (2026). https://doi.org/10.3390/su18073331
16. Du, J., Li, H., Zhang, W., Pan, W., Yin, J.: Airport Taxiway–gate joint scheduling problem: a multi-objective optimization approach based on a spatiotemporal graph. Aerospace. **13**(4), 384 (2026). https://doi.org/10.3390/aerospace13040384
17. Zhang, C., Juraschek, M., Herrmann, C.: Deep reinforcement learning-based dynamic scheduling for resilient and sustainable manufacturing: a systematic review. J. Manufact. Syst. **77**, 962–989 (2024)
18. Liu, H., Li, S., Fan, W., Ip, W.-H., Yung, K.-L.: Adaptive dynamic programming with reinforcement learning on optimization of flight departure scheduling. Aerospace. **11**(9), 754 (2024). https://doi.org/10.3390/aerospace
19. Rakhmangulov, A., Osintsev, N., Mishkurov, P.: Spatio-Temporal Graphs in Transportation: Challenges, Optimization, and Prospects. Multidisciplinary Digital Publishing Institute (MDPI) (Apr. 01, 2025). https://doi.org/10.3390/systems13040263
20. Zhang, Y., Yue, M., Wang, J., Yoo, S.: Multi-agent graph-attention deep reinforcement learning for post-contingency grid emergency voltage control. IEEE Trans. Neural Netw. Learn. Syst. **35**(3), 3340–3350 (2024). https://doi.org/10.1109/TNNLS.2023.3341334
21. Choo, B., Kim, D.K.: Applying transformer-based dynamic-sequence techniques to transit data analysis †. Eng. Proc. **102**(1), 12 (2025). https://doi.org/10.3390/engproc2025102012
22. Chen, Y., Li, C., Zhao, X.: TSGformer: a unified temporal–spatial graph transformer with adaptive cross-scale modeling for multivariate time series. Systems. **13**(8) (2025). https://doi.org/10.3390/systems13080688

23. Mkrtchian, G., Gorodnichev, M.: Patch-based transformer–graph framework (PTSTG) for traffic forecasting in transportation systems. Appl. Sci. **15**(19), 10468 (2025). https://doi.org/10.3390/app151910468
24. Xiong, Z., Wang, Y., Tian, Y., Liu, L., Zhu, S.: RoPT: route-planning model with transformer. Appl. Sci. (Switz.). **15**(9) (2025). https://doi.org/10.3390/app15094914
25. Gao, Q., Schweidtmann, A.M.: Deep reinforcement learning for process design: review and perspective. Curr. Opin. Chem. Eng. **44**, 101012 (2024). https://doi.org/10.1016/j.coche.2024.101012
26. Liu, Z., Huang, L., Gao, Z., Luo, M., Hosseinalipour, S., Dai, H.: GA-DRL: Graph Neural Network-Augmented Deep Reinforcement Learning for DAG Task Scheduling over Dynamic Vehicular Clouds. [Online]. http://arxiv.org/abs/2307.00777 (Jul. 2024)
27. Wang, P., Wang, K., Song, Y., Wang, X.: AutoLDT: a lightweight spatio-temporal decoupling transformer framework with AutoML method for time series classification. Sci. Rep. **14**(1), 29801 (2024). https://doi.org/10.1038/s41598-024-81000-1
28. Song, L., Li, Y., Xu, J.: Dynamic job-shop scheduling based on transformer and deep reinforcement learning. Processes. **11**(12), 3434 (2023)
29. Xu, S., Li, Y., Li, Q.: A deep reinforcement learning method based on a transformer model for the flexible job shop scheduling problem. Electronics (Switzerland). **13**(18) (2024). https://doi.org/10.3390/electronics13183696
30. Elrod, M., et al.: Graph based deep reinforcement learning aided by transformers for multi-agent cooperation. In: 2025 IEEE International Conference on Communications Workshops (ICC Workshops), vol. 2025, pp. 415–420. https://doi.org/10.1109/ICCWorkshops67674.2025.11162376
31. Dong, Z., et al.: A reinforcement learning-based approach for solving multi-agent job shop scheduling problem. Int. J. Prod. Res. **63**(10), 3512–3537 (2025). https://doi.org/10.1080/00207543.2024.2423807

Leveraging Weather Data and Predictive Analytics to Enhance Special Perils Insurance Uptake in Malaysia

Yau Lai Fung[1], Suhaida Halamy[2], Noorsidi Aizuddin Mat Noor[3], and Aslina Baharum[4](✉)

[1] School of Computing and Artificial Intelligence, Faculty of Engineering and Technology, Sunway University, Petaling Jaya, Selangor, Malaysia

[2] Faculty of Information Science, Universiti Teknologi MARA, Cawangan Sarawak, Samarahan, Malaysia

[3] Mass Appraisal, Housing and Planning Research Group, Centre for Real Estate Department, Faculty of Built Environment and Surveying, Universiti Teknologi Malaysia, Johor Bahru, Johor, Malaysia

[4] The Design School, Faculty of Innovation & Technology, Taylor's University, Subang Jaya, Selangor, Malaysia

aslina.baharum@taylors.edu.my

Abstract. The increasing frequency and severity of climate-related disasters have heightened the importance of adequate insurance protection against natural hazards. Despite the growing exposure of Malaysian motorists to floods and extreme weather events, the adoption rate of Special Perils insurance coverage remains relatively low. This study investigates the relationship between weather conditions and consumer decisions to purchase Special Perils add-on coverage in motor insurance policies. Drawing upon secondary data obtained from a Malaysian insurance agency and historical weather records, the research integrates meteorological information with customer and vehicle characteristics to identify factors influencing insurance up-take. A data science framework based on predictive analytics and machine learning is proposed to forecast consumer adoption behaviour. The study further evaluates the potential of weather-informed marketing strategies in increasing awareness and adoption of Special Perils cover-age. Findings from the literature indicate that weather data, demographic characteristics, driving behaviour, and insurance-related factors collectively influence purchasing decisions. The study contributes to the emerging intersection of climate analytics, insurance innovation, and consumer behaviour by proposing a predictive framework capable of supporting targeted marketing and risk management strategies within the Malaysian insurance industry.

Keywords: Special Perils Insurance · Predictive Analytics · Weather Data · Consumer Behaviour · Machine Learning

N. Mohd Sharef et al. (Eds.): CAMP 2026, CCIS 3093, pp. 196–211, 2027.
https://doi.org/10.1007/978-981-92-4302-0_15

1 Introduction

Climate change has significantly increased the occurrence of extreme weather events worldwide, leading to more frequent and intense natural hazards [1, 2]. In Malaysia, floods, heavy rainfall, landslides, and storms have become more frequent and severe due to changing climate patterns and increasing rainfall intensity, resulting in substantial economic losses and damage to vehicles and property, as evidenced by major flood disasters such as those in 2021. Although motor insurance policies provide protection against standard risks, many losses associated with natural disasters are only covered through optional Special Perils add-ons. Despite increasing climate-related risks, adoption rates of Special Perils coverage remain relatively low among Malaysian motorists. This gap suggests a disconnect between risk exposure and insurance purchasing behaviour, a phenomenon widely observed in natural hazard insurance markets where demand is influenced by risk perception, experience, and willingness to pay [2, 3]. Understanding the factors that influence customers to purchase additional protection has therefore become an important concern for insurance providers.

Recent developments in data science, machine learning, and predictive analytics offer new opportunities for insurers to better understand customer behaviour and improve risk assessment. Predictive analytics enables insurers to analyse historical and environmental data to forecast patterns of risk exposure and insurance demand [1, 4]. Simultaneously, advances in meteorological data availability, including long-term rainfall and climate datasets, provide valuable insights into environmental conditions that may influence consumer perceptions of risk. This study explores the integration of weather data and predictive analytics to forecast the adoption of Special Perils coverage. Specifically, it examines whether heavy rainfall events influence customer purchasing decisions and proposes a data-driven framework for improving insurance marketing and risk assessment. The objectives of this study are:

1. To investigate the relationship between heavy rainfall events and insurance purchasing behaviour.
2. To develop a predictive framework capable of forecasting Special Perils adoption.
3. To provide recommendations for insurance companies to enhance coverage uptake through data-driven strategies.

2 Related Work

Special Perils insurance is an optional extension to motor insurance policies that covers losses caused by natural disasters such as floods, landslides, and storms. As climate-related risks continue to increase, particularly in Malaysia where heavy rainfall and monsoon flooding are common, this coverage has become increasingly important [1, 5]. However, despite growing exposure to risk, adoption remains relatively low. This reflects a broader issue in natural hazard insurance markets, where uptake is often lower than expected given the level of risk [2]. Several factors contribute to this low adoption, including limited awareness, insufficient understanding of climate-related risks, affordability concerns, and lack of exposure to disaster-related information. Previous studies show that insurance purchasing behaviour is strongly influenced by risk perception, financial literacy, trust in insurers, and past experiences with losses [2, 3]. In developing economies, these behavioural and psychological factors play an even more significant role in shaping decisions to purchase additional coverage [6]. Understanding these determinants is therefore essential for insurers aiming to improve Special Perils coverage uptake.

Weather conditions are also known to influence consumer behaviour across various industries. Research shows that factors such as rainfall, temperature, and sunlight affect purchasing decisions, spending patterns, and mobility by influencing mood and perception of risk [7, 8]. The concept of weather marketing has emerged to describe the use of meteorological data to tailor products and promotional strategies based on prevailing weather conditions [9]. This approach allows businesses to deliver more relevant and timely messages to consumers. In the insurance context, extreme weather events such as heavy rainfall and flooding can increase awareness of potential risks and financial losses. As a result, consumers may become more willing to purchase additional protection. Studies suggest that exposure to weather-related risks and past disaster experiences can influence both initial and repeat insurance purchasing behaviour. Furthermore, rainfall intensity is closely linked to increased insurance claims, reinforcing the connection between weather conditions and financial risk. Integrating weather data into insurance marketing strategies therefore offers strong potential to improve Special Perils adoption by enhancing risk awareness and encouraging proactive decision-making.

The insurance industry is increasingly adopting data science and predictive analytics to support decision-making, including risk assessment, fraud detection, and customer segmentation. Predictive models use historical data and machine learning techniques to identify patterns and forecast future outcomes [1, 10]. Common methods include Logistic Regression, Decision Trees, Random Forests, Support Vector Machines, Gradient Boosting, and Deep Learning models, which have demonstrated strong predictive performance in various insurance applications [1, 11]. Predictive analytics provides several benefits, such as improved risk assessment, better customer segmentation, more targeted marketing strategies, and increased operational efficiency [4, 12]. The integration of external data sources, particularly weather data, further enhances these models by incorporating environmental risk indicators. Studies show that weather-related factors are strongly associated with insurance claims and financial losses, highlighting their importance in predictive modelling [11, 13]. Including weather variables also helps explain customer behaviour, particularly in relation to risk perception and insurance purchasing decisions.

Despite these advances, existing research has largely focused on the impact of weather on insurance claims and risk modelling rather than on customer adoption behaviour. There is also limited research within the Malaysian context that examines how weather conditions influence insurance purchasing decisions. While Malaysia is highly exposed to flood risks, most studies focus on disaster impacts and policy responses rather than individual insurance uptake. Furthermore, although predictive analytics is widely used in insurance, limited attention has been given to integrating weather data into machine learning models to predict insurance adoption. This represents a clear research gap. Therefore, this study addresses this limitation by developing a predictive framework that combines customer characteristics and weather variables, particularly heavy rainfall, to better understand and forecast Special Perils insurance uptake.

3 Methodology

The methodology follows a structured process consisting of several stages, as illustrated in Fig. 1. The approach aligns with the **Cross-Industry Standard Process for Data Mining (CRISP-DM)**, a widely recognized framework for guiding data analytics and machine learning projects. By following this systematic framework, the study ensures a rigorous and organized process for data collection, preparation, model development, and performance evaluation. The methodology supports the systematic development of predictive models while enhancing the reliability and reproducibility of the research findings.

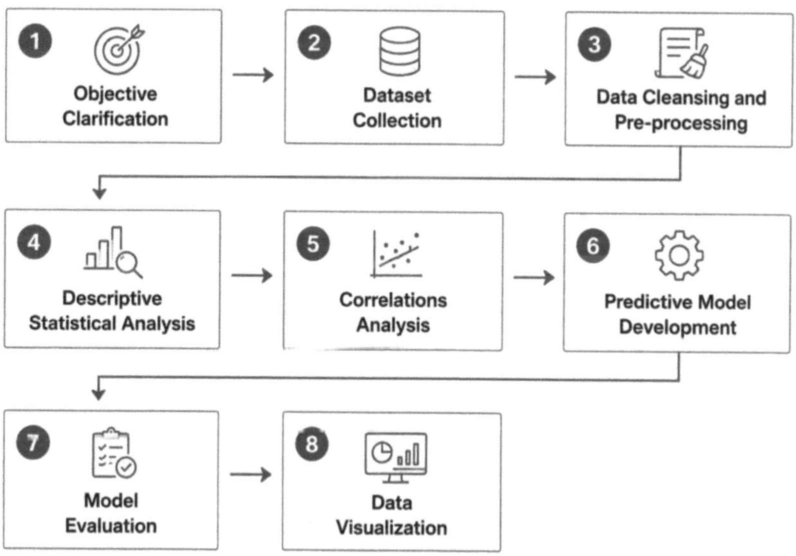

Fig. 1. Research Design

Several preprocessing procedures were conducted to ensure data quality. The dataset has 10 variables and encompasses 1655 records of vehicle insurance transactions. After the initial data cleaning and pre-processing, the dataset of 9 variables and 1,639 entries pertaining to customer and vehicle information will be used for further analysis and modelling.

This iterative process of data cleaning, preprocessing, EDA or during model fine-tuning ensures that the final dataset is optimized for the subsequent modelling phase, leading to more accurate and reliable predictions of Special Perils addon uptake. The models are evaluated using a comprehensive set of metrics, including accuracy, precision, recall, F1-score, and AUC-ROC. These metrics provide a holistic view of the model's performance, considering both its ability to correctly classify instances and its ability to distinguish between the two classes (Special Perils uptake or not). Scikit-learn provides various metrics (accuracy, precision, recall, F1-score, AUC-ROC) to evaluate the performance of the models, allowing for a comprehensive assessment of their predictive power and generalization ability. To evaluate the model's performance, a combination of metrics will be used (see Table 1):

Table 1. Model Evaluation Metrics

Metrics	Description
Accuracy	The overall proportion of correct predictions (both true positives and true negatives).
Precision	The ratio of correctly predicted positive observations (customers who adopted Special Perils and experienced heavy rain during purchase) to the total predicted positive observations.
Recall	The ratio of correctly predicted positive observations to all observations in the actual positive class (all customers who purchased Special Perils).
F1 Score	The harmonic means of precision and recall, providing a balanced measure that considers both the model's ability to avoid false positives and its ability to capture true positives.
AUC ROC	Area Under the Receiver Operating Characteristic curve, measuring the model's ability to distinguish between customers who purchased Special Perils and those who did not, across different classification thresholds.

The final model selection will rely on a thorough assessment of these indicators, including not only their predictive efficacy but also their interpretability. The study aims to provide actionable insights to stakeholders in the insurance sector, making interpretability a crucial goal. Through a comprehensive analysis of the strengths and weaknesses of each model based on several criteria, the objective to determine the best appropriate model for forecasting the adoption of Special Perils during instances of heavy rain weather in Malaysia should be achieved.

Beyond developing a predictive model for Special Perils adoption, this study introduces a research component through a comparative and enhanced modeling framework. The novelty of the study lies in the integration of temporal weather-related features with customer and vehicle characteristics, combined with a feature-selection and model-optimization process to improve predictive performance. Specifically, the study develops an enhanced predictive model by incorporating engineered variables derived from purchase timing and heavy-rain conditions, followed by hyperparameter optimization using cross-validation. The performance of the enhanced model is then compared against several widely used baseline classifiers, including Logistic Regression, Decision Tree, Random Forest, and Gradient Boosting. The research aims to answer the following question:

Can an enhanced machine-learning framework that integrates weather-related feature engineering and model optimization improve the prediction of Special Perils uptake compared with conventional classification models?

To ensure a rigorous comparison, all models are trained and evaluated using the same preprocessed dataset and data-splitting strategy. Their performance is assessed using Accuracy, Precision, Recall, F1-score, and AUC-ROC. Statistical comparison of model performance may also be conducted using repeated cross-validation results to determine whether the enhanced approach provides a meaningful improvement over the baseline models. By introducing feature engineering, model optimization, and comparative evaluation, the study contributes not only a practical insurance prediction system but also empirical evidence on methodological improvements for weather-aware insurance analytics, thereby strengthening its research contribution and publication readiness.

4 Results and Analysis

Insurance firms may develop more focused efforts to encourage the purchase of Special Perils coverage and modify their policies to better meet the changing demands of consumers in response to shifting weather conditions by examining these patterns and determining their underlying causes. Table 2 presents a concise overview of the significant findings derived from the dataset variables.

Table 3 presents the descriptive statistics for numerical variables, providing a detailed overview of their distribution and central tendencies.

Based on Table 4, the average age of customers in the dataset is 33 years, with ages ranging from 17 to 79. The standard deviation of 11 indicates a moderate spread in customer ages. The average age of insured vehicles is 7 years, ranging from 0 (new vehicles) to 20 years. The standard deviation of 4 suggests a narrow spread in vehicle ages, with most vehicles falling within a few years of the average. The average vehicle capacity is 1557 cc, with values ranging from 989 cc to 3471 cc. The standard deviation of 340 indicates a wide range of engine sizes in the insured vehicles. The average premium paid is RM1679, ranging from RM378 to RM9089. The standard deviation of 750 suggests a substantial variation in premium amounts, due to differences in coverage types, vehicle values, and other factors.

Table 5 offers a visualization analysis of numerical variables using bar charts and boxplots to illustrate patterns and relationships within the data.

Meanwhile, the adoption of Special Perils was minimal in October and November but saw a surge in December, aligning with the month of the most frequent heavy rains. Table 6 summarize the key insights for target variables.

In this study, correlation between target variable of Special Perils with other numerical variables are review with correlation Matrix Heatmap and stacked bar. Meanwhile the target variable with categorical variables such as gender and drive location are review with stacked bar and Chi-squared Test. Tables 7, 8, 9, and 10 provide a summary of key insights for the correlation between target variables and other variables.

The location more prone to heavy rain can be reviewed with the contingency table. Notably, Kuala Lumpur area has the highest proportion of customers purchasing Special Perils coverage, while several locations have no customers purchasing this coverage.

Table 2. Summary of Data Variables' Key insights

Dataset Variables	Key insights
Customer Age	The average customer age is 33 years (range: 17–79 years), with the highest concentration of customers between 20 and 40 years old. Older customers represent a smaller but notable proportion of the dataset.
Vehicle Age	The average age of insured vehicles is 7 years, ranging from 0 (new vehicles) to 20 years. Most vehicles being new (0–5 years old).
Vehicle Capacity	The average vehicle capacity is 1,557 cc, with considerable variation among vehicles. The distribution suggests two distinct vehicle categories, which may represent different customer preferences and insurance needs.
Premium	The average premium paid is RM1,679. Premium values show considerable variation, reflecting differences in vehicle types and insurance coverage. Most premiums are below RM2,000, while a small number of higher premiums are associated with luxury cars.
Gender	The data reveals a disparity in gender distribution among customers, with men being the majority.
Heavy Rain	Around 22% of customers experienced heavy rain during their motor insurance purchase. Heavy rainfall became more frequent from October onwards, peaking in mid-December, which suggests an increased exposure to severe weather risks across different states.
Special Perils	Special Perils coverage was adopted by only 16% of customers. The uptake increased significantly in mid-December, coinciding with peak rainfall periods, indicating a possible relationship between heavy rain events and customer demand for additional protection.
Drive Location	The majority of motor insurance customers were located in Selangor, Johor, Kuala Lumpur, Perak, and Negeri Sembilan. The concentration of customers in these states suggests potential target areas for future marketing efforts.
Purchase Data	he number of policies sold increased steadily from October to December, indicating stronger demand for motor insurance at the end of the year. This trend may be influenced by seasonal factors and promotional activities.

Table 11 provides a comparison of four classification models: Logistic Regression, Decision Tree, Random Forest and XGBoost. The models are assessed using measures such as accuracy, precision, recall, F1 score, and AUC-ROC.

All models have high precision but low recall, suggesting that they are conservative in predicting potential customers who would be interested in the add-on and tend to favour the majority class, particularly good at predicting when customers will not purchase Special Perils. This is a classic sign of class imbalance. Considering the objective of evaluating the impact of heavy rains weather on the adoption of Special Perils, and the need to strike a compromise between accuracy and the capacity to identify new consumers, the Random Forest model is the most appropriate option for this research. The model's excellent accuracy, balanced precision, and recall, and the greatest F1 score

Table 3. Numerical Variables Descriptive Statistic

Numerical Variables				
	Customer Age	Vehicle Age	Vehicle Capacity	Premium
Count	1639	1639	1639	1639
Mean	33	7	1557	1679
Std	11	4	340	750
Min	17	0	989	378
25%	24	4	1339	1212
50%	31	7	1497	1521
75%	40	10	1786	1922
Max	79	20	3471	9089

make it a comprehensive choice for forecasting the uptake of Special Perils in the setting of severe rains in Malaysia. Random Forest boasts the highest accuracy among the four models, correctly classifying 90% of instances. It offers a good balance between precision (0.91) and recall (0.46), indicating a reasonable ability to avoid false positives while still capturing a substantial portion of true positives. This metric combines precision and recall, and Random Forest's score of 0.61 is the highest, suggesting it is the best model for this specific task. Its AUC-ROC score is also relatively high, signifying a good overall ability to discriminate between customers who will and will not adopt Special Perils coverage. The confusion matrix shows that the model correctly predicts 404 true negatives and 24 true positives. However, it also has 5 false positives and 64 false negatives. The mean cross-validation accuracy of 0.8963 suggests that the model generalizes well to unseen data. The classification report indicates that the model has high precision (0.86 for class 0 and 0.83 for class 1) but lower recall (0.99 for class 0 and 0.27 for class 1), especially for the minority class (Special Perils uptake). The results suggest that while the model is good at avoiding recommending the add-on to customers who would not buy it, it could be better at identifying those who would. This information can be valuable for insurers in refining their marketing strategies and targeting efforts.

The Random Forest model demonstrates good overall accuracy and is particularly good at predicting when customers will not purchase Special Perils (high precision). However, it struggles to identify all potential customers who would be interested in the add-on (lower recall for the minority class). This could be due to the class imbalance in the dataset, where the number of customers who did not purchase the add-on is much larger than those who did. To enhance the model's capacity in identifying prospective customers for Special Perils, it would be beneficial to gather a more extensive dataset, particularly one that includes a greater number of cases where Special Perils are considered, to improve the model's learning capabilities. Besides, investigating other functionalities that might be pertinent to the forecasting assignment, such as the historical weather data of the customer's area or their past records of insurance claims. Alternatively, optimise the model's performance by fine-tuning its hyperparameters, with a particular focus on

Table 4. Numerical Variable Visualization Analysis

Numerical Variables	Histogram & Boxplot	Visualization analysis & Insights
Customer Age	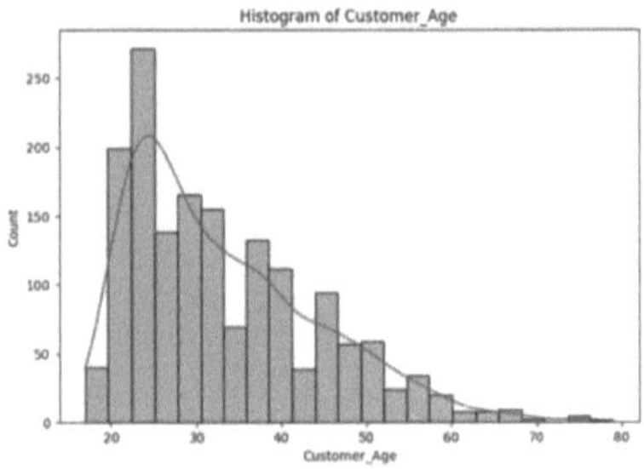 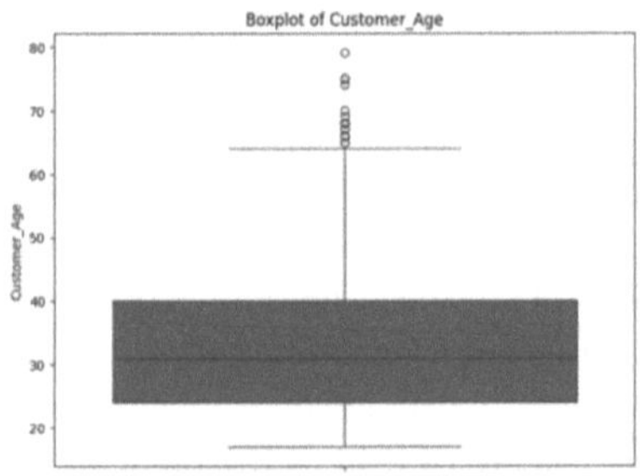	Histogram display skewness of 0.94 indicates a moderate right skew. Majority of customers falling between 20 and 40 years old. There is a smaller but noticeable group of elderly customers.
Vehicle Age	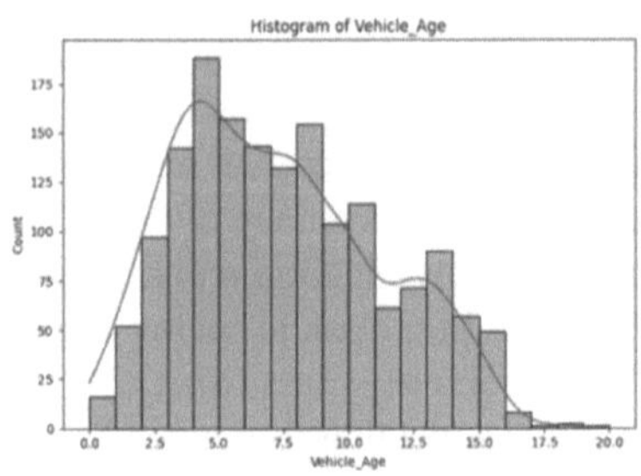 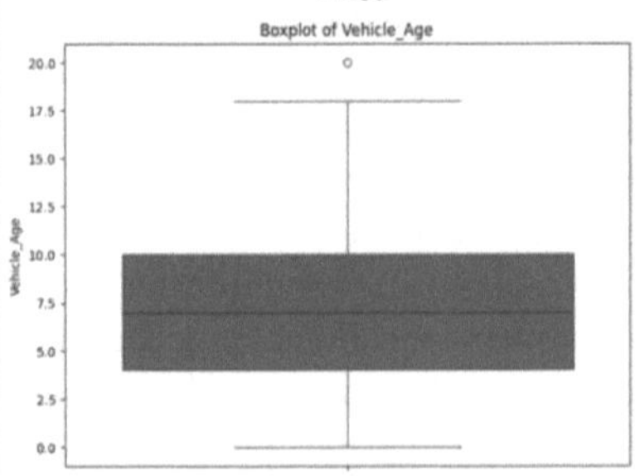	Histogram display skewness of 0.39 indicates a mild right skew. Most vehicles being new (0–5 years old). There is a long tail indicating some older vehicles.

(*continued*)

Table 4. (*continued*)

Numerical Variables	Histogram & Boxplot	Visualization analysis & Insights
Vehicle Capacity	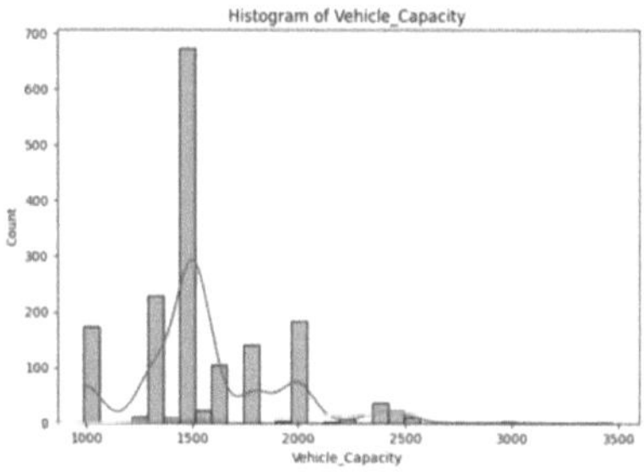 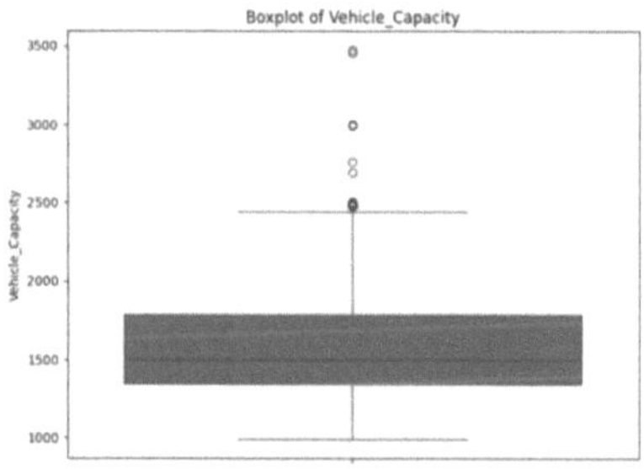	Histogram display skewness of 0.92 indicates a moderate right skew. The distribution is bimodal, with peaks around 1300 cc and 1800 cc. Insights: This suggests two distinct categories of vehicle capacity. This could represent different customer segments with varying preferences and needs.
Premium	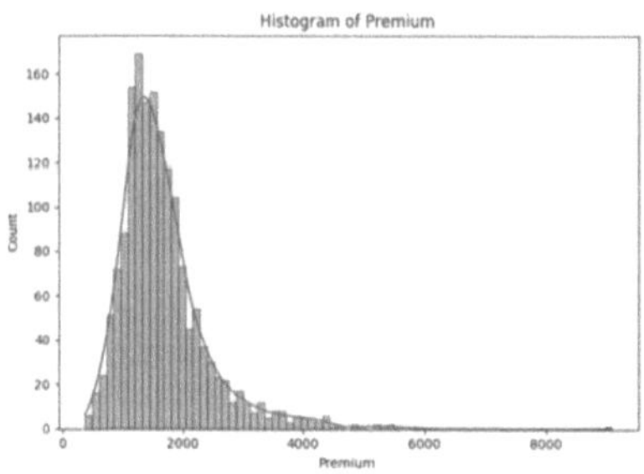 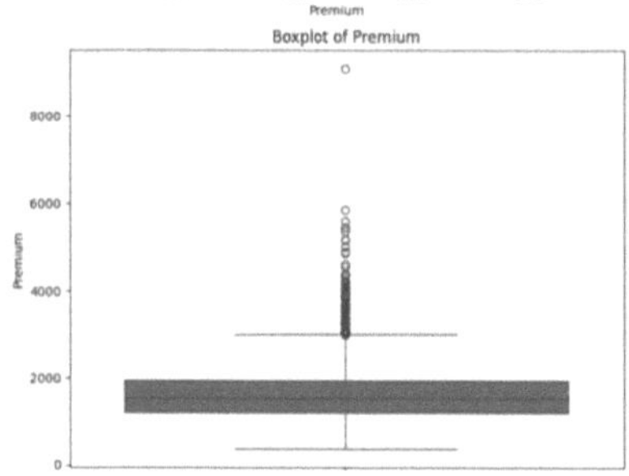	Histogram display skewness of 2.21 indicates a substantial right skew.26 Most premiums clustered below RM2000. There are a few outliers with remarkably high premiums. When referring to the dataset owner, high premiums often pertain to insurance for luxury cars.

Table 5. Categorical Variables Visualization Analysis

Categorical Variables	Bar Chart	Visualization analysis vs Purchase Date
Purchase Date		**Insight:** Motor insurance sales increased from October to December, peaking in December, which suggests higher year-end demand.
Heavy Rain	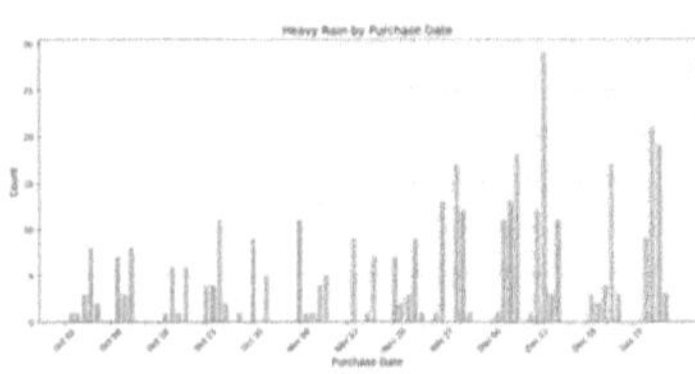	**Insight:** Heavy rainfall peaked in mid-December, increasing the risk of severe weather-related hazards for customers in different states.
Special Perils	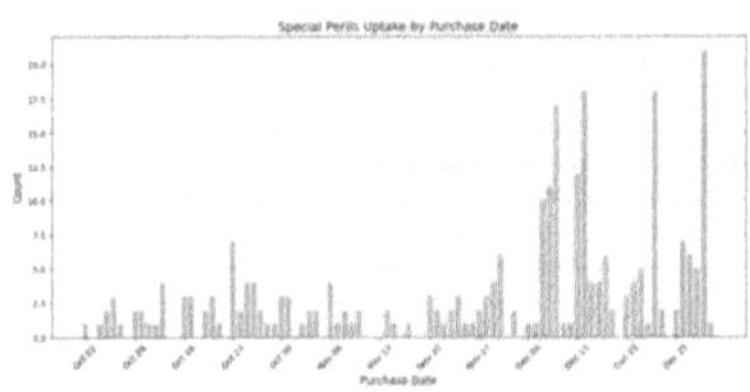	**Insight:** Special Perils uptake increased during periods of heavy rainfall, indicating that weather conditions may influence customer purchasing decisions.

the minority class will enhance the accuracy and effectiveness of the predictive model for identifying prospective customer for Special Perils add-on insurance, it is crucial to focus on five specific aspects.

The findings support the development of weather-informed marketing initiatives. For example, insurers may:

- Launch targeted campaigns during periods of heavy rainfall.
- Provide location-specific risk alerts.
- Offer customized policy bundles.
- Deliver personalized recommendations based on predicted risk exposure.
- Such approaches can simultaneously improve customer protection and business performance.

Table 6. Summary Key insight for Target Variables vs Variables Correlation

Variables	Key insights
Customer Age	Correlation Matrix 0.19 indicates that customer age has little relationships with the acquisition of Special Perils coverage.
Vehicle Capacity	Correlation Matrix 0.14 indicates that customer age has little relationships with the acquisition of Special Perils coverage
Vehicle Age	Correlation Matrix −0.16 indicates that customer age has little relationships with the acquisition of Special Perils coverage
Premium	Correlation Matrix 0.23 has more substantial determinant in deciding to get this coverage
Heavy Rains	Correlation Matrix 0.30 indicate the most substantial determinant compared to other numerical variable in deciding to get this coverage. Heavy rain is the strongest factor associated with Special Perils uptake, with adoption increasing during periods of heavy rainfall.
Gender	Gender does not appear to significantly influence the customer's decision to opt for Special Perils coverage.
Drive Location	Chi-squared test confirms a relationship between driving location and Special Perils uptake. The probability of obtaining Special Perils policy is dependent upon the customer's driving location

Table 7. Correlation Matrix Heatmap

Target Variables Special Perils Uptake with Numerical Variables Correlation	Visualization analysis
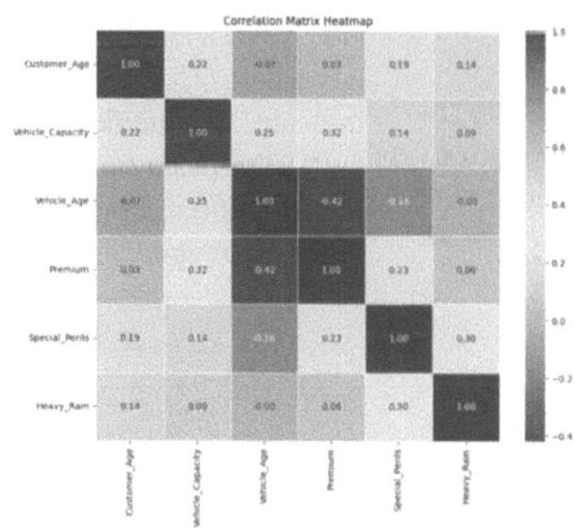	The correlation study, using Special Perils as the primary dependent variable, indicates that individual and insured vehicle variables such as customer age (0.19), vehicle capacity (0.14) and vehicle age (−0.16), have little relationships with the acquisition of Special Perils coverage. On the other hand, the impact of heavy rain (0.30) and premium (0.23) is a more substantial determinant in deciding to get this particular coverage. The results indicate that customer age, vehicle capacity, and vehicle age have limited influence on Special Perils uptake. However, heavy rain and premium amount are more important factors. The likelihood of purchasing Special Perils coverage increases during periods of heavy rainfall and among customers with higher insurance premiums.

Table 8. Target Variables Special Perils Uptake by Categorical variable occurrences

Target Variables Special Perils	Visualization	Visualization analysis
Heavy Rain occurrence	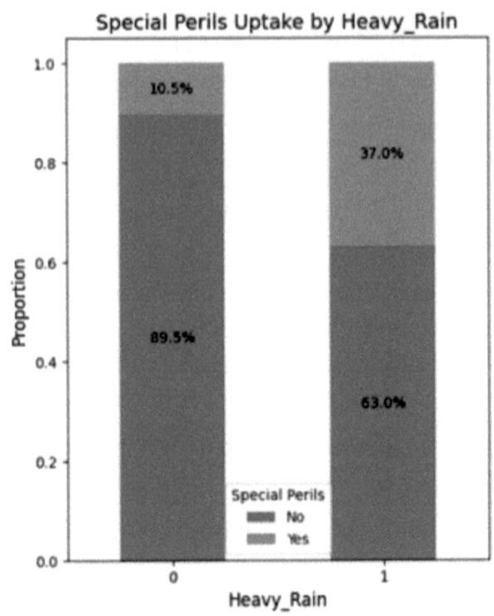	The chart indicates that Special Perils uptake increases during periods of heavy rain. Customers exposed to heavy rainfall were more likely to purchase the coverage, supporting the findings of the correlation analysis. This suggests that weather conditions are an important factor in predicting Special Perils adoption.
Gender	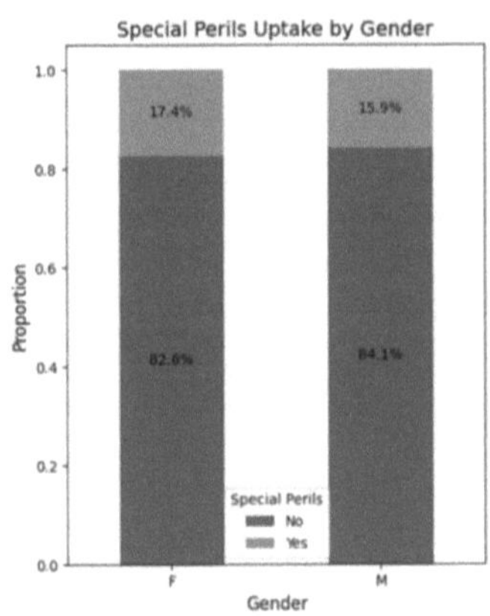	Although male customers made up a larger proportion of the dataset, Special Perils uptake was only slightly higher among female customers (17.4%) than male customers (15.9%). This indicates that gender is not a significant factor in the adoption of Special Perils coverage.

Table 9. Chi-squared Test value

Chi-squared test	Value	Analysis
Chi-squared statistic	330.16218203906755	The results of the chi-squared test indicate a statistically significant association between Drive Location and the purchase of Special Perils coverage. As the p-value is well below 0.05, the null hypothesis was rejected. **Insight:** The probability of purchasing Special Perils coverage differs across driving locations, suggesting that location-related factors may influence customer decisions.
P-value	3.792297671729711e-64	

Table 10. Contingency Table - Drive Location vs Special Perils Uptake

Contingency Table - Drive Location vs Special Perils Uptake			
Drive Location	Special Peril	Special Peril	% Special Perils uptake over
(State)	Uptake (No)	Uptake (Yes)	Total Motor insurance purchased
Johor	275	19	6%
Kedah	88	0	0%
Kelantan	30	1	3%
Kuala Lumpur	90	123	58%
Melaka	93	24	21%
Negeri Sembilan	124	13	9%
Other	13	2	13%
Pahang	46	4	8%
Perak	124	18	13%
Pulau Pinang	41	3	7%
Selangor	423	62	13%
Terengganu	23	0	0%

Table 11. Predictive Model Result Evaluation Comparison

	Accuracy	Precision	Recall	F1 Score	AUC ROC
Logistic Regression	0.89	0.80	0.44	0.57	0.85
Decision Tree	0.84	0.65	0.15	0.25	0.77
Random Forest	0.90	0.91	0.46	0.61	0.81
XGBoost	0.89	0.90	0.42	0.57	0.83

5 Conclusion

This study explored the use of weather data and predictive analytics to improve the uptake of Special Perils insurance in Malaysia. The findings show that heavy rainfall events can increase consumers' perception of risk and influence their decision to purchase additional insurance coverage. By integrating weather data with customer and vehicle information, insurers can develop predictive models to better forecast adoption behaviour, improve customer targeting, and enhance marketing and risk management strategies. However, several challenges remain, including variability in weather data, difficulty in establishing direct causal relationships, data quality limitations, and concerns over model interpretability and data privacy. Future research should focus on using long-term and real-time data, incorporating behavioural factors, and applying explainable AI techniques. Overall, combining weather analytics with machine learning offers strong potential for innovation, helping insurers improve customer protection while strengthening business performance in the face of increasing climate risks.

References

1. Kanchetti, D.: Climate change and insurance: using predictive analytics to navigate emerging risks. ESP J. Eng. Technol. Advanc. **1**(1), 184–194 (2021)
2. Mushonga, F.B., Mishi, S.: Natural hazard insurance demand: a systematic review. Jàmbá: J. Disaster Risk Stud. **14**(1), a1223 (2022)
3. Wagner, K.R.H.: Adaptation and adverse selection in markets for natural disaster insurance. Am. Econ. J. Econ. Pol. **14**(3), 380–421 (2022)
4. Haldar, P., Mandal, D.K., Gupta, U.: Data Science for Green Insurance: a Predictive Analytics Approach to Climate Risk Assessment. IGI Global (2026)
5. NAHRIM: The impact of climate change on hydro-climate of Malaysia based on IPCC AR5 (2021)
6. Hott, C., Tran, T.X.: Natural catastrophes and insurance in a developing economy: new theoretical and empirical evidence. Environ. Dev. Econ. **30**, 402–420 (2025)
7. Murray, K.B., Di Muro, F., Finn, A., Popkowski Leszczyc, P.: The effect of weather on consumer spending. J. Retail. Consum. Serv. **17**, 512–520 (2010)
8. Piłat-Borcuch, M.: Weather and purchasing decisions in behavioural economics research. Humanit. Soc. Sci. Rev. (2025)
9. Chen, X., Zhao, Y.: Climate risk and insurance purchasing behaviour. Front. Sustain. Food Syst. (2024)

10. Olaniyan, D.: Enhancing multi-step stock price forecasting with social media sentiment and engagement metrics. Master's thesis, University of Denver (2025)
11. Malote, A.: Applying Predictive Analytics to Climate-Related Insurance Risk Assessment. University of the Witwatersrand (2022)
12. Mahankal, M., Misal, S., Deshpande, M., Lande, G.: Investigating the impact of climate change on insurance models and products. Int. J. Res. Fin. Manag. **8**(2), 208–210 (2025)
13. Shi, Y.: Impacts of Extreme Weather on Insurance Claims. Norwegian School of Economics (NHH) (2025)

Performance Comparison of Jump Point Search and A* for Grid-Based Campus Pathfinding

Muhammad Izmy Shah Aries Iskandar[1], Azyan Yusra Kapi[2(✉)], Azrina Suhaimi[2], and Mohamad Faizal Ab Jabal[2]

[1] Faculty of Computer and Mathematical Sciences, Universiti Teknologi MARA, Terengganu Branch Kuala Terengganu Campus, Kuala Terengganu, Terengganu, Malaysia
2024514507@student.uitm.edu.my

[2] Faculty of Computer and Mathematical Sciences, Universiti Teknologi MARA, Johor Branch Pasir Gudang Campus, Pasir Gudang, Johor, Malaysia
{azyanyusra,azrin253,m.faizal}@uitm.edu.my

Abstract. Route planning for campus navigation requires pathfinding algorithms that can generate feasible routes while maintaining acceptable execution time and memory usage. This study compares Jump Point Search (JPS) and A* for grid-based pathfinding in a campus environment. The campus area was represented as a 100×100 grid consisting of traversable areas and obstacles. Both algorithms were tested using selected route scenarios and compared in terms of latency, memory usage, and route distance. The results indicate different performance behaviours between JPS and A* within the selected campus route scenarios. A* achieved lower latency in the 100×100 grid, while JPS showed lower memory usage in selected medium- and long-distance route cases. Both algorithms produced comparable route distances, suggesting that route quality was generally preserved. In the farthest route case, JPS used 15.23 KB of memory compared with 17.60 KB for A*, although A* achieved lower latency. These findings suggest that JPS may offer memory-related advantages in some campus pathfinding scenarios, while A* remains more efficient in terms of execution time for the tested small-grid environment. Future work should include larger grid sizes, additional route scenarios, and repeated experiments to further validate the performance trends.

Keywords: Campus navigation · Jump Point Search · A* algorithm · Grid-based route planning · Pathfinding performance · Memory usage

1 Introduction

Pathfinding is commonly used in navigation tasks, where the goal is to identify a feasible or optimal route from a starting point to a target destination. Previous research has indicated that the performance of pathfinding can be influenced by the features of the navigation environment, such as the complexity of the terrain and the way the search space is represented [1].

University campuses can be difficult to navigate, particularly for freshmen, new staff, and visitors who are unfamiliar with the arrangement of buildings, walkways, offices,

N. Mohd Sharef et al. (Eds.): CAMP 2026, CCIS 3093, pp. 212–226, 2027.
https://doi.org/10.1007/978-981-92-4302-0_16

classrooms, and facilities. Users frequently encounter challenges such as confusion, delay, and tension when attempting to reach their goal at any university. When there is no proper signboard, the problem of orienting becomes more problematic. Previously, challenges of orientation were noted in relation to first-year students and visitors [2]. Additionally, navigating indoors using GPS can be challenging due to weak signals or the complete lack of GPS availability in such environments [3].

Pathfinding algorithms play a crucial role in navigation applications as they assist in identifying the route from the starting point to the destination. Dijkstra's algorithm serves a referential purpose in this work, as it is among the earliest algorithms developed for addressing the shortest path problem on weighted directed graphs that do not have negative edge weights [4]. Although it is optimal, it is not goal-directed, which means that the algorithm might explore many vertices before reaching the destination node. A*, however, addresses this issue by utilising a heuristic function to estimate the cost from the current vertex to the goal [5]. Therefore, the A* algorithm is more appropriate to serve as the foundational algorithm for comparing the performance of JPS. It is important to note that in grid-like environments, the process will examine neighbouring vertices in an iterative manner, which may result in unnecessary operations if multiple symmetric routes are present.

JPS is an enhancement of A* specifically designed for grid-based pathfinding problems. It is not necessary to evaluate each and every neighbour node when implementing the JPS algorithm since it utilises symmetry breaking principles which help in omitting unnecessary evaluation and determining the vital nodes referred to as jump points [6]. In essence, it becomes possible for the algorithm to avoid excessive node evaluations without compromising on the resulting quality of the path generated by the algorithm. In this case, it is therefore appropriate for JPS to be assessed in campus navigation since campus maps can be visualised in terms of grid-based representations made up of passable and impassable areas.

Although JPS has potential for reducing unnecessary node evaluations in grid-based pathfinding, its performance in campus route planning still requires further investigation. Previous studies have applied JPS in domains such as robotics, pipe routing, and dynamic path planning. However, existing campus navigation studies have mostly focused on system interfaces, augmented reality, GPS, QR codes, or mobile-based systems [7, 8]. Therefore, a practical comparison between JPS and A* is useful for examining the trade-off between execution time and memory usage in a campus-scale grid environment.

In this context, this study compares the performance of JPS and A* in a grid-based campus pathfinding environment. The campus area is represented as a 100×100 grid consisting of walkable regions and obstacles to illustrate selected campus navigation scenarios. The contribution of this study lies in the practical implementation and performance comparison of JPS and A* using selected route scenarios. Rather than proposing a new pathfinding algorithm, this work provides empirical insight into the behaviour of both algorithms in terms of latency, memory usage, and route distance on a 100×100 campus grid.

2 Related Work

This section reviews related work on campus navigation systems, grid-based pathfinding, and JPS and A* algorithms. The review highlights that many campus navigation studies focus on user assistance and interface design, while fewer examine algorithm performance.

2.1 Campus Navigation Systems

Campus navigation has gained significance as an area of research because university settings normally contain a number of buildings, paths, classroom blocks, offices, and indoor spaces that can be difficult to find. The problem is more prevalent among students and employees who are fresh in the university setting, along with visitors. Earlier literature indicates that complicated building designs and environments, lack of sufficient signage, and an unfamiliar environment are responsible for confusion and delay among users [2, 9].

While navigational apps in commercial sectors are commonly adopted by individuals for general purposes of traveling, these applications may fail to adequately assist in the process of navigating within the premises of a campus, especially when navigating indoors. While navigational applications using GPS technology work quite effectively outdoors, the signal of such devices becomes weaker indoors, meaning that their usefulness is reduced considerably when looking to navigate to specific rooms or other facilities [3].

Various research works have introduced digital approaches for better campus navigation. Reference in [10] introduced a web-based campus navigation system with mobile augmented reality intervention. Similarly, SINS_AR, an augmented reality-based smart indoor navigation system, was introduced by [8]. Other researchers have used approaches based on QR codes and web-based techniques, mobile navigation systems, GPS, and IoT [7, 9].

This demonstrates that navigational aids for campus environments have moved from traditional static mapping and directional aids to more advanced digital, mobile, web-enabled, and even augmented reality methods. Nonetheless, most studies conducted so far concentrate more on aspects such as the user interface, visualisation, and presentation of information. A limited number of studies evaluate the effectiveness of the pathfinding method used. The pathfinding algorithm needs to be able to accurately and efficiently calculate routes to be regarded as a useful tool in a web-based or mobile-based campus navigation system.

2.2 Grid-Based Pathfinding Algorithms

Pathfinding algorithms are widely used in artificial intelligence, robotics, games, and navigation systems to determine an effective route from a starting point to a destination [11]. In navigation applications, the selection of an algorithm influences the overall efficiency of the navigation system and includes aspects such as optimality of the computed path, processing time, memory allocation, and adaptability to various environments [12]. Popular path finding algorithms involve Dijkstra's algorithm and A* algorithm, which have been widely employed as benchmark solutions in route planning tasks. Recent research conducted by [13] introduced Incremental Line Search as an optimisation method for grid-based pathfinding. In simple terms, the search domain is confined to a corridor whose bounds fluctuate depending on the situation, resulting in reduced computational complexity.

A* is often used in grid-based route planning because it combines the travelled cost with a heuristic estimate to guide the search toward the goal [5]. However, despite having a heuristic component, A* can still explore adjacent nodes incrementally. With grid maps having free spaces or similar pathway sequences, there might be redundant node explorations that will occur. Therefore, optimisation methods such as JPS are useful to consider for reducing unnecessary search operations in grid-based route planning.

In this case, the application of grid representation for the analysis of route planning in a university campus is helpful since the physical environment may be represented by structured cells. In this case, each cell may represent walkable or non-walkable zones. Walkable cells represent pathways, corridors, open spaces, or pedestrian ways. Non-walkable cells, on the other hand, represent structures like buildings, barriers, and other objects. In this way, a pathfinder may be used to develop a route from the start point to the destination within the environment.

Nevertheless, despite the efficiency of A* as compared to Dijkstra's algorithm, this search strategy involves the examination of neighbouring nodes in steps. This makes this algorithm have redundant node expansions in cases where there is a presence of several symmetrical paths in the environment. This problem has triggered the development of more efficient grid-based searching algorithms including JPS.

2.3 JPS and A* Algorithm

Recent advances in grid-based pathfinding have expanded the JPS idea to weighted grid environments. For instance, [14] developed Weighted Jump Point Search (JPSW), which uses symmetry-breaking and pruning algorithms to increase pathfinding performance on weighted grid maps. JPS is an optimisation of A* for uniform-cost grid maps [6]. One of the basic objectives of JPS is to reduce the number of unnecessary nodes that will be pruned symmetrically. Unlike A* search, which examines all neighbour nodes, JPS skips irrelevant nodes and continues in other directions until it encounters some relevant nodes, also known as jump points. Jump points can occur when JPS reaches the target point, encounters an obstacle, or finds forced neighbours [6].

A* remains one of the most popular path-finding algorithms due to its simplicity, effectiveness, and ability to offer ideal paths when the proper circumstances are satisfied. Some recent research has focused on enhancing the performance of the A* algorithm for use in path planning. [15] created a new version of the A* algorithm for AGV path planning that employs three-neighbourhood search and an efficient evaluation function method. When working with a grid map, A* keeps extending a significant number of nodes, which makes it harder to find the shortest path. JPS addresses this issue by determining the necessary nodes while eliminating those that are unnecessary, leading to an optimal path.

Previous studies have applied JPS and its variants in several pathfinding domains. [16] applied JPS in environments with temporal obstacles, while [17] adapted JPS for pipe auto-routing. [18] proposed a path planning and collision avoidance method for distributed multi-robot systems in complex dynamic environments, incorporating a bidirectional alternating JPS-based global planner. Reference in [19] combined JPS with Ant Colony Optimisation for mobile robot path planning. The reviewed studies indicate that JPS has mostly been tested in grid-based planning problems outside campus navigation, such as robot movement, pipe routing, and temporal-obstacle settings. Therefore, it is necessary to explore the performance of JPS in a campus route planning scenario.

However, it does not guarantee that JPS will always take less time for execution in every setting. The recursion mechanism of jumps made by JPS and the mandatory neighbour check may add to the computation time required by the algorithm. Thus, in some settings, the performance improvement offered by JPS would be determined by factors such as the size of the environment, the density of obstacles in the grid, the path length, and the complexity of the environment itself. When dealing with small grids, A* would offer better performance due to its simplified node expansions.

2.4 Positioning of the Present Study

Several prior studies on campus navigation have considered different digital approaches, such as websites, mobile applications, QR codes, IoT, GPS, and augmented reality. These studies highlight the need to provide better wayfinding support for students, staff, and visitors in complex campus environments. Nevertheless, most of these studies focus on navigation support, interface design, or visualisation rather than the performance of the route generation process.

Regarding route calculation techniques, previous studies have adopted traditional pathfinding algorithms such as A* and Dijkstra. In addition, JPS has been applied effectively in other path planning domains, including robotics, pipe routing, and dynamic obstacle environments. However, fewer studies discuss how the choice of pathfinding algorithm affects route generation performance in a campus navigation setting. For this reason, this paper compares JPS and A* using the same 100×100 campus grid. The two algorithms are compared based on three metrics: route distance, latency, and memory usage.

3 Methodology

This section explains how JPS and A* were implemented and compared in a grid-based campus pathfinding environment. The methodology includes the representation of the campus map as a grid, the implementation of both algorithms, and the experimental procedure used to measure route distance, latency, and memory usage. This study focuses on algorithm performance comparison rather than the development of a complete navigation system.

3.1 Research Design

This study compares the performance of JPS and A* for route planning on the UiTM Kuala Terengganu campus. The objective is to examine the trade-off between execution time, memory usage, and route distance when both algorithms are applied to selected campus route scenarios. The study was conducted in three main phases: preliminary analysis, design and implementation, and testing and evaluation.

3.2 Grid Representation

This experiment involved a campus map from UiTM Kuala Terengganu. The researchers used OpenStreetMap to identify campus landmarks such as buildings, offices, recreation areas, pathways, passageways, and shortcuts. The campus layout was then manually encoded into a 100×100 grid representation to support pathfinding computation.

The grid representation used binary values, where 0 indicates a walkable cell and 1 indicates an obstacle. This allows the algorithms to distinguish between accessible and restricted areas during route computation. Each selected campus location was mapped to a corresponding grid coordinate. When a user selected a starting point and destination, these coordinates were used as input for the pathfinding algorithm.

The algorithms used these coordinates to conduct their searches on the grid and generate valid routes without crossing obstacle cells. The use of the grid makes it possible to conduct a comparative study on the performance of JPS and A* since both operate on similar settings. The grid made use of eight directions for the movement process which included moving in straight lines as well as diagonally. Therefore, the octile distance heuristic was used to estimate the cost from the current node to the destination. The same movement rule and heuristic function were applied to both JPS and A* to ensure a fair comparison.

3.3 Implementation of A* Algorithm

A* was implemented as the baseline algorithm for comparison. In grid-based route planning, A* evaluates neighbouring cells and selects the next node based on the estimated total cost:

$$f(n) = g(n) + h(n)$$

where g(n) is the cost from the start node to the current node, and h(n) is the heuristic estimate from the current node to the destination. The algorithm continues selecting nodes with the lowest estimated cost until the destination is reached or no valid path is found.

A* was selected as the comparison algorithm because it is widely used in pathfinding and serves as the basis for JPS. Comparing JPS with A* therefore allows the study to evaluate whether the jump-based optimisation provides practical benefits in the campus route planning context.

3.4 Implementation of JPS

JPS was implemented as the main algorithm evaluated in this study. Unlike A*, where neighbouring nodes are visited one-by-one, in JPS, a jumping procedure is performed, which allows skipping unnecessary nodes in grid maps. In JPS, the process starts with loading the grid of the campus map and verifying the start and goal nodes. Then, an open list containing the start node is initialised, and the node with the smallest cost is selected.

After selecting the node, the jump procedure is called recursively in different movement directions. The process goes on until either the goal node, an obstacle or a forced neighbour is encountered. Whenever a successful jump is obtained, it is stored into the open list with updated cost values. This process is repeated until the goal is reached or the open list gets empty. The implementation covered the processes of initialising the open list, selecting the node with the lowest cost, performing recursive jumping, detecting obstacles and forced neighbours, and generating the final path.

3.5 Experimental Evaluation

The experimental evaluation was conducted using six selected route scenarios on the 100×100 campus grid. The selected routes were intended to represent different navigation distances, including short-distance, medium-distance, and long-distance campus paths. For each route scenario, the same starting point and destination were used for both JPS and A* to ensure a fair comparison.

The algorithms were compared based on three metrics: latency, memory usage, and route distance. Latency was used to measure the execution time required to generate a route, while memory usage was recorded to compare the computational resource requirements of both algorithms. Route distance was included to verify whether both algorithms produced comparable path lengths for the same route scenario.

Since the evaluation was based on selected route scenarios and repeated executions were not conducted, the results are presented as descriptive findings rather than statistical performance measurements. Therefore, mean and standard deviation analysis for repeated runs is not reported in this study. This scope is appropriate for examining the practical behaviour of JPS and A* within the tested campus grid, while larger-scale and repeated experiments are recommended for future work.

4 Experimental Setup

The following subsections describe the experimental setting used to assess the performance of JPS and A*. The experimental setup consists of the prototype environment, test case selection, evaluation metrics, and the testing method. The JPS and A* algorithms were evaluated on identical campus grids, starting points, destinations, hardware, and software.

4.1 System Environment

As shown in Fig. 1, the prototype was implemented as a web-based campus route planning system. The backend was developed using Python, while the frontend interface used Leaflet.js for map visualisation. Through the interface, users can select the starting point, destination, and pathfinding algorithm, either JPS, A*, or both algorithms for comparison.

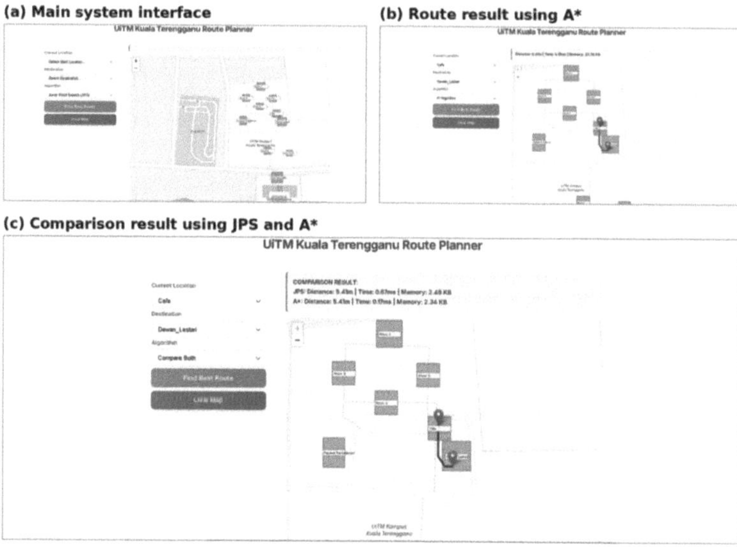

Fig. 1. Prototype interface and example output of web-based campus route planner: (a) main interface of the system, (b) output result when applying A* search algorithm, and (c) comparison mode output result when applying JPS and A* algorithms.

From the image shown in Fig. 1(a), it can be seen that the user is supposed to provide the starting point, destination, and algorithm in order to obtain the result. Fig. 1(b) gives an example of the output route obtained by using A* algorithm, while Fig. 1(c) presents the output result of the comparison between JPS and A* algorithms. Values provided in Fig. 1 above are some of the example outputs obtained from the prototype.

These experiments were carried out using Dell Latitude E7470 laptop, having an Intel Core i7-6600U processor, 8 GB RAM, and 476 GB storage capacity. The software used includes Windows 10 Pro, Visual Studio Code, and Python. Specifications have been provided because latency and memory consumption could be dependent on the software and hardware environment.

4.2 Test Cases

The experimental setup was designed to evaluate the performance of JPS and A* using the same campus grid and route conditions. The UiTM Kuala Terengganu campus map was represented as a 100 × 100 grid, where walkable areas and obstacles were encoded to support pathfinding computation. Six test cases were selected to represent different route characteristics based on distance level as shown in Table 1.

The test cases were grouped into three levels: close, medium, and far. Each level consisted of two route cases.

4.3 Evaluation Metrics

JPS and A* algorithms were analysed based on three criteria including validity of the path and comparison of route distance, latency, and memory requirements. Path validity

Table 1. Selected campus route test cases based on distance level.

Case	Route description	Distance	Distance level
Case 1	Kolej Kerawang to Block E	~8.25 m	Close
Case 2	Kolej Kerawang to Cafe	~21.66 m	Medium
Case 3	Block A to Block D	~7.41 m	Close
Case 4	Pejabat to Surau	~27.66 m	Far
Case 5	Dewan Lestari to Block C	~18.66 m	Medium
Case 6	Block B to Kolej Kerawang	~35.07 m	Far

was used to determine whether the generated route could reach the destination without crossing obstacle cells. As both algorithms were developed for use in uniform-cost grids, they should have comparable results when applied to similar grid structures.

Latency is the amount of time taken by the algorithm to generate the route in milliseconds (ms). Memory usage is the maximum memory utilised in generating the route and is measured in kilobytes (KB). The latency and memory have been chosen to test the performance of the algorithms in addition to determining their validity. Latency and memory usage were gathered using the integration of the back end in the prototype. In this regard, the back end received the start point, destination, and algorithm selected and then performed the pathfinding algorithm. After that, the generated route, along with the corresponding latency and memory usage, was sent back to the user interface.

4.4 Experimental Procedure

In the experiment, both JPS and A* were tested on the same 100 × 100 campus grid using six selected route scenarios. For each case, the same starting point and destination were used for both algorithms. The latency, memory usage, and route distance produced by each algorithm were recorded and compared.

The objective of this experiment is to examine the relative performance behaviour of JPS and A* in a selected campus grid environment. The comparison focuses on whether the jump and forced-neighbour mechanisms in JPS provide practical advantages in terms of memory usage, and whether these mechanisms introduce additional latency in a small-grid setting.

5 Results and Discussion

This section discusses the results obtained from testing JPS and A* on the six selected route cases. The discussion focuses on three aspects: path validity and route distance, latency, and memory usage. The results are used to analyse the trade-off between execution time and memory efficiency in the tested campus grid.

5.1 Path Validity and Route Distance Comparison

During the initial testing, focus was given to the validity of the paths and the comparison of route distances. In all six tests, both the JPS and A* algorithms successfully produced valid paths to reach the goal destination from the selected starting point. In all six tests conducted within the 100 × 100 campus grid setting, the route distances produced by JPS and A* were comparable or nearly the same.

Path quality was assessed by comparing the route distances generated by JPS and A* for the same start and destination points. Similar route distances indicate that both algorithms produced comparable valid routes under the tested grid conditions. After confirming path validity, the comparison focused on computational efficiency using latency and memory usage. Table 2 summarises the performance comparison between JPS and A* across the six selected route cases.

The results show that both algorithms were able to generate valid routes for all six selected route cases. This indicates that JPS and A* can both produce feasible campus routes in the tested grid-based environment. However, the findings should be interpreted within the scope of the selected 100 × 100 campus grid and the six route scenarios used in this study. The two algorithms showed different performance patterns: A* achieved lower latency in all tested cases, while JPS showed lower memory usage in selected medium-distance and long-distance route cases. Therefore, the results are presented as descriptive evidence of algorithm behaviour in the tested campus setting rather than as a general benchmark for all grid-based pathfinding environments. The detailed performance values are presented in Table 2.

Table 2. Performance comparison between JPS and A*

Case	Distance level	Route description	Algorithm	Distance (m)	Latency (ms)	Memory usage (KB)	Path status
Case 1	Close	Kolej Kerawang to Block E	JPS	~8.25	1.99	5.55	Comparable / valid
			A*	~8.24	0.52	5.36	Comparable / valid
Case 2	Medium	Kolej Kerawang to Cafe	JPS	~21.66	3.40	8.36	Comparable / valid
			A*	~21.66	1.78	8.51	Comparable / valid
Case 3	Close	Block A to Block D	JPS	~7.41	0.92	2.98	Comparable / valid
			A*	~7.41	0.64	3.33	Comparable / valid
Case 4	Far	Pejabat to Surau	JPS	~27.66	6.67	10.13	Comparable / valid

(continued)

Table 2. (*continued*)

Case	Distance level	Route description	Algorithm	Distance (m)	Latency (ms)	Memory usage (KB)	Path status
			A*	~27.66	2.30	13.95	Comparable / valid
Case 5	Medium	Dewan Lestari to Block C	JPS	~18.66	3.71	6.78	Comparable / valid
			A*	~18.66	1.73	8.58	Comparable / valid
Case 6	Far	Block B to Kolej Kerawang	JPS	~35.07	11.56	15.23	Comparable / valid
			A*	~35.07	3.06	17.60	Comparable / valid

5.2 Latency Comparison

Figure 2 presents the latency comparison between JPS and A* across the six route cases.

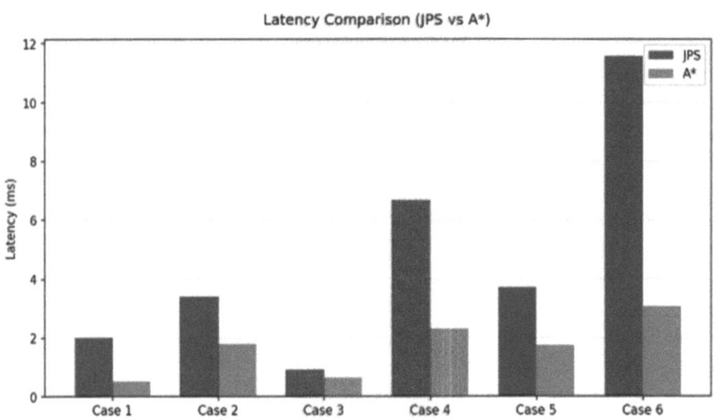

Fig. 2. Latency comparison between JPS and A* across six campus route cases.

Based on the results, A* consistently recorded lower latency than JPS in all six test cases. For example, in Case 1, A* recorded 0.52 ms compared with 1.99 ms for JPS. In Case 6, which represents a far route case, A* recorded 3.06 ms, while JPS recorded 11.56 ms.

The findings suggest that A* achieved better latency performance than JPS within the tested 100 × 100 grid environment. One possible reason is that the grid size used in this study was relatively small, allowing A* to compute routes efficiently using its direct node expansion process. In contrast, JPS involves additional procedures such as lookahead operations, jump point detection, and forced neighbour detection, which may introduce processing overhead in smaller grid environments.

Thus, based on the latency findings, JPS did not demonstrate faster execution than A* in the selected route scenarios. In the tested 100 × 100 campus grid, A* was more favourable when execution time was the main performance consideration.

5.3 Memory Usage Comparison

Figure 3 presents the peak memory usage comparison between JPS and A* across the six route cases. In contrast to the latency results, JPS showed better memory efficiency in most test cases. For example, in Case 3, JPS used 2.98 KB, while A* used 3.33 KB. In Case 4, JPS used 10.13 KB, while A* used 13.95 KB. Similarly, in Case 6, JPS consumed 15.23 KB, compared to A*'s 17.60 KB.

These results suggest that JPS may provide memory-related advantages in some medium-distance and long-distance route cases. This can be explained by the fact that JPS omits unnecessary intermediate nodes that do not need to be included on the open list. Therefore, JPS conserves memory since fewer nodes need to be stored during the search process. As previously noted, the difference in memory consumption in cases with similar routes is minimal. In Case 1, JPS used 5.55 KB, whereas A* utilised only 5.36 KB.

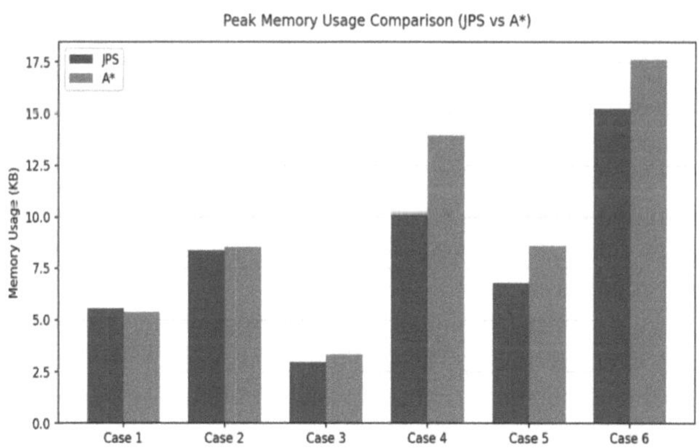

Fig. 3. Peak memory usage of JPS and A* across six selected campus route cases.

5.4 Discussion

The analysis indicates a trade-off between latency and memory usage. A* achieved lower latency in all six route cases, making it more suitable when faster route computation is the main priority. This may be due to the relatively small 100 × 100 grid used in this study, where the direct node expansion strategy of A* can compute routes efficiently without substantial overhead.

In comparison, JPS showed lower memory usage in most cases, particularly in the medium-distance and long-distance route cases. This suggests that JPS may provide memory-related advantages when the route involves more cells to be explored. The jump search strategy used in JPS reduces unnecessary intermediate node evaluations, which can reduce the number of nodes stored during the search process.

Overall, the results suggest that JPS should not be considered universally faster than A*. In the tested 100 × 100 campus grid, A* performed better in terms of latency, while JPS showed lower memory usage in selected route cases. Therefore, the choice of algorithm depends on the priority of the application. If execution time is the main concern, A* is more suitable for the tested campus grid. However, if memory usage is a key consideration, JPS may be useful, especially for medium-distance and long-distance route scenarios.

It should be noted that this study is limited to six selected route scenarios on a 100 × 100 campus grid. Therefore, the findings should not be generalised as a broad performance benchmark for JPS and A* across all grid-based environments. In addition, repeated executions and statistical analysis were not conducted in this study. Future work should extend the evaluation using larger grid sizes, additional randomised start–destination pairs, repeated experimental runs, and statistical measures such as mean and standard deviation to further validate the observed performance trends.

6 Conclusion

This study presented a performance comparison of JPS and A* for grid-based campus route planning. The campus environment was represented using a 100 × 100 grid, and both algorithms were compared using six selected route scenarios in terms of latency, memory usage, and route distance. The results showed that both algorithms were capable of generating valid routes in the tested grid environment. However, the two algorithms demonstrated different performance behaviours. A* achieved lower latency across all selected route cases, while JPS showed lower memory usage in selected medium-distance and long-distance route cases.

The findings suggest that JPS cannot be generally concluded to be faster than A* in a small campus grid environment. The additional processing involved in the recursive jump process and forced-neighbour checking may introduce overhead in the tested 100 × 100 grid. Nevertheless, the lower memory usage observed in some route cases suggests that JPS may still be useful when memory efficiency is prioritised, especially in larger or more detailed campus grid maps. The main contribution of this study is the practical performance comparison of JPS and A* for campus-scale grid-based pathfinding.

This study has several limitations. First, the campus map was manually represented as a grid, which may limit scalability when larger or more complex maps are used. Second, the evaluation was conducted using only six selected route scenarios, without repeated executions or statistical analysis. Third, the system assumes a static campus environment and does not consider real-time conditions such as route closures, construction, rain, or flooding.

Future work should extend the evaluation using larger grid sizes, additional randomised start–destination pairs, repeated experimental runs, and statistical measures such as mean and standard deviation. Further improvements may include automated map construction using OpenStreetMap or building plans, real-time route reporting, accessibility-aware routing for stairs and ramps, and code-level optimisation of the JPS implementation. This direction is supported by previous work on JPS code optimisation [20], corridor-based search and incremental line search [13], and path quality indicators such as turning points, path smoothness, and suitability for physical navigation [21].

References

1. Kapi, A.Y., Sunar, M.S., Algfoor, Z.A.: Summary of pathfinding in off-road environment. In: 2020 6th International Conference on Interactive Digital Media (ICIDM), pp. 1–4 (2020)
2. Abiodun, F.: Technical Report on: Navigation the University: Experience of Disorientation and Wayfinding of CEA and IMES Students at University of Cologne (2024)
3. Tahir, R., Krogstie, J.: Impact of navigation aid and spatial ability skills on wayfinding performance and workload in indoor-outdoor campus navigation: challenges and design. Appl. Sci. **13** (2023). https://doi.org/10.3390/app13179508
4. Dijkstra, E.W.: A note on two problems in connexion with graphs. Numer Math (Heidelb). **1**, 269–271 (1959). https://doi.org/10.1007/BF01386390
5. Hart, P.E., Nilsson, N.J., Raphael, B.: A formal basis for the heuristic determination of minimum cost paths. IEEE Trans. Syst. Sci. Cybern. **4**, 100–107 (1968). https://doi.org/10.1109/TSSC.1968.300136
6. Harabor, D., Grastien, A.: Online graph pruning for pathfinding on grid maps. In: Proceedings of the AAAI Conference on Artificial Intelligence, pp. 1114–1119 (2011)
7. Saradha, K.R., Sakthi, S., Dakshatha, et al.: Smart AR navigation: enhancing campus wayfinding with augmented reality. In: 2025 International Conference on Computing and Communication Technologies (ICCCT), pp. 1–7. IEEE (2025)
8. Alkady, Y., Rizk, R., Alsekait, D.M., et al.: SINS_AR: an efficient smart indoor navigation system based on augmented reality. IEEE Access. **12**, 109171–109183 (2024). https://doi.org/10.1109/ACCESS.2024.3439357
9. Chaitra, K., Lavanya, K.H., Amrutha, S., et al.: Campus navigation system using QR code and web technologies. Int. J. Creat. Res. Thoughts (IJCRT). **13**, c575–c582 (2025)
10. Nordin, N., Markom, M.A., Suhaimi, F.A., Ishak, S.: A web-based campus navigation system with mobile augmented reality intervention. J. Phys. Conf. Ser. **1997**, 012038 (2021). https://doi.org/10.1088/1742-6596/1997/1/012038
11. Pardede, S.L., Athallah, F.R., Huda, Y.N., Zain, F.D.: A review of pathfinding in game development. CEPAT: J. Comput. Eng.: Progr. Appl. Technol. **1**, 47 (2022). https://doi.org/10.25124/CEPAT.V1I01.4863
12. Lawande, S.R., Jasmine, G., Anbarasi, J., Izhar, L.I.: A systematic review and analysis of intelligence-based pathfinding algorithms in the field of video games. Appl. Sci. **12**, 5499 (2022). https://doi.org/10.3390/app12115499
13. Elshahed, A., Bin Majahar Ali, M.K., Mohamed, A.S.A., et al.: Efficient pathfinding on grid maps: comparative analysis of classical algorithms and incremental line search. IEEE Access. **13**, 98473–98484 (2025). https://doi.org/10.1109/ACCESS.2025.3575168
14. Carlson, M., Moghadam, S.K., Harabor, D.D., et al.: Optimal pathfinding on weighted grid maps. Proc. AAAI Conf. Artif. Intell. **37**, 12373–12380 (2023). https://doi.org/10.1609/aaai.v37i10.26458
15. Guo, Y., Tong, X., Huang, L., et al.: Research on AGV based on improved A-star algorithm. Sci. Rep. **16**, 3629 (2026). https://doi.org/10.1038/s41598-025-33653-9

16. Hu, S., Harabor, D.D., Gange, G., et al.: Jump point search with temporal obstacles. In: Proceedings of the International Conference on Automated Planning and Scheduling, pp. 184–191 (2021)
17. Min, J.G., Ruy, W.S., Park, C.S.: Faster pipe auto-routing using improved jump point search. Int. J. Nav. Archit. Ocean Eng. **12**, 596–604 (2020). https://doi.org/10.1016/j.ijnaoe.2020.07.004
18. Yang, Z., Li, J., Yang, L., et al.: Path planning and collision avoidance methods for distributed multi-robot systems in complex dynamic environments. Math. Biosci. Eng. **20**, 145–178 (2023). https://doi.org/10.3934/mbe.2023008
19. Chen, T., Chen, S., Zhang, K., et al.: A jump point search improved ant colony hybrid optimization algorithm for path planning of mobile robot. Int. J. Adv. Robot. Syst. **19** (2022). https://doi.org/10.1177/17298806221127953
20. Kapi, A.Y., Sunar, M.S., Algfoor, Z.A.: A comparison of pathfinding algorithm for code optimization on grid maps. Int. J. Adv. Comput. Sci. Appl. **13** (2022). https://doi.org/10.14569/IJACSA.2022.0131247
21. Wang, Y., Gao, J., Song, W., et al.: A path planning algorithm for mobile robots combining bidirectional scanning and sampling with virtual obstacles. Intell. Serv. Robot. **19**, 11 (2025). https://doi.org/10.1007/s11370-025-00658-2

Information Retrieval

CF-RAG: Cross-Farm Agricultural Decision Intelligence with Counterfactual Retrieval-Augmented Generation and Adaptive Knowledge Transfer

Xing Yunhao[1] and Nurfadhlina Mohd Sharef[1,2](✉)

[1] Faculty of Computer Science and Information Technology, Universiti Putra Malaysia, Serdang, Selangor, Malaysia
nurfadhlina@upm.edu.my

[2] Institute of Mathematical Research, Universiti Putra Malaysia, Serdang, Selangor, Malaysia

Abstract. The use of large language models (LLMs) for advisory purposes in agriculture is growing, but their usage for intervention decision-making is limited by hallucinations, weak numerical reasoning, and limited ability to adapt across farms with different environmental conditions. Works on retrieval-augmented generation (RAG) improve knowledge relevance, but did not much validate intervention recommendations suitability. This is vital in agriculture since inappropriate interventions may directly affect crop performance. This paper proposes CF-RAG as a cross-farm agricultural decision intelligence framework where counterfactual RAG is combined with adaptive knowledge transfer to provide safer advice on interventions. Source-domain agricultural records were transformed into reusable intervention cases through rule-based knowledge engineering, physiological simulation, and vector-based semantic indexing. We used a publicly available lettuce environment dataset as the source domain, and a Batavia lettuce hydroponic dataset as the target environment. Results show that the proposed framework achieved 86.7% intervention suitability with 100% safety compliance which outperform baseline approaches. The results illustrate the efficacy of incorporating retrieval grounding, adaptive knowledge transfer, and counterfactual validation into an agricultural application that can assist agricultural decision-making in cross-farm and data-limited settings.

Keywords: Agricultural decision intelligence · Retrieval-augmented generation · Counterfactual validation · Cross-farm knowledge transfer · Intervention recommendation · Large language models

1 Introduction

Artificial intelligence (AI) is increasingly influencing modern farming and agriculture applications, which include crop monitoring and disease detection, yield prediction, resource management, and operational decision support (Kamilaris & Prenafeta-Boldú, 2018; Talaviya et al., 2020; Benos et al., 2021). Most recently, large language models

N. Mohd Sharef et al. (Eds.): CAMP 2026, CCIS 3093, pp. 229–246, 2027.
https://doi.org/10.1007/978-981-92-4302-0_17

(LLMs) have emerged under a new era in agricultural advisory that enables systems to comprehend domain knowledge, communicate using natural language, and make context-specific recommendations.

New applications are pointing to a promising future. Recent research on row crop production found AI-generated decisions among other matters like planting, fertilisation, irrigation, insecticide, and cover crop management decisions to be logical, practical, operationally feasible, and faster than manual decision processes (Chamara et al., 2026). Also, technologies such as AgroLLM and current multimodal retrieval-based agriculture advisory platforms show us the possibility of implementing generative AI with agricultural knowledge support (Ravindran et al., 2026; Liu et al., 2026).

Generated recommendations provide agricultural recommendations that may directly impact a crop's physiology, performance, and use of resources so they must be biologically safe. However, the practical utility of LLMs in intervention decision-making is still limited (Ji et al., 2023). This is because LLMs normally generate answers from what they learned during training through internal learned patterns, associations and general prior knowledge. This left room for ungrounded generation because that knowledge may be incomplete, outdated, inaccurate and invented (hallucinated).

Retrieval-augmented generation (RAG) has taken off as a promising mechanism for enhancing grounding by augmenting generative reasoning with knowledge retrieval (Lewis et al., 2020). Retrieval-grounded solutions have improved the relevance of advisory outputs in agriculture, allowing recommendations to draw from both previous sources of information and domain knowledge repositories (Ravindran et al., 2026; Liu et al., 2026).

But traditional RAG is still mainly a retrieval and contextualisation mechanism, not a decision-making platform. Knowledge retrieval can enhance the informational relevance but does not specify whether the recommended intervention is likely to be a beneficial, safe, or operationally feasible solution under the current environmental state. As a consequence, the existing agricultural LLM systems are largely based on advisory or reasoning tools without operational decision-support mechanisms.

Another issue relates to the adaptability across farms. Farm systems differ greatly in terms of environmental conditions, cultivation practices and infrastructure, crop varieties, and operational constraints (Verdouw et al., 2021). Also, local historical data may not be sufficient for reliable farm-specific intelligent decision models for many farms, especially for smaller or newer farms. This results in practical incentive to cross-farm learning, that is learning in one context might inform the decision-making in another.

Transfer learning has been studied widely in agriculture to solve such data limitations, especially in crop classification, disease recognition, and yield prediction (Kamilaris & Prenafeta-Boldú, 2018; Zhang et al., 2025). However much of this work largely centres around the knowledge transfer in the context of a technical modelling approach for predictive adaptation. In fact, the decisions made in these interventions are more contextually sensitive. What appears to be most effective for one farm does not necessarily work for another, even within what may be similar production conditions. Consequently, the mismatch between the generalisation of models and the applicability of operational decisions remains a chronic one.

Counterfactual reasoning offers an interesting path as decisions regarding intervention in agriculture are inherently contingent on what-if exploration. It is standard practice of farmers and practitioners to assess implicit alternatives (e.g., modifying nutrient concentration, timing of irrigation or environmental conditions) which will usually lead to better crop outcomes. But operational decision support is about figuring out those alternatives before they occur.

The existing counterfactual applications in agriculture have generally been for explanation (Guidotti, 2024), retrospective review, or policy-thinking, rather than for immediately-in-time recommendation of an intervention. Thus, the combined utility of knowledge-grounded advisory generation and adaptive cross-farm learning with intervention-level counterfactual validation is relatively low. This limitation becomes more relevant where experimentation itself is not feasible, would be expensive or operationally risky.

In controlled-environment agriculture, an inappropriate intervention could weaken crop conditions and cause inappropriate resource consumption (Shamshiri et al., 2018b). Such repeated testing of alternative interventions on production crops is highly unlikely in a commercial context. More fundamentally, the outcomes of intervention are counterfactual. After one action has been undertaken, other unrealised outcomes cannot be observed directly under identical circumstances. This is precisely why simulation is a convenient and scientifically defensible tool for a systematic assessment of intervention in advance of applying it to the market. Simulation has become an increasingly practical mechanism for evaluating agricultural intervention scenarios when direct real-world experimentation is costly, disruptive, or operationally constrained (Verdouw et al., 2021).

Drawing upon these issues, we introduce CF-RAG: Cross-Farm Agricultural Decision Intelligence via Counterfactual Retrieval-Augmented Generation paired with Adaptive Knowledge Transfer. This framework turns source data in agriculture into a reusable decision knowledge base, recovers context-relevant past intervention analogues, and evaluates the generated recommendation via a counterfactual simulation mechanism before the final recommendation. In this work, adaptive knowledge transfer does not refer to parameter transfer or model adaptation, but the contextual reuse of intervention experiences from one agricultural environment to support recommendation generation in another.

The framework transforms source-domain environmental observations into a reusable intervention knowledge repository through rule-based knowledge mapping, where physiological constraints and simulated crop response models are used to enrich raw records with diagnostic and intervention-relevant contextual information. These structured cases are then embedded and stored in a vector database for retrieval during decision generation. The counterfactual reasoning is operationalised through simulation. When a recommendation is generated, the system evaluates the likely outcome if that intervention were to be applied under the current crop conditions. Because these alternative outcomes cannot be directly observed in reality without implementing the action, simulation provides a practical mechanism for intervention assessment. A simple recommendation refinement mechanism is included to allow unsuitable recommendations to be revised following validation feedback.

The paper offers three contributions. First, it presents an integrated agricultural decision intelligence framework that integrates retrieval grounding, adaptive knowledge transfer, and counterfactual intervention validation for operational recommendation support. Second, it provides a simulation-based strategy for the assessment of safety for intervention in situations in which continued real-world experimentation is impractical or operationally risky. Third, it demonstrates empirical performance using direct LLM generation and traditional RAG baselines in a cross-farm agriculture advisory ecosystem, showing enhanced utility and safety of decision.

The rest of this paper is structured as follows. Section 2 presents related studies involving agricultural retrieval-augmented advisory systems, counterfactual reasoning and adaptive cross-farm decision support systems. The proposed methodology is described in Sect. 3. Section 4 presents the findings and their discussion and in Sect. 5 conclude.

2 Related Work

2.1 Agricultural Advisory Systems Augmented with Retrieval

New opportunities in the field of agricultural advisory arise from access to interactive knowledge, contextual rationale and natural language-modeled decision support through large language models. Recent works have shown generative AI can enable practical agricultural reasoning, especially in advisory practices, where recommendations are based on domain knowledge and operational context.

For example, AI-generated decisions for planting, fertilisation, irrigation, insecticide application and cover crop management were considered practical, logical and operationally executable, with responses significantly faster than traditional manual decision processes as determined by a more recent case study using ChatGPT applied to row crop production (Chamara et al., 2026). Likewise, AgroLLM showed that it is possible to bridge farmers with the agricultural best practices through knowledge-based large language model interactions intended to enhance accessibility and knowledge transfer (Ravindran et al., 2026).

The integration of RAG has taken up this trend further, more grounding is achieved through external knowledge retrieval instead of internal model memory (Lewis et al., 2020). As a result, this method is beginning to be implemented in agriculture-specific implementations. (Liu et al., 2026) for instance introduced a multimodal system to answer agricultural questions in a multimodal approach based on retrieval systems that combined retrieval strategies with structured reasoning elements to enhance its domain awareness and quality of response. These findings imply more mature status of agricultural knowledge-grounded agri-analysis based advisory systems. Still, the vast majority of the existing systems are mainly advisory in nature. Instead, their aim is not to validate whether an intervention is operationally feasible under the system's current environmental situation, but to collect information, to elucidate decisions made or to facilitate access to knowledge. The contextual relevance on itself is inadequate in intervention-sensitive fields, such as agriculture. The recommendation may be likely to be agronomically practicable in broad but untenable in particular with respect to the physiological environment of the current crop. This reveals a significant limitation of the current agricultural LLM advisory systems.

2.2 Transfer of Knowledge from Cross-Farm Adaptation

Agricultural environments are extremely heterogeneous and vary widely in climate, cultivation, operating practices and infrastructures, and availability of resources (Verdouw et al., 2021). Therefore, intelligent systems implemented on one farm environment might be less effective at another one. The development of farm-specific models for this challenges become more crucial in instances where there are few local historical data available.

Transfer learning has been utilized extensively in agriculture to alleviate data scarcity by transferring learning from one domain to the other by learning from a single context for specific performance characteristics. Applications can be found in crop disease recognition, crop classification, crop yield prediction, and remote sensing analysis (Kamilaris & Prenafeta-Boldú, 2018). Recent attempts also showed advanced applications in prediction of crop yield using cross-scale transfer learning, and the value of knowledge reuse within agricultural contexts have not only been appreciated but also proven (Zhang et al., 2025).

But even with such advancements, the majority of transfer learning approaches still tend towards prediction-driven approaches. The focus tends to be on bettering the accuracy of models using feature adaptation, representation transfer or model fine-tuning. Although helpful, such efforts are not directly directed at the decision making on intervention based on their effectiveness, as operation of the approach not only relies on prognostic generalisation but context also. Agricultural decisions are context-dependent and those interventions that are effective in one context may not always be transferable elsewhere. Current technical transfer systems consequently are still insufficient to solve the practical issue of adaptive decision support on entire farms. This indicates that forward transfer from prediction transfer to adaptive decision intelligence entails an approach where past experience can be used for recycling, being re-interpreted and contextually adopted, rather than being transferred in a fixed approach.

2.3 Counterfactual Reasoning for Decision Support

Counterfactual reasoning has become increasingly relevant in AI decision-support because it enables what-if evaluation of alternative actions rather than merely predicting likely outcomes. Unlike conventional predictive models, which estimate what is likely to happen based on observed patterns, counterfactual reasoning supports reasoning about hypothetical interventions and their possible consequences, making it particularly useful for decision-making under uncertainty (Guidotti, 2022; Pearl, 2009). This is highly relevant in agriculture, where many operational decisions are inherently intervention-oriented. Questions such as whether increasing nutrient concentration, adjusting irrigation timing, or modifying environmental conditions would improve crop performance require evaluation of alternative actions rather than prediction alone.

However, counterfactual reasoning in agriculture has been more focused on interpretation and intervention effect analysis than real-time operational decision support

(Guidotti, 2024; Tsoumas et al., 2025). Some adaptive inference approaches have been explored more broadly in machine learning, but iterative refinement of agricultural intervention recommendations remains underexplored. This gap is important because unsuitable interventions can directly affect crop physiology, productivity, and resource efficiency (Shamshiri et al., 2018a, b). In practical agricultural settings, repeatedly testing multiple intervention alternatives on production crops is often infeasible due to cost, operational disruption, and biological risk. More fundamentally, intervention outcomes are counterfactual by nature: once a particular action is taken, the outcomes of alternative unrealised actions cannot be directly observed under the same conditions. Therefore, simulation provides a practical mechanism for evaluating intervention suitability before implementation, particularly in agricultural settings where direct experimentation may be costly, operationally disruptive, or risky to crop performance (Shamshiri et al., 2018b; Verdouw et al., 2021).

Table 1 Comparison of existing works and the proposed solution

Capability	AgroLLM (Ravindran et al., 2026)	Agricultural RAG Systems (Liu et al., 2026)	CF-RAG
Knowledge-grounded recommendation	✓	✓	✓
Retrieval-augmented generation	✓	✓	✓
Cross-farm knowledge reuse	✗	✗	✓
Intervention-level validation	✗	✗	✓
Counterfactual reasoning	✗	Limited	✓
Physiological safety assessment	✗	✗	✓
Recommendation refinement	✗	✗	✓

Table 1 highlights that current agricultural advisory systems are mainly concerned with knowledge access and recommendation generation. Although retrieval mechanisms help to gain contextual grounding, they do not explicitly consider whether a recommended intervention is safe or likely to improve the current crop condition. CF-RAG extends retrieval-based advice to include cross-farm knowledge re-use, physiological safety assessment, and counterfactual intervention validation prior to recommendation deployment.

3 Methodology

3.1 Framework Overview

CF-RAG is a cross-farm agricultural decision intelligence framework which combines retrieval-augmented generation, adaptive knowledge transfer, counterfactual validation, and iterative refinement to facilitate the recommendation of interventions. The framework aims to assist in agricultural decision-making where farm-specific historical information is scarce and where unconstrained language modelling creates unsuitable recommendations. In contrast to typical retrieval-augmented advisory systems which mainly retrieve relevant information to be used for response generation, CF-RAG offers an explicit intervention validation layer. Candidate recommendations are not taken solely with reference to semantic relevance, but are evaluated using projected intervention outcomes before final recommendation. There are five key aspects of the framework:

(i) building source-domain knowledge,
(ii) contextual retrieval from the source,
(iii) recommendation generation,
(iv) counterfactual intervention validation,
(v) recommendation refinement based on validation feedback.

The workflow, in summary, starts with processing traditional source-domain agricultural experiences into decision knowledge that can be reused. The system collects relevant prior intervention analogues when environmental conditions for the target farm are presented, proposes an initial recommendation via an LLM, uses counterfactual simulations to evaluate projected outcomes, and subsequently chooses to accept or refine the recommendation in light of validation results. Adaptive knowledge transfer in the current study is defined as the contextual reappraisal and reinterpretation of source-domain intervention experiences as they inform and enable decision-making in a new farm context, and is not a direct transfer of model parameters or predictive representations.

3.2 Source and Target Dataset

To evaluate cross-farm decision adaptation, two independent datasets representing different agricultural contexts are used in the present study. The source domain is the publicly available lettuce environmental dataset from Kaggle that contains historical crop growth observations and environmental data suitable for intervention-oriented knowledge extraction. The dataset comprises approximately 3,000 records from 70 lettuce plants and covers daily environmental measurements of temperature, humidity, total dissolved solids (TDS), pH, and crop growth progression. The dataset provides high quality, high-resolution information which can easily be analyzed under the conditions of data science. This data is the actual experiential source of knowledge used to derive prior patterns of intervention-response.

The target domain is based on a Batavia lettuce hydroponic dataset that was gathered under controlled-environment agriculture for hydroponic conditions. Compared with this source dataset, the target environment is more organized and farm-specific. Relevant overlapping environmental data attributes between both datasets were matched to enable knowledge recycling and reuse among the different domains. Variables not present consistently within datasets were removed for comparability.

The usage of source and target datasets represents a natural cross-farm adaptation scenario, in which the historical experiences of one environment for decision making is applied to the decision making of a setting without local decision history. For example, a source-domain record with temperature = 27.5 °C and humidity = 50% would be diagnosed as heat stress, assigned the intervention COOLING, and stored as a reusable intervention case. During retrieval, similar target farm conditions can therefore reuse previously successful intervention experiences even when originating from different farms.

3.3 Source-Domain Knowledge Construction

CF-RAG builds upon a reusable decision knowledge repository by converting source-domain agricultural observations in the first stage. For effective retrieval-based decision support, raw environmental observations alone cannot be enough, as they do not explicitly reflect intervention implications. As a result, source records were transformed into structured knowledge examples that represent environmental conditions, intervention situations, and expected patterns of crop response. This reframes the source dataset as a decision-based knowledge pool instead of an observational dataset in the traditional sense. The information instances were encoded in vector representations using Sentence-BERT and indexed in ChromaDB for semantic similarity retrieval during inference. It helps the system to retrieve previous cases which are contextually similar to the target farm condition as observed now.

Source-domain environmental records were transformed into structured intervention cases using a rule-based knowledge engineering process. Gaussian response functions were first used to estimate crop response to deviations from optimal environmental conditions, while Liebig's Law of the Minimum was applied to identify the dominant limiting factor affecting crop performance. Each record was subsequently enriched with diagnostic information, inferred intervention categories, expected crop response, and textual case descriptions. The resulting intervention cases were stored as reusable knowledge instances, encoded with Sentence-BERT embeddings, and indexed in ChromaDB for semantic retrieval.

3.3.1 Cross-Farm Rules Knowledge Engineering

The source dataset contained approximately 3,000 environmental observations but lacked explicit intervention records and daily crop growth measurements. Each source-domain record was subsequently transformed into a structured intervention case consisting of environmental parameters, growth stage information, limiting factor diagnosis, recommended intervention, simulated crop response and context description for retrieval. Agricultural experiences learned from the source domain is converted into a reusable knowledge base with this flow: Source Farm Records → Environmental Parameters → Gaussian Suitability Scores → Liebig Limiting Factor Detection → Diagnosis

Generation → Intervention Assignment → Textual Case Construction → Sentence-BERT Embedding → ChromaDB Knowledge Repository. This strategy enables experiential knowledge reuse when transfer is not treated as an adaptation problem for the model, when the experience of the intervention from the source domain becomes context referenced in decision making for the target recommendation generation.

The optimal setpoint μ represents the centre of the recommended operating range for Batavia lettuce, while σ controls the sensitivity of crop response to deviations from the optimum. Larger σ values indicate higher tolerance to environmental variation, whereas smaller values indicate greater sensitivity.

To transform these observations into reusable intervention knowledge, a rule-based knowledge engineering process was developed. First, environmental observations were evaluated against the optimal physiological requirements of Batavia lettuce. For each environmental parameter, Gaussian response functions were used to estimate crop suitability relative to the optimal operating condition (Table 2). Individual suitability scores were then combined using Liebig's Law of the Minimum, where the most limiting environmental factor determined the overall crop response. Thresholds were derived from recommended Batavia lettuce physiological operating ranges reported in controlled-environment agriculture studies (Shamshiri et al., 2018a, b). Based on the identified limiting factor, diagnostic labels and intervention recommendations were automatically assigned. The resulting cases formed the intervention knowledge repository used by the retrieval module.

Table 2 Rule-Based Mapping between Environmental Conditions and Interventions

Condition	Diagnosis	Recommended Intervention
Temperature > 22 °C	Heat stress	COOLING
Temperature < 18 °C	Low temperature stress	HEATING
Humidity <60%	Water deficit risk	HUMIDITY_UP
Humidity >80%	Excess humidity risk	HUMIDITY_DOWN
pH < 5.5	Acidic nutrient solution	pH_UP
pH > 6.5	Alkaline nutrient solution	pH_DOWN
TDS < 800 ppm	Nutrient deficiency	TDS_UP
TDS > 1200 ppm	Nutrient excess	TDS_DOWN
All parameters within range	Optimal condition	NONE

3.4 Retrieval-Augmented Recommendation Generation

Upon displaying a target farm state, CF-RAG first constructs a representation for a query using current environmental parameters (temperature, pH, TDS, and humidity). Such a query is sent to the vector repository, which retrieves the top-k most relevant instances of prior knowledge using semantic similarity. Here, however, k = 3 yielded enough contextual grounding and also kept the retrieval noise to a minimum. The retrieved intervention analogues are added to the prompt context provided to the large language model.

Sentence-BERT embeddings were leveraged to encode the source-domain intervention cases as well as the target farm queries. Retrieval was achieved using cosine similarity from the ChromaDB vector repository. Early experiments suggested that a retrieval size of $k = 3$ was sufficient to ground knowledge with context but also reduced irrelevant retrieval noise. Hence, the recommendation generation module was fed the three most similar intervention cases.

Gemini 2.5 Flash is used as the recommendation generation engine. Unlike direct LLM prompting, which is wholly done based on knowledge of internal models that the system has already trained on, retrieval grounding allows the system to generate recommendations based on agricultural experiences relevant to the current operational environment. Nonetheless, retrieval grounding does not provide enough evidence to warrant intervention suitability. CF-RAG introduces the explicit counterfactual validation layer for this reason.

Table 3 Prompt Structure Used for Recommendation Generation

Component	Description
Current Target State	Temperature, humidity, pH, TDS
Retrieved Cases	Top-3 intervention analogues
Expert Constraints	Batavia lettuce physiological ranges
Task Instruction	Generate intervention recommendation

The recommendation prompt (Table 3) was composed of the present target farm condition, intervention analogues retrieved, physiological operating constraints (optimal range according to (Shamshiri et al., 2018a)), and an instruction requesting an actionable intervention recommendation. This structure guaranteed that generated outputs are anchored in both historical intervention experience and crop-specific operational constraints. Table 4 shows the optimal environmental parameters used in counterfactual validation.

Table 4 Optimal Environmental Parameters Used in Counterfactual Validation

Variable	Symbol (μ)	Optimal Range	Selected Optimal Value
Temperature (°C)	μtemp	18–22 °C	20 °C
pH	μpH	5.5–6.5	6.0
TDS (ppm)	μtds	800–1200	1000
Humidity (%)	μhum	60–80%	70%

3.5 Counterfactual Intervention Validation

CF-RAG's key differentiating element is the counterfactual intervention validation module. After the initial recommendation is created, the system simulates the environmental impact of the proposed intervention before recommendation acceptance. This represents the in vivo nature of agricultural decision-making, as intervention decisions are inherently anchored around what-if thinking.

A Gaussian response-based simulation model was applied as an instrumental surrogate system to estimate crop response of the intervention scenarios. This simulation is not intended to reproduce a mechanistic crop growth model by nature, but to perform a computationally feasible assessment of the efficiency of an intervention by modifying the environmental response assumed on data from the target environment.

The validation stage assesses whether the predicted post-intervention environmental condition is in accordance with predetermined physiological safety limits for Batavia lettuce. Any recommendations outside these restriction parameters are considered unsafe. In addition to validating their safety, the system performs a projected usefulness assessment of the intervention to assess whether recommendations are safe and operationally meaningful. This mechanism of counterfactual validity transforms recommendation generation from a purely generative procedure into an assessment process.

The counterfactual validation module approximates the crop response, R under different intervention scenarios using Gaussian response functions. For environmental variable x, it is computed as (Eq. 1):

$$R(\mathrm{x}) = \exp\left(-\frac{(x-\mu)^2}{(2\sigma)^2}\right) \tag{1}$$

$$\sigma = \frac{Upper - Lower}{2} \tag{2}$$

where μ is the optimal environmental value and σ represents the acceptable tolerance around that optimum. Higher response values correspond to the environment being closer to the preferred physiological range.

The intervention suitability assessment compares the projected crop condition before and after the recommended intervention. Following Liebig's Law of the Minimum, the overall suitability score is determined by the most limiting environmental factor. To evaluate a recommended intervention, the suitability score was computed before and after applying the simulated intervention. An intervention was considered suitable when the projected post-intervention suitability score exceeded the original suitability score, indicating an improvement in the crop's physiological condition. Otherwise, the intervention was classified as unsuitable and subjected to recommendation refinement.

$$S_{pre} = \min\left(R_{temp}, R_{hum}, R_{pH}, R_{tds}\right) \tag{3}$$

$$S_{post} = \min\left(R'_{temp}, R'_{hum}, R'_{pH}, R'_{tds}\right) \tag{4}$$

$$Intervention\ Suitability = \begin{cases} 1 & S_{post} > S_{pre} \\ 0 & otherwise \end{cases} \tag{5}$$

3.6 Recommendation Refinement Mechanism

To improve operational robustness, CF-RAG includes a recommendation refinement mechanism that revises unsuitable recommendations based on validation feedback. If a generated recommendation fails safety compliance or is assessed as unsuitable under the projected crop condition, the system does not immediately return the output. Instead, simulation feedback is translated into corrective prompt guidance and returned to the recommendation generation process. This creates a simple iterative refinement loop that allows weaker recommendations to be revised before final acceptance.

This mechanism helps reduce the effect of imperfect historical retrieval matches or generative reasoning variability, thereby improving the practical reliability of the decision-support process without requiring model retraining.

3.7 Experimental Design and Evaluation

The CF-RAG Framework was evaluated using a scenario-based experimental methodology based on the Batavia lettuce target dataset which contained 30 daily environmental observations. From these, 15 representative intervention scenarios were selected to cover diverse operational crop conditions, including nutrient imbalance, pH stress, humidity stress, temperature-related stress, and mixed-condition cases. Selection focused on scenarios requiring meaningful intervention assessment, while avoiding repetitive near-identical conditions. For each scenario, an intervention recommendation had to correspond to a specific crop state. Three configurations were evaluated:

- Baseline 1: Direct LLM
  Gemini 2.5 Flash created intervention recommendations directly from the target crop condition without reference to retrieval grounding, expert restrictions, or evaluation of intervention validation.
- Baseline 2: Traditional RAG.
  Recommendations were based upon retrieval-grounded contextual support from the source-domain knowledge repository that contains historical intervention-response cases, without intervention validation.
- Proposed CF-RAG.
  A fully integrated framework combined retrieval grounding, expert physiological restrictions, and counterfactual simulation-based validation of the interventions before recommending them.

In all three cases, recommendations were generated independently for each scenario, leading to 45 evaluated recommendation instances for the comparative experiment. The quality of recommendations was assessed according to the three criteria:

(i) Safety Compliance, which assesses whether generated recommendations remain within predefined physiological operating boundaries for Batavia lettuce. Unsafe recommendations included interventions that would move crop conditions into harmful ranges, such as excessive nutrient concentration, unsuitable pH, or dangerous temperature levels. The score was calculated as:

$$\textit{Safety Compliance} = \frac{\#\textit{Safe Recommendation}}{\textit{Total Recommendation}} \times 100 \tag{6}$$

(ii) Intervention Suitability, which assesses whether the projected post-intervention condition indicates that the recommendation is appropriate for improving the observed crop state. This was evaluated through the counterfactual simulation module, which projected the likely environmental outcome if the recommended intervention were applied. A recommendation was considered suitable if the projected condition demonstrated measurable improvement over the original crop state. The score was calculated as:

$$Intervention\ Suitability = \frac{\#Suitable\ Intervention}{Total\ Recommendation} \times 100 \tag{7}$$

(iii) Contextual Alignment, which assessed whether the generated recommendation appropriately reflected retrieved agricultural knowledge and the surrounding environmental context, rather than unsupported generative outputs. This was particularly relevant for retrieval-based configurations. Contextual alignment was manually assessed by verifying whether the recommendation was consistent with the retrieved intervention cases and associated environmental conditions. The score was calculated as:

$$Contextual\ alignment = \frac{\#Contextually\ Aligned\ Recommendation}{Total\ Recommendation} \times 100 \tag{8}$$

Two experiments were conducted. The first was a comparative performance evaluation across the three complete configurations to assess overall recommendation quality. The second was a component-level analysis to examine how expert rule integration and counterfactual validation contributed to performance improvement within the proposed framework.

To clarify the workflow, a representative intervention scenario from the evaluation set is shown in Table 5.

Table 5 Example intervention scenario

Scenario Element	Description
Target Crop Condition	Temperature = 24.6 °C; Humidity = 68.2%; TDS = 812 ppm; pH = 5.2 (acidic condition requiring corrective intervention)
Direct LLM Recommendation	*"Increase nutrient concentration to improve crop growth."*
Assessment	The recommendation appeared plausible but failed to address the actual limiting factor, which was the acidic nutrient solution. Increasing TDS alone would not correct the pH imbalance and could worsen the crop condition.
Traditional RAG Recommendation	*"Current environmental conditions appear acceptable. Continue monitoring."*

(*continued*)

Table 5 (*continued*)

Scenario Element	Description
Assessment	Retrieval grounding improved contextual relevance by referencing prior cases, but the recommendation still failed to identify the immediate intervention required for pH correction.
CF-RAG Recommendation	*"Apply pH correction (PH_UP) to restore the nutrient solution toward the acceptable physiological range."*
Counterfactual Validation Outcome	The simulation projected improved crop suitability under the corrected pH condition, and the recommendation was accepted.
Interpretation	This example shows that retrieval grounding alone may still produce contextually plausible but operationally unsuitable recommendations. Counterfactual validation provides an additional safeguard by assessing likely intervention suitability before implementation.

4 Results and Discussion

There are two experiments that have been conducted. Section 4.1 presents the results from comparing the whole systems while Sect. 4.2 progressively adds design components to isolate what causes the performance gains.

4.1 Comparative Performance of Recommendation Approaches

The comparative evaluation examined whether the proposed CF-RAG framework improves intervention recommendation quality compared with conventional generative advisory approaches. Table 6 summarises the comparative results across the three evaluated configurations.

Table 6 Comparative performance in safety, intervention suitability and contextual consistency

Method	Safety (%)	Intervention suitability (%)	Contextual Consistency
Direct LLM	46.7	40.0	Low
Conventional RAG	73.3	66.7	Moderate
CF-RAG	100.0	86.7	High

Figure 1 highlight significant discrepancies among the tested configurations. In a few scenarios, direct LLM generation generated recommendations that were logically plausible at surface level, but discrepancies arose when comparing recommendations to the physiological operating requirements of the crop. This indicates the known limitation of unconstrained large language models wherein such answers can seem coherent but not strongly based on operational decision-making.

The retrieval grounding using traditional RAG increased contextual relevance because the recommendation engine could tap into prior agricultural knowledge instead of having to draw solely from its internal language model associations. Even with that, retrieval support alone proved still inadequate in determining the suitability of the intervention. This enabled recommendations to still be contextual while suggesting actions that had not been operationally suitable under the current crop condition.

The CF-RAG framework achieved the best performance, with an acceptable safety compliance level of 100% and intervention suitability at 86.7%. These data indicate that retrieval grounding is significantly facilitated when combined with explicit intervention validation. Instead of considering recommendation generation as the final result, CF-RAG adds another layer of assessment, which determines whether a proposed intervention is most likely to be safe and beneficial before being adopted.

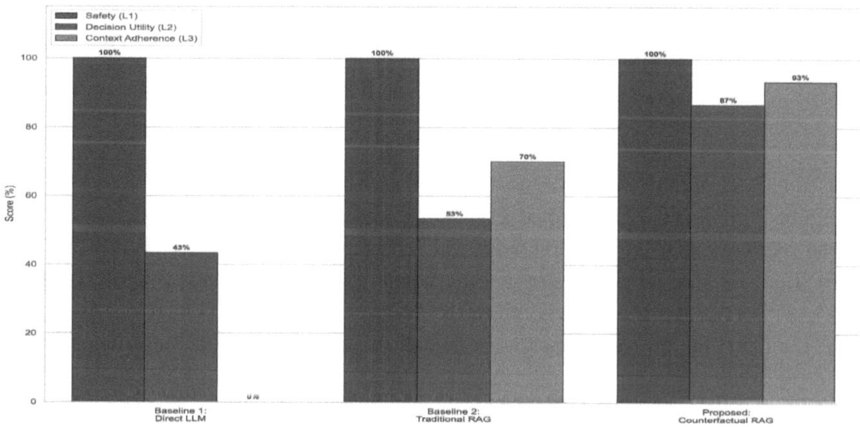

Fig. 1 Comparative performance of Direct LLM, Traditional RAG, and CF-RAG across safety (red bar), intervention suitability (blue bar), and contextual adherence (green bar).

The improvement also indicates that recommendation quality in agricultural decision support should not be judged solely by linguistic plausibility or informational relevance. Operational usefulness requires explicit assessment of intervention consequences.

4.2 Component Contribution Analysis Experiment

While Sect. 4.1 compared the overall performance of complete recommendation approaches, this section addressed whether major components of CF-RAG contributed to performance improvement overall. The 15 intervention cases used in the comparison experiments were replicated using three configurations:

- Traditional RAG based on retrieval-grounded recommendation generation without explicit intervention constraints or validation;

- RAG + Expert Rules, in which physiological operating constraints on Batavia lettuce were given to guide recommendation generation;
- CF-RAG with full support, and with validation of counterfactual interventions included.

The results as shown in Table 7 indicate that each new component increased the quality of recommendation. Introducing expert physiological constraints increased Intervention Suitability for this study from 53.3% to 73.3%, signaling that explicit knowledge on agricultural operating mechanisms reduced inappropriate advice. The inclusion of counterfactual validation brought suitability of interventions to 86.7%. It has been suggested that retrieval grounding alone can potentially increase contextual relevance, but further intervention validation offers additional protection from contextually plausible but operationally weak indications.

Table 7 Component analysis experiments result

Configuration	Intervention Suitability (%)
Traditional RAG	53.3
RAG + Expert Rules	73.3
CF-RAG	86.7

Figure 2 indicate that retrieval grounding alone is insufficient for reliable intervention recommendation. While contextual retrieval improves access to relevant prior information, substantial gains emerge only when domain knowledge is explicitly structured and intervention recommendations are subjected to counterfactual validation. The results confirm that the counterfactual verification mechanism contributes meaningfully to filtering weak decisions and improving the practical reliability of the framework.

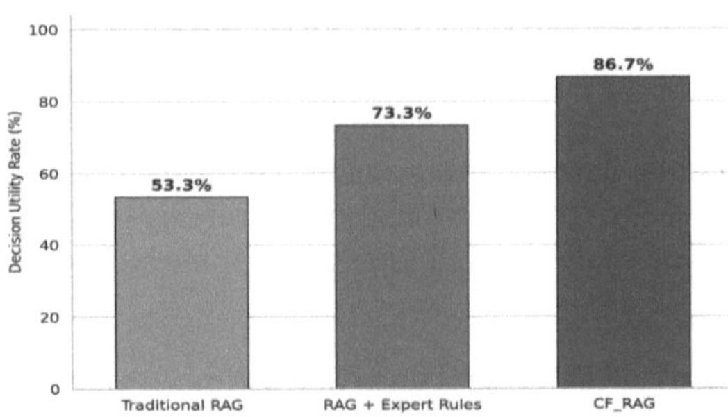

Fig. 2 Component analysis results using intervention suitability

The component analysis helps understand the influence of each framework element in context with respect to contribution. When expert physiological constraints were implemented, the suitability of intervention increased by 20 percentage points, from 53.3% to 73.3%. Indeed, explicit agricultural operating knowledge appears to reduce the contextually viable but agronomically inappropriate recommendations. Counterfactual validation added another 13.4 percentage points increase in the increase of appropriate application from 73.3% to 86.7%. These results suggest that retrieval grounding alone enhances contextual significance, and that subsequent validation of the intervention would afford an extra protection against inappropriate operational recommendations.

5 Conclusion

This paper presents a cross-farm agricultural model designed to increase the reliability of agricultural intervention recommendation in cross-farm settings. Although there are recent developments in large language models and retrieval-based advisory systems as agricultural knowledge support, there are still some challenges to use these approaches for the decision making of strategic intervention to make them practicable due to the weak contextual adjustment, low degree of quality control, and lack of explicit mechanism to judge if the proposed intervention is expected to be useful prior to implementation.

In order to overcome these shortcomings, the present work provides three contributions:

(i) holistic cross-farm decision intelligence framework integrated in the analysis
(ii) validation of counterfactual interventions by simulation
(iii) empirical evaluation against baselines.

The proposed work implemented rule-based knowledge transformation followed by vector-based semantic retrieval. Experimental results showed clear improvement over baseline methods. The proposed CF-RAG achieved 86.7% intervention suitability, compared to 43.3% for Direct LLM and 53.3% for Traditional RAG. Component analysis additionally demonstrated that expert knowledge integration significantly elevated intervention suitability, at the same time counterfactual validation enhanced intervention reliability and diminished ineffective recommendations. This highlights the need for agricultural decision intelligence which extends beyond context-aware recommendation generation since reliable intervention support will have to explicitly assess the likely impact of an intervention before action is taken.

This study is still exploratory, with limitations across several aspects. The evaluation was performed in a controlled hydroponic lettuce environment with a simplified simulation mechanism and more generalisability would be enhanced through broader validation across crop types, production environments, and richer agronomic simulation models. A total of 15 representative intervention scenarios from 30 target environmental observations were used to carry out the current evaluation. Since the study was exploratory and only a few evaluation scenarios were performed, formal statistical significance testing was not done. While adequate for proof-of-concept evaluation, broader, more varied datasets in more diverse crops, environments, and production scenarios would provide more robust evidence of generalizability.

Subsequent work may scale the framework to support adaptive agricultural advisory systems in real-time, including the adaptation to diverse farming environments at a larger operational scale. Statistical validation of observed performance differences will be further validated through more rigorous analyses of the observed performance changes with larger datasets in future studies. Additional effort should also be devoted to comparing CF-RAG against emerging agricultural retrieval-enhanced advisory systems and multimodal agricultural foundation models to further establish comparative effectiveness across diverse agricultural decision-support settings in future work.

References

Chamara, N., et al.: Can generative AI make farming decisions? Current status and future pathways – a case study in row crop production with ChatGPT. Artif. Intell. Agric. **16**(2), 837–853 (2026). https://doi.org/10.1016/j.aiia.2026.03.008

Guidotti, R.: Counterfactual explanations and algorithmic recourse for machine learning: a review. ACM Comput. Surv. **55**(5), 1–37 (2022). https://doi.org/10.1145/3524056

Guidotti, R.: Counterfactual explanations and how to find them: literature review and benchmarking. Data Min. Knowl. Disc. **38**(5), 2770–2824 (2024). https://doi.org/10.1007/s10618-022-00831-6

Kamilaris, A., Prenafeta-Boldú, F.X.: Deep learning in agriculture: a survey. Comput. Electron. Agric. **147**, 70–90 (2018)

Lewis, P., et al.: Retrieval-augmented generation for knowledge-intensive NLP tasks. Adv. Neural Inf. Proces. Syst. **33**, 9459–9474 (2020) https://proceedings.neurips.cc/paper/2020/hash/6b493230205f780e1bc26945df7481e5-Abstract.html

Liu, B. et al.: An intelligent multi-modal Q&A system for agriculture combining APGM, PBTCS, and RAG. Smart Agric. Technol. **13**, 101829 (2026). https://doi.org/10.1016/j.atech.2026.101829

Pearl, J.: Causality: Models, Reasoning, and Inference, 2nd edn. Cambridge University Press (2009)

Ravindran, D.J.S., Skarga-Bandurova, I., V, S., Awais, M., S, M.: AgroLLM: connecting farmers and agricultural practices through large language models for enhanced knowledge transfer and practical application. AgriEngineering. **8**(1) (2026). https://doi.org/10.3390/agriengineering8010038

Shamshiri, R., et al.: Advances in greenhouse automation and controlled environment agriculture: a transition to plant factories and urban agriculture. Int. J. Agric. Biol. Eng. **11**(1), 1–22 (2018b) https://ijabe.org/index.php/ijabe/article/view/3210

Shamshiri, R.R., Jones, J.W., Thorp, K.R., et al.: Review of optimal temperature, humidity, and vapour pressure deficit for microclimate evaluation and control in greenhouse cultivation of tomato. Int. Agrophys. **32**(2), 287–302 (2018a). https://doi.org/10.1515/intag-2017-0005

Tsoumas, I., et al.: Leveraging causality and explainability in digital agriculture. Environ. Data Sci. **4**, e23 (2025). https://doi.org/10.1017/eds.2025.14

Verdouw, C.N., Tekinerdogan, B., Beulens, A., Wolfert, S.: Digital twins in smart farming. Agric. Syst. **189**, 103046 (2021). https://doi.org/10.1016/j.agsy.2020.103046

Zhang, J., et al.: Transfer learning for improved crop yield predictions in a cross-scale pathway: a case study for Brazilian national soybean. Int. J. Appl. Earth Obs. Geoinf. **145**, 104981 (2025). https://doi.org/10.1016/j.jag.2025.104981

Improving Factual Grounding in Text-Image Plant Disease Diagnosis Using Multimodal Graph-Enhanced RAG

Nurul Sharefa Haneem Shahri and Nurul Amelina Nasharuddin(✉)

Faculty of Computer Science and Information Technology, Universiti Putra Malaysia, Serdang, Malaysia
gs75334@student.upm.edu.my, nurulamelina@upm.edu.my

Abstract. Deep learning, vision-language models, and most recently retrieval-augmented generation (RAG) has significantly improved on plant disease diagnosis. On the other hand, practical diagnosis is not only a disease label. Farmers or agricultural experts often need to be aware of the reasons behind a suspected disease, the symptom evidence supports it, the potential severity of the condition and the treatment that is appropriate for the particular crop. However, a lot of RAG systems still depend on flat passage retrieval. Which may overlook the connections between crop, symptom, disease, pathogen, severity and treatment. To overcome this problem, graph-enhanced retrieval is promising as it can return linked evidence rather than discrete text chucks. Twenty studies on plant disease detection, RAG, hallucination, GraphRAG, multimodal and evaluation on RAG are examined for this review. A PRISMA review approach was used to guide paper identification, screening, and selection, followed by thematic analysis across the selected studies. According to the review, RAG aids in explanation and treatment recommendation, while current image-based models are helpful for visual recognition. But the literature also demonstrated that these fields are still not closely related. While multimodal plant diagnostic studies sometimes concentrate on categorisation without strong external knowledge grounding, the majority of agriculture GraphRAG studies continue to be text-based. Thus, the primary gap found in this review is the limited integration of multimodal plant disease diagnosis, graph-enhanced retrieval, factual grounding and hallucination evaluation in a single cohesive approach. Its main contributions are to bring these areas together and highlight the need for more connected research direction.

Keywords: Plant Disease Diagnosis · Retrieval-Augmented Generation · Graph-Enhanced Retrieval

1 Introduction

Up until today, the major problem in the agriculture world is plant diseases and pests because they can reduce yield, lower crop quality, and lead to increase the cost of crop protection. Identifying the disease from image is not the only challenge. In real field

N. Mohd Sharef et al. (Eds.): CAMP 2026, CCIS 3093, pp. 247–262, 2027.
https://doi.org/10.1007/978-981-92-4302-0_18

conditions, symptoms that look similar such as leaf spots or yellowing may come from different causes, including pest damage, nutrient imbalance, pathogens, or environmental stress. Recent studies show that deep learning has improved plant disease detection, but a number of issues persist in the literature, particularly real-field variability, dataset bias, lighting changes, background noise, occlusion and poor generalization outside controlled dataset [1–3].

For numerous years, classification has been the main focus of plant disease artificial intelligence (AI). This is because the model receives a plant image and predicts a disease class. Although it is a limited form of diagnosis, this is still useful. In practice, a user would tend to follow up with inquiries. For example, whether the crop species change the interpretation, another stress element could be the cause of the symptom or what treatment is suitable. These questions cannot be addressed by disease labels alone. As proof by Ding et al. [4], their study shifts to discussion from simple recognition to simple prescription suggestion. It shows that diagnosing plant disease should help with management choices, not only identifying the disease.

Large Language Models (LLMs) is the program that can explain results in natural language, making them useful when it comes to developing agricultural AI systems. However, generated answers that seem fluent are not always a correct one. LLMs may produce hallucinations, where the response sounds confident but there is no available evidence and make it unsupported, fabricated, or not consistent [5, 6]. In plant disease diagnosis, this becomes risky because an incorrect recommendation can delay disease control, waste chemicals, or lead to unsuitable treatment. For this reason, factual grounding becomes central. In this review, factual grounding means that the generated diagnosis and recommendation are supported by reliable plant disease evidence, including crop type, visible symptoms, likely disease or pest, pathogen information, severity, and treatment knowledge. For example, Crop GraphRAG [7] shows that retrieving connected crop disease and pest knowledge can improve answer traceability, which is important for ensuring that a diagnosis is supported by evidence. In this review, a grounded answer is not only one that sounds correct, but one that can be linked back to relevant disease symptoms, causes, severity, and treatment information.

RAG is one way to improve grounding because it lets the model retrieve external knowledge before answering [8, 9]. However, standard RAG often retrieves text passages based on lexical or semantic similarity and this approach is not always enough for diagnosis. This is because a plant disease answer may require several connected pieces of information, and the connections are sometimes more important than the individual sentence that was retrieved. This is where graph-enhanced retrieval becomes relevant. Crop GraphRAG [7], the main anchor paper in this review, already shows that graph-based retrieval can improve crop disease and pest question answering. The remaining question is how this idea can be extended to a multimodal setting, where the user provides both an image and a textual symptom description.

2 Review Scope and Search Strategy

This review was conducted to identify the various methods in the literature on multimodal graph-enhanced RAG for plant disease diagnosis. Furthermore, this review aims to identify the gaps in the research literature. The selected papers were chosen due to each paper plays a clear role in building the argument. Some papers explain the limitations of plant disease detection, some introduce plant disease RAG, some define hallucination and RAG evaluation, and others support graph-based or multimodal retrieval.

Most papers were selected from the period 2023 to 2026. However, Lewis et al. [8] is an exception as they introduced the original RAG framework and is still necessary for explaining the baseline concept. Most of the sources reviewed came from indexed journals and recognised academic publishers, such as Springer, Elsevier, ACM, IEEE, Frontiers, MDPI, and Nature Scientific Reports. Multiple combinations of search keywords are applied for locating the papers, covering both the agricultural diagnosis aspect and the technical dimension of the study. Some of the keywords are "plant disease diagnosis RAG," "crop disease GraphRAG," "plant disease large language model," "multimodal plant disease diagnosis," "hallucination mitigation RAG," "knowledge graph retrieval augmented generation," "RAG factual grounding," and "RAG evaluation."

The inclusion criteria were: (i) the paper pertains to plant disease diagnosis, RAG, hallucination, GraphRAG, multimodal learning, or RAG evaluation; (ii) the paper offers methods, findings, or evaluative concepts that can enhance the proposed study; (iii) the source is recent when feasible; and (iv) the work can be utilized critically, rather than merely as background information. Papers were excluded if they were too general, focused on agricultural issues outside plant disease diagnosis, did not explain their methods clearly, or were not closely aligned with the direction of this review.

Based on Table 1, this review used a PRISMA method to make the paper selection process more transparent and reproducible. PRISMA is widely used in systematic reviews because it provides a structured way to report how studies are identified, screened, and included [10]. However, a full PRISMA protocol was not applied because the aim of this review is to provide a focused thematic analysis rather than a complete systematic review or meta-analysis. Therefore, the PRISMA flow was adapted to summarise the search, screening, and selection process while still allowing critical discussion of the selected studies according to the research scope.

Table 1. PRISMA paper selection summary

Selection stage	Estimated number	Explanation
Records identified	55	Papers found through database and manual reference searching using RAG, GraphRAG, plant disease, hallucination, and multimodal diagnosis keywords.
Records screened	38	Titles and abstracts were checked for relevance to plant disease AI, retrieval, factuality, or multimodal diagnosis.
Full papers assessed	28	Papers with closer relevance were read in more detail, especially method, dataset, contribution, and limitation sections.
Papers included	20	Final papers selected because they directly support the literature review themes and proposed research gap.

3 Overview of the Selected Literature

The twenty selected papers form a layered foundation rather than one single line of work. The primary anchor is Crop GraphRAG [7] as it is a direct application of graph-enhanced RAG to crop disease and pest question answering. Chat Demeter [11] and the pomegranate RAG system [12] show that RAG is already used in plant disease diagnosis, but mostly as a support layer after image-based detection. PlantIF [13] is useful for the multimodal side because it uses graph learning to fuse image and text information for plant disease diagnosis, but it does not retrieve external knowledge. The remaining papers help provide the problem more broadly, which are plant detection limitations [1–3] and also LLM or vision-language models (VLMs) in plant stress decision support [14]. In this context, VLMs refer to models that can process both visual and textual information, such as analysing a diseased leaf image together with a written symptom description. Other reviewed works cover standard RAG [8], RAG surveys and hallucination studies [5, 6, 9, 15], graph retrieval [16–18], and evaluation methods [19, 20]. See Table 2.

Table 2. Selected papers grouped by role in the review

Role in review	Papers	How they are used
Main agricultural anchors	[7, 11–13]	Used to show what has already been done in plant disease RAG, Crop GraphRAG, and multimodal graph learning.
Plant disease detection context	[1–4]	Used to explain why image-only diagnosis is still incomplete, especially for real-field and treatment-oriented use.
RAG and hallucination foundation	[5, 6, 8, 9, 15, 21]	Used to define standard RAG, factual grounding, hallucination, and why retrieval design affects answer reliability.
Graph retrieval and KG reasoning	[16–18]	Used to support the argument that relational retrieval and multi-hop reasoning are needed for diagnosis.
Multimodal and evaluation support	[14, 19, 20]	Used to justify multimodal inputs and evaluation measures such as faithfulness, answer relevance, and hallucination assessment.

4 Literature Review Matrix

The literature review matrix in Table 3 was prepared to organise the twenty selected papers in a clearer way. Instead of only listing the papers, the matrix summarises each study based on its purpose, methodology, key findings, and implications for this review.

This helps show how each paper contributes to the topic, whether through plant disease detection, RAG, GraphRAG, hallucination analysis, multimodal diagnosis, or evaluation methods.

5 Results

5.1 Plant Disease Detection

Studies on the identification of plant disease have technically improved. As reviewed by Shafay et al. [1], Nyawose et al. [2], and Wang et al. [3], numerous CNN, Vision Transformer, attention-based and hybrid models that work well on plant disease picture datasets. The studies are important because they show that visual recognition is no longer novel. The discipline has advanced from simple classification models to increasingly intricate architecture that can handle various imaging conditions.

Table 3. Literature Review Matrix

Author	Purpose	Methodology	Key Findings	Implications
Wu et al. [7]	To build a crop pest and disease Q&A system that can give more grounded answers than normal text retrieval.	The authors developed Crop GraphRAG using a crop disease and pest knowledge graph. Instead of retrieving only text passages, the system also retrieves related graph neighbours and summarises the evidence before generating the answer.	The graph-based approach performed better than BM25 + LLM, Naive RAG, and DeepSeek-R1 in accuracy, relevance, coverage, and traceability. The results suggest that structure matters when the question depends on connected agricultural facts.	The work is still mainly text-based, so there is room to extend the idea to plant disease diagnosis that uses both leaf images and text descriptions.
Zhang et al. [11]	To make plant disease diagnosis more interactive by combining image detection with an LLM-based response system.	Chat Demeter uses a multi-agent design. A CNN-transformer model handles disease detection, while retrieval and conversational agents help produce responses for users.	The system shows that plant disease AI can be more useful when it does not stop at a disease label. It can also support explanation and user interaction.	The work is useful for showing the value of RAG in plant disease diagnosis, but the retrieval part is not strongly structured. It does not fully model relations such as crop, symptom, disease, pathogen, severity, and treatment.
Revathi and Agash [12]	To diagnose pomegranate disease, estimate severity, and provide treatment suggestions.	The study combines deep learning for image-based disease classification and severity estimation with a RAG-based LLM that retrieves treatment information.	It gives a practical pipeline from disease detection to treatment recommendation. This is helpful because farmers usually need advice, not only a class label.	The study is directly relevant, but it is limited to pomegranate and uses standard RAG rather than graph-based retrieval. The proposed research can build on this by supporting wider diagnosis reasoning.

(continued)

Table 3. (*continued*)

Author	Purpose	Methodology	Key Findings	Implications
Wu et al. [13]	To improve plant disease diagnosis by using both visual symptoms and text descriptions.	PlantIF applies multimodal semantic fusion with graph learning. Image features and symptom text are connected through graph-based interaction instead of being simply combined.	The fusion of image and text improved diagnosis performance compared with single-modality methods. This suggests that plant disease symptoms are better understood when visual and textual cues are considered together.	This supports the multimodal part of the proposed study. Still, PlantIF is mainly a diagnosis classifier. It does not retrieve external disease knowledge or generate grounded treatment explanations.
Musazade et al. [14]	To review how LLMs and VLMs can be used in plant stress phenotyping and decision support.	The paper reviews recent work on visual recognition, contextual reasoning, text mining, semantic annotation, and RAG-related plant stress applications.	The review argues that plant stress analysis is moving toward multimodal and knowledge-supported systems, not only image recognition.	This paper helps justify the broader direction of the proposed study. It does not provide a system to follow directly, but it gives good support for combining VLMs, LLMs, and retrieval.
Ding et al. [4]	To recommend plant disease prescriptions from existing disease and treatment records.	The authors use sentence embedding retrieval to match disease descriptions with relevant plant disease prescription records.	The retrieval approach worked better than keyword matching for finding treatment recommendations. It shows that treatment retrieval is already important in plant disease decision support.	The limitation is that the retrieval is still flat and similarity based. A graph-based approach may help reduce wrong matches when symptoms overlap but diseases or treatments differ.
Shafay et al. [1]	To review recent progress and remaining issues in plant disease detection.	The paper discusses CNNs, attention models, transformers, datasets, and practical issues in plant disease detection.	Many detection models perform well in experiments, but real field conditions remain difficult. Problems such as overlapping symptoms, lighting variation, and limited explainability are still common.	Image classification alone is not enough as a diagnosis system. It also should also give explanation on the answer they generate especially when disease have similar looking symptoms
Nyawose et al. [2]	To review machine learning and deep learning work in plant disease detection.	The authors analyse model types, datasets, preprocessing methods, and evaluation practices used in plant disease detection studies.	A common pattern is that models achieve strong results on curated datasets, but their performance may not transfer well to real-world farm images.	This paper shows why a plant disease diagnosis system needs more than high classification accuracy in controlled datasets.

(*continued*)

Table 3. *(continued)*

Author	Purpose	Methodology	Key Findings	Implications
Wang et al. [3]	To summarise deep learning applications for detecting plant diseases and pests.	The review covers image-based and remote-sensing-based methods, including CNN and transformer approaches.	The reviewed works show clear progress in detection performance. However, most systems are still designed around recognising disease or pest categories, not reasoning about what to do next.	The paper supports the argument that current plant disease AI is detection heavy. The proposed study can fill part of this gap by focusing on grounded diagnosis and recommendation.
Lewis et al. [8]	To introduce the original Retrieval-Augmented Generation framework for knowledge-intensive NLP tasks.	The framework combines dense retrieval with a sequence-to-sequence generator, so the model can use retrieved documents during generation.	RAG improved performance on knowledge-intensive tasks compared with models that depend only on internal parameters.	This is the foundation for the standard RAG baseline. The proposed study extends this idea by adding graph retrieval and multimodal input.
Huang and Huang [9]	To review how retrieval-augmented text generation is used with large language models and why it matters for reducing unsupported generation.	The paper surveys RAG from the retrieval process, covering pre-retrieval, retrieval, post-retrieval, and generation stages. It also discusses evaluation methods and open challenges.	RAG can improve the reliability of LLM outputs by bringing in external information, but the quality of the answer still depends on how the system retrieves, filters, and uses that information.	This supports the use of standard RAG as a baseline in the proposed study. It also helps explain why standard RAG may still be insufficient for plant disease diagnosis when the answer needs connected evidence such as crop, symptom, disease, severity, and treatment.
Zhang et al. [15]	To review why hallucination still happens in RAG-based LLM systems and how it can be reduced.	The review analyses hallucination sources across retrieval, generation, and verification stages.	The paper shows that retrieval can reduce hallucination, but it can also introduce noise if the retrieved context is irrelevant, incomplete, or poorly aligned with the question.	This supports the proposed study's focus on stronger retrieval quality. For plant disease diagnosis, wrong or incomplete context could lead to unsafe treatment advice.
Liu et al. [21]	To study how RAG improves LLM applications in biomedical settings.	The authors conduct a systematic review and meta-analysis of RAG use in biomedical and clinical tasks.	RAG often improves factual accuracy in high-stakes domains, but the benefits depend on good knowledge sources, careful system design, and proper evaluation.	Although the domain is healthcare, the lesson is relevant. Plant disease diagnosis also involves decision support, so grounding and evaluation should be treated seriously.
Peng et al. [16]	To review graph-based RAG approaches and explain how they differ from standard RAG.	The survey covers graph construction, graph indexing, subgraph retrieval, graph-guided generation, and evaluation strategies.	GraphRAG is useful when questions require relational or multi-hop reasoning. It can retrieve connected facts instead of isolated passages.	This paper gives theoretical support for using graph retrieval in the proposed framework, especially for crop–symptom–disease–treatment relations.

(continued)

Table 3. (*continued*)

Author	Purpose	Methodology	Key Findings	Implications
Jia et al. [17]	To improve RAG by making the retrieval process more aware of document or knowledge structure.	StructRAG uses a structure-aware retrieval framework rather than treating all information as isolated text chunks.	The study shows that structure-aware retrieval can support better answers when the task depends on organised or connected information.	This helps support the proposed use of graph-enhanced retrieval. For plant disease diagnosis, disease knowledge is naturally structured around crop, symptom, cause, severity, and treatment.
Sun et al. [18]	To make LLM reasoning more interpretable by connecting it with knowledge graph traversal.	Think-on-Graph lets an LLM explore entities and relations step by step in a knowledge graph until it reaches an answer.	The approach improves multi-hop reasoning and gives clearer reasoning paths compared with black-box generation.	This is useful for diagnosis-oriented graph retrieval. A plant disease answer may need to connect the crop type thru the observed symptom, then relate it to possible disease, pathogen and suitable treatment
Ji et al. [5]	To define and organise the problem of hallucination in natural language generation.	The authors review hallucination types, causes, detection methods, and mitigation techniques across several generation tasks.	The paper distinguishes hallucination that contradicts a source from hallucination that adds unsupported information. This distinction is important for evaluating generated answers.	It helps define what "hallucination" means in this research. In plant disease diagnosis, unsupported treatment advice can be treated as a serious hallucination issue.
Anh-Hoang et al. [6]	To examine hallucination in LLMs and the effect of prompting strategies.	The paper reviews and analyses how different prompting approaches influence hallucination behaviour.	Prompting can change hallucination rates, but it does not fully solve factual accuracy problems, especially in specialised domains.	This supports the argument that prompt engineering alone is not enough. The proposed system needs external evidence and structured grounding.
Es et al. [19]	To provide a practical way to evaluate RAG systems.	Retrieval Augmented Generation Assessment (RAGAs) evaluates RAG outputs using measures such as faithfulness, answer relevance, context precision, and context recall.	The framework makes it easier to check whether an answer is actually supported by the retrieved context, instead of only checking if it sounds correct.	These metrics are suitable for evaluating the proposed system, especially factual grounding and answer relevance.
Niu et al. [20]	To create a benchmark for studying hallucination in RAG-generated outputs.	RAGTruth uses annotated RAG outputs with hallucination labels, including fabrication, contradiction, and unverifiable content.	The study shows that hallucination remains a problem even when systems use retrieval. It also gives a clearer way to label different kinds of errors.	This paper can guide the hallucination evaluation design. The benchmark itself is not agricultural, so the proposed study may need plant disease-specific evaluation examples.

Despite this, the flaw is clear. The system is not necessarily prepared for use in the real world even if it has a high accuracy score on a benchmark. Most datasets are collected under controlled conditions, while field images have shadows, poor camera quality, overlapping leaves, mixed symptoms or background noise. The possibility of distinct biological explanations for apparently identical symptoms is also another issue. For instance, water stress, pest damage, nutrient deficiency, can all cause yellowing disease. This is why classification alone is not sufficient to evaluate practical plant disease diagnosis systems.

The bigger limitation is that most detection systems only output the predicted class. They do not explain the reasoning how to move from crop type to symptoms evidence, potential cause, severity and treatment. This is important as plant disease diagnosis is a decision-support task, not only recognition task. Sentence embedding retrieval for prescription advice is one way that Ding et al. [4] partially address this. This is more in line with actual agricultural decision support. Nevertheless, semantic similarity still remains the basis for retrieval, not on explicit relationship between crop, symptoms, disease, pathogen and treatment. In other words, it assists in retrieving advice, but it does not have the complete control over the advice follows a valid diagnostic chain or not.

5.2 RAG in Plant Disease Diagnosis

RAG was originally proposed for knowledge-intensive natural language processing (NLP) tasks by retrieving external passages and using them during generation [8]. In its basic form, RAG reduces dependence on a language model's internal memory. The RAG survey in [9] later described that RAG is not just a simple step of retrieving documents before generating an answer. The process can involve several stages, including preparing the query, retrieving relevant content, filtering or refining the retrieved material, and then using it during generation. This is useful for the present study because it shows that the reliability of a RAG system depends on the whole pipeline, not only on the presence of external documents. For plant disease diagnosis, this plays a role because the answer may need several connected pieces of evidence, such as crop type, symptom, possible disease, severity, and treatment.

Recent plant disease systems show the usefulness of this approach. Chat Demeter [11], one of the tools, integrates image-based plant disease diagnosis with RAG-enhanced knowledge retrieval in a multi-agent system. While Revathi and Agash [12] combine disease detection, severity estimation, and a RAG-based LLM for pomegranate treatment remedies. These systems are important because they begin to generate more useful diagnostic responses and go beyond simple classification.

However, their limitations are also important for identifying the remaining gap. In these systems, RAG often works as a text support layer after disease detection. The model detects or estimates the disease, then retrieves related information. This pipeline is practical, but it may not explicitly represent diagnostic relations. The system may not be able to determine whether the treatment is linked to the correct crop, symptom pattern, severity level, or pathogen even though a retrieved paragraph mentions a disease and its treatment. Additionally, the pomegranate system is crop-specific, which makes generalisation more limited. As a result, existing plant disease RAG work supports the direction, but it does not entirely resolve structured factual grounding.

5.3 Hallucination and Factual Grounding

Hallucination is not only a problem of "wrong answers". Ji et al. [5] distinguish hallucination that contradicts a source from hallucination that adds unsupported information. Anh-Hoang et al. [6] further show that prompting strategies and model behaviour can affect hallucination rates. This is important because a plant disease system may sound confident even when the answer is not sufficiently supported by retrieved evidence.

RAG is often presented as a solution to hallucination, but the literature is more cautious. According to Zhang et al. [15] hallucination can still happen in RAG if the retrieved passages are irrelevant, incomplete, noisy, or incorrectly used by the generator. A similar argument for high-stakes decision support is made by Liu et al. [21]. Despite their work in biomedicine, their argument is relevant to agriculture because RAG design must consider knowledge quality, retrieval quality, and evaluation into account. A system may retrieve a correct paragraph about a pesticide but still apply it to the wrong crop or severity condition if the evidence chain is not explicit.

For this reason, factual grounding should be evaluated at more than one level. It is not enough to check whether an answer sounds relevant. A grounded plant disease answer should be traceable to retrieved evidence, and the retrieved evidence should match the user input. RAGAs [19] is useful here because it evaluates faithfulness, answer relevance, context precision, and context recall. RAGTruth [20] also provides a way to think about hallucination annotation in RAG outputs. These evaluation papers are not agriculture-specific, but they are practical for designing the assessment of the proposed framework.

5.4 Graph-Enhanced Retrieval

A plant disease diagnosis is naturally relational. Crop, symptom, disease, pest or pathogen, severity, and treatment must all be connected by the system. Although similar material can be retrieved using standard passage retrieval, the relationships between these entities are not always preserved. This is the main reason graph-enhanced retrieval is relevant. Peng et al. [16] describe GraphRAG as a direction that employs graph structures to improve retrieval and generation, particularly when the problem calls for multi-hop reasoning. In addition, Edge et al. [17] demonstrate that graph-based retrieval can support more coherent answers for complex queries by organising data beyond discrete chunks.

Crop GraphRAG [7] is the strongest evidence from the agricultural domain. It creates a crop disease and pest knowledge graph, retrieves adjacency subgraphs, summarises passages, and then produces answers. In comparison to BM25 + LLM, naive RAG and other baselines, the article reports better accuracy, coverage, relevance, and traceability. This demonstrated how structured graph retrieval can improve agricultural Q&A, which is closely related to the planned study. However, the system is still mainly text-based. It does not start from a user-provided leaf image, and it does not completely address multimodal diagnosis which requires that visual symptoms aligned with textual and graph evidence.

Think-on-Graph [18] provides another useful idea, which is that LLM reasoning can be guided through graph traversal. For plant diagnosis, this can develop into diagnosis-oriented graph retrieval. Instead of retrieving random paragraphs about disease, the system could retrieve connected evidence that links the crop type, observed symptoms, possible disease, pathogen, severity level, and suitable treatment. This type of path is easier to inspect and can lessen the chance that the model incorporates irrelevant information and is simpler to examine. The limitation is that the majority of graph reasoning techniques are still evaluated on general-domain knowledge graphs, so their direct use in plant disease diagnosis needs adaptation.

5.5 Multimodal Diagnosis of Plan Disease

Plant diagnosis is not purely textual. In practice, a user might upload a picture of a leaf together with a description of the symptom, crop type, location, or disease stage. Plant stress phenotyping can be supported by LLMs and vision-language models through recognition, reasoning, semantic annotation, and integration of external knowledge, according to Musazade et al. [14]. This supports the idea that image and text should not be treated independently by future plant diagnosis systems.

PlantIF [13] is especially important because it uses multimodal semantic interactive fusion through graph learning for plant disease diagnosis. The study highlights that image features and text descriptions can be fused more effectively when graph learning is used to model cross-modal relationships. This is similar to the planned research in spirit, because both methods emphasize graph structure. The difference is that PlantIF is primarily a diagnostic classification framework. It does not produce treatment suggestions through RAG, obtain external plant disease knowledge, or assess the factual foundation of generated responses.

The pattern across the literature is therefore uneven. Although they frequently lack external knowledge retrieval, multimodal systems are improving image-text alignment. While RAG systems frequently operate primarily with text, they do retrieve external knowledge. GraphRAG systems improve relational retrieval, but current agricultural examples are still not fully multimodal. This creates a clear space for a framework that connects all three, which are image evidence, user text, and diagnosis-oriented graph retrieval.

6 Comparative Analysis

Tables 4 and 5 compare the methods used across the reviewed studies. The comparison is helpful because the papers do not all move one single technical path. Some are centred on image classification, while others use RAG, graph-based retrieval, multimodal fusion, or evaluation on hallucination. By placing them side by side, it becomes clearer as it shows which studies detect plant diseases and which ones move closer to grounded diagnosis and recommendation.

Table 4. Comparisons on problem and context

Group of studies	Main problem addressed	What they do well	What remains weak
Plant disease detection reviews and models [1–3]	Recognising disease or pest patterns from plant images.	Show strong progress in CNN, ViT, attention, and hybrid visual models.	Mostly focus on classification; limited reasoning, explanation, or treatment grounding.
Plant disease RAG systems [11, 12]	Generating more useful diagnosis or treatment responses using retrieved knowledge.	Connect visual diagnosis with natural-language recommendation.	RAG is usually flat passage retrieval and often applied after visual detection.
GraphRAG and KG reasoning [7, 16–18]	Retrieving structured and connected evidence for complex questions.	Support relational and multi-hop reasoning better than isolated chunks.	Agricultural GraphRAG is still mainly text-based; multimodal alignment is limited.
Multimodal plant diagnosis [13, 14]	Combining image and text information for better plant stress or disease diagnosis.	Improve image-text fusion and show the value of visual-semantic alignment.	Usually lacks external domain knowledge retrieval and grounded generation.
Hallucination and RAG evaluation [5, 6, 15, 19, 20]	Measuring and reducing unsupported generated answers.	Provide useful definitions and metrics for faithfulness and hallucination.	Most benchmarks are domain-general, not designed for plant disease diagnosis.

Table 5. Comparisons on methodology

Approach	Typical method	Strength	Limitation for this research problem
Image classification	CNN, ViT, attention networks, hybrid DL models.	Efficient for recognising visible disease patterns.	Cannot by itself justify treatment or manage ambiguous symptom causes.
Standard RAG	Embedding-based passage retrieval followed by LLM generation.	Improves access to external knowledge and reduces dependence on model memory.	May retrieve fragmented or semantically similar but biologically wrong passages.

(*continued*)

Table 5. (*continued*)

Approach	Typical method	Strength	Limitation for this research problem
GraphRAG	KG construction, subgraph retrieval, graph-to-text context, LLM generation.	Preserves relationships and supports multi-hop evidence retrieval.	Needs a well-designed schema and may be harder to build and evaluate.
Multimodal fusion	Image-text encoders, graph learning, cross-modal alignment.	Connects visual symptoms with textual descriptions.	Often stops at classification and does not retrieve external treatment knowledge.
Evaluation frameworks	RAGAs, hallucination annotation, expert judgement.	Measures faithfulness, relevance, and context use.	Domain-specific plant disease evaluation sets still need to be built or adapted.

For Table 6, it focuses on the evaluation elements of the analysis studies. This comparison is crucial because success is measured differently in various research. Plant disease detection studies frequently rely on accuracy, memory, precision or F1-score, whereas RAG research typically uses metrics like answer relevance, context precision, faithfulness or rate of hallucination. When these evaluation methods are viewed together, it gives the result that classification accuracy is not the only way to measure factual grounding. A trustworthy approach for diagnosing plant disease should also be assessed according to how well the retrieved evidence supports its explanation and recommendations.

Table 6. Comparisons on data and experimental setup

Paper / group	Data or setting	Evaluation focus	Issue for proposed study
Crop GraphRAG [7]	Crop disease and pest knowledge graph with Q&A evaluation.	Accuracy, coverage, relevance, traceability.	Strong anchor but does not evaluate image-based diagnosis.
Chat Demeter [11]	Plant images with agricultural knowledge retrieval.	Interactive diagnosis and user-oriented response.	No explicit graph retrieval or hallucination-focused evaluation.
Revathi and Agash [12]	Pomegranate disease images and remedy corpus.	Disease detection, severity, treatment remedy.	Single crop and flat RAG limit generalisation.
PlantIF [13]	Large multimodal plant disease image-text data.	Diagnostic classification accuracy and F1.	No external RAG and no generated recommendation grounding.

(*continued*)

Table 6. (*continued*)

Paper / group	Data or setting	Evaluation focus	Issue for proposed study
RAGAs / RAGTruth [19, 20]	General RAG evaluation and hallucination datasets.	Faithfulness, context quality, hallucination labels.	Needs adaptation to plant disease terms and expert validation.

7 Discussion

The literature suggests several consistent findings. First, plant disease detection is technically mature visual classification, but it is still not equivalent to diagnosis. The label itself is not enough as the practical user needs an explanation and a management recommendation. Second, RAG is becoming a useful layer for agriculture decision support because it can retrieve external knowledge and generate natural language responses. Next, vanilla RAG is still weak if the answer depends on relations among multiple entities. Fourth, graph retrieval can support traceability and multi-hop reasoning, but the most robust agricultural example, Crop GraphRAG [7], is text-based. Finally, multimodal plant diagnosis is evolving rapidly but is drifting from external knowledge grounding.

The literature also has some tensions. Image-based methods are good at recognition but weak at explanation. Although RAG-based systems are capable of generating more comprehensive responses, inadequate retrieval can deceive them. Multimodal models can align image and text features, but they may still experience hallucinations if the explanation they generate is not tied to reliable knowledge. These tensions are not contradictions in a negative sense rather than they show that each method solves one part of the problem.

The main gap is therefore quite specific where the current plant disease RAG and multimodal diagnosis studies have not sufficiently addressed factual grounding in text-image plant disease diagnosis through diagnosis-oriented graph retrieval. Existing systems either classify images, retrieve flat passages, perform text-only GraphRAG, or fuse image-text features without external knowledge grounding. What is missing is an integrated framework where image evidence and user symptom descriptions are mapped to a structured graph of crop, symptom, disease, pathogen, severity, and treatment relationships before the LLM generates the answer.

This gap is important because plant disease diagnosis often needs multi-hop reasoning. A symptom can be shared by more than one disease. One treatment may be valid for one crop but not another. A severe case may require a different response than an early case. A grounded system should therefore retrieve the correct relationship chain and not just a paragraph that seems semantically similar. Therefore, the main motivation for the proposed study is to focus on multimodal graph-enhance RAG instead of vanilla RAG alone.

8 Conclusion

The reviewed literature shows that plant disease diagnosis is moving from image classification toward more interactive and knowledge-supported decision systems. While deep learning is still helpful for visual recognition, it is not a complete solution for treatment prescription and reasoning. Flat passage retrieval can nevertheless result in weak grounding when the answer depends on multiple interconnected disease relationships, even while RAG enhances access to external knowledge. Crop GraphRAG [7] is the most convincing agricultural proof that GraphRAG provides a more robust retrieval structure. However, multimodal plant diagnosis systems are frequently not based on external knowledge, and agricultural GraphRAG work is still primarily text-based. This review therefore supports a research direction that combines image-text input, diagnosis-oriented graph retrieval, and RAG-based generation to improve factual grounding and reduce hallucination in plant disease diagnosis.

From the gap identified in this review, future work may further explore how multimodal plant disease diagnosis can be combined with graph-enhanced RAG. This direction is worth considering because image-based diagnosis alone may miss important contextual details, while standard RAG may retrieve information that is relevant but not strongly connected. A future system could link leaf images and user symptom descriptions with structured plant disease knowledge, such as crop type, symptom, disease, pathogen, severity, and treatment. Evaluation should also move beyond classification accuracy by including factual grounding, answer relevance, traceability, and hallucination reduction.

References

1. Shafay, M., et al.: Recent advances in plant disease detection: challenges and opportunities. Plant Methods. **21**(1), 140–140 (2025). https://doi.org/10.1186/s13007-025-01450-0
2. Nyawose, T., Maswanganyi, R.C., Khumalo, P.: A review on the detection of plant disease using machine learning and deep learning approaches. J. Imaging. **11**(10), 326–326 (2025). https://doi.org/10.3390/jimaging11100326
3. Wang, S., et al.: Advances in deep learning applications for plant disease and Pest detection: a review. Remote Sens. **17**(4), 698–698 (2025). https://doi.org/10.3390/rs17040698
4. Ding, J., Qiao, Y., Zhang, L.: Plant disease prescription recommendation based on electronic medical records and sentence embedding retrieval. Plant Methods. **19**(1) (2023). https://doi.org/10.1186/s13007-023-01070-6
5. Ji, Z., et al.: Survey of hallucination in natural language generation. ACM Comput. Surv. **55**(12) (2022). https://doi.org/10.1145/3571730
6. Anh-Hoang, D., Tran, V., Nguyen, L.-M.: Survey and analysis of hallucinations in large language models: attribution to prompting strategies or model behavior. Front. Artif. Intell. **8**, 1622292 (2025). https://doi.org/10.3389/frai.2025.1622292
7. Wu, H., Xie, N., Wang, X., Fan, J., Li, Y., Meng, Z.: Crop GraphRAG: pest and disease knowledge base Q&A system for sustainable crop protection. Front. Plant Sci. **16**, 1696872 (2026). https://doi.org/10.3389/fpls.2025.1696872
8. Lewis, P., et al.: Retrieval-augmented generation for knowledge-intensive NLP tasks. Neural Inf. Process. Syst. **33**, 9459–9474 (2020)

9. Huang, Y., Huang, J.X.: A survey on retrieval-augmented text generation for large language models. ACM Comput. Surv. **58**, 1–38 (2026). https://doi.org/10.1145/3805774
10. Page, M.J., et al.: The PRISMA 2020 statement: an updated guideline for reporting systematic reviews. BMJ. **372**, Art. no. n71 (2021). https://doi.org/10.1136/bmj.n71
11. Zhang, S.: Chat Demeter: a multi-agent system for plant disease diagnosis integrating CNN-transformer models. Front. Plant Sci. **16**, 1695227 (2026). https://doi.org/10.3389/fpls.2025.1695227
12. Revathi, A.R., Agash, A.A.: Pomegranate disease diagnosis with severity estimation and treatment remedies using deep learning and RAG-based LLM. Sci. Rep. **15**(1), 41278 (2025). https://doi.org/10.1038/s41598-025-25160-8
13. Wu, X., et al.: PlantIF: multimodal semantic interactive fusion via graph learning for plant disease diagnosis. Plant Phenomics, 100132 (2025). https://doi.org/10.1016/j.plaphe.2025.100132
14. Musazade, E., Mrisho, I.I., Gao, J., Feng, X.: Integration of LLMs and VLMs in plant stress phenotyping: from trait recognition to decision support. Plant Phenomics, 100161 (2025). https://doi.org/10.1016/j.plaphe.2025.100161
15. Zhang, W., Zhang, J.: Hallucination mitigation for retrieval-augmented large language models: a review. Mathematics. **13**(5), 856 (2025). https://doi.org/10.3390/math13050856
16. Peng, B., et al.: Graph retrieval-augmented generation: a survey. ACM Trans. Inf. Syst. (2025). https://doi.org/10.1145/3777378
17. Jia, R., Zhang, B., Méndez, S.J.R., Omran, P.G.: StructRAG: structure-aware RAG framework with scholarly knowledge graph for diverse question answering. Companion Proc. ACM Web Conf. **2025**, 2567–2573 (2025). https://doi.org/10.1145/3701716.3717819
18. Sun, J. et al.: Think-on-graph: deep and responsible reasoning of large language model with knowledge graph. In: Proceeding of the International Conference on Learning Representations (ICLR) (2024)
19. Es, S., James, J., Espinosa Anke, L., Schockaert, S.: RAGAs: automated evaluation of retrieval augmented generation. ACLWeb. https://aclanthology.org/2024.eacl-demo.16/ (01 Mar 2024)
20. Niu, C., et al.: RAGTruth: a hallucination corpus for developing trustworthy retrieval-augmented language models, pp. 10862–10878, (2024) https://doi.org/10.18653/v1/2024.acl-long.585.
21. Liu, S., McCoy, A.B., Wright, A.: Improving large language model applications in biomedicine with retrieval-augmented generation: a systematic review, meta-analysis, and clinical development guidelines. J. Am. Med. Inform. Assoc. **32**, 605–615 (2025). https://doi.org/10.1093/jamia/ocaf008

Invariant Fish Image Representation Using Extended Local Directional Pattern Momentgram

Mas Rina Mustaffa(✉), Noorul Shuhadah Osman, Shyamala Doraisamy, and Hizmawati Madzin

Multimedia Department, Faculty of Computer Science and Information Technology, Universiti Putra Malaysia, Serdang, Selangor, Malaysia
MasRina@upm.edu.my

Abstract. Fish species retrieval systems have gained considerable attention due to their importance in marine biodiversity observation, fisheries management, and underwater environmental monitoring. However, underwater fish images are often captured under unconstrained conditions, resulting in variations in position, scale, orientation, illumination, and background complexity. These challenges highlight the importance of developing invariant feature representation methods for fish-oriented CBIR systems. Previous studies have shown that Local Directional Pattern (LDP) is capable of effectively extracting local shape and texture information while maintaining robustness against noise. Nevertheless, conventional LDP lacks invariance to basic geometric transformations. Therefore, this study proposes an invariant feature representation approach by applying 2D Dual-Tree Complex Wavelet Transform (2D-DTCWT) on momentgrams generated using Zernike Moment Invariants (ZMInv), followed by Local Directional Pattern (LDP) extraction. The proposed method integrates global invariant shape representation with wavelet-based local structural encoding and directional texture features to improve fish image retrieval performance under varying underwater imaging conditions. Experiments were conducted using the Fish4Knowledge mask image dataset containing 27,370 fish images. Retrieval performance was evaluated using Mean Average Precision (MAP), Average 11-Point Precision-Recall, and Two-tailed paired t-test analysis. The proposed ZMInvLDP method achieved a MAP value of 91.3%, outperforming several benchmark methods including Fusion Method, ZMInvLBP, EZLH, and IZLH. Statistical analysis further confirmed that the improvement achieved by the proposed method is significant. The experimental results demonstrate that the proposed invariant descriptor is effective for fish species image retrieval in natural underwater environments.

Keywords: Content-Based Image Retrieval · Fish Species Image Retrieval · Local Directional Pattern · Zernike Moment Invariants · Invariant Feature Representation

N. Mohd Sharef et al. (Eds.): CAMP 2026, CCIS 3093, pp. 263–277, 2027.
https://doi.org/10.1007/978-981-92-4302-0_19

1 Introduction

The study of marine biodiversity and fish species has become increasingly important for supporting sustainable fisheries management, underwater ecosystem preservation, and aquatic resource monitoring. As a result, Content-based Image Retrieval (CBIR) techniques have been extensively investigated for automatic fish image representation and retrieval using visual descriptors including shape, texture, and colour information.

In fish-oriented CBIR systems, feature extraction remains a major challenge because underwater fish images are often captured under unconstrained conditions with variations in position, orientation, scale, illumination, and background complexity. These variations make invariant feature representation an important requirement for accurate fish image retrieval.

Previous studies have shown that shape descriptors are highly effective for fish image retrieval due to the distinctive morphological characteristics of fish species. Local Directional Pattern (LDP) is capable of extracting local texture and shape information while maintaining robustness against noise. However, conventional LDP does not possess invariance to basic geometric transformations such as rotation, scaling, and translation.

Therefore, this study proposes an invariant fish image retrieval descriptor by first generating momentgram representations using Zernike Moment Invariants (ZMInv), transforming the momentgrams using 2D Dual-Tree Complex Wavelet Transform (2D-DTCWT) to obtain wavelet-based shape encoded images, and subsequently extracting Local Directional Pattern (LDP) features. This combination aims to preserve invariant global shape characteristics while enhancing local structural and directional information for fish image retrieval. The proposed method integrates global invariant shape representation with wavelet-enhanced local directional texture features to improve retrieval performance under varying underwater imaging conditions. Experiments conducted using the Fish4Knowledge mask image dataset demonstrate that the proposed method achieves superior retrieval performance compared to benchmark descriptors.

2 Related Works

Marine biodiversity monitoring and fish species analysis have attracted increasing attention due to their importance in fisheries management, underwater ecological conservation, and marine resource sustainability. This has encouraged the development of computer vision and CBIR approaches for automatic analysis and retrieval of fish images.

Recent advances in deep learning have significantly improved underwater fish analysis. CNN-based approaches have shown strong capability in extracting discriminative underwater image features. However, many deep learning methods require large, annotated datasets and high computational resources, limiting their suitability for lightweight retrieval systems. Furthermore, several existing studies focus mainly on classification rather than image retrieval.

Although deep learning methods have demonstrated strong performance, handcrafted descriptors are still valuable for CBIR applications because they require lower computational cost and perform effectively on limited datasets. In fish image retrieval, shape descriptors are particularly effective because fish species can often be differentiated through their body contour and morphological characteristics. Invariant descriptors are also important because underwater fish images commonly exhibit variations in rotation, scaling, translation, illumination, and viewing angle.

Earlier fish-oriented CBIR systems mainly focused on morphological and texture representations. Zheng et al. [1] proposed a CBIR system for fish recognition using morphological and ontology-based representations. Hu et al. [2] presented a fish recognition approach based on Multi-class Support Vector Machine (SVM) utilising colour and texture descriptors, whereas Yao et al. [3] employed a combination of clustering-based segmentation and morphological processing for fish image boundary extraction. However, these methods either utilised limited shape information or depended heavily on controlled datasets and parameter settings.

The performance of CBIR systems is strongly influenced by the discriminative capability of the selected visual descriptors. Shape descriptors are generally considered more discriminative for fish species retrieval because fish morphology differs significantly across species. Singh and Sharma [4] compared several region-based shape descriptors and reported that Zernike Moments (ZM) achieved superior retrieval performance due to their robustness in representing global shape information.

Papakostas and Macedonia [5] proposed Moment-Based Local Binary Patterns (ZMInvLBP), where Local Binary Pattern (LBP) was computed on momentgrams generated using Zernike Moment Invariants (ZMInv). The method improved rotational invariance but focused mainly on global shape representation. Singh [6] later combined Hough Transform and Zernike Moments to improve image retrieval performance, although the method suffered from high computational complexity.

Goyal and Walia [7] investigated several local descriptors integrated with ZM and demonstrated that ZM fused with Local Directional Pattern (LDP)-based descriptors achieved higher retrieval accuracy than other fusion approaches. They subsequently improved LDP using weighted image patches and wavelet-based shape encoding [8]. Although the proposed descriptors improved retrieval performance, the experiments mainly involved clean grayscale datasets with simple object boundaries. Farokhi et al. [9] also investigated the fusion of ZM with Hermite Kernels (HKs) and reported good computational efficiency for image representation.

Local texture representation is also valuable for underwater fish analysis because fish images commonly exhibit distinctive scale, fin, and surface texture patterns. Haralick and Shanmugam [10] introduced the Gray Level Co-occurrence Matrix (GLCM), while He and Wang [11] proposed Texture Unit (TU) and Texture Spectrum methods for texture analysis. Ojala et al. [12] proposed Local Binary Pattern (LBP), which is computationally efficient and robust against monotonic illumination variation. To improve the robustness of LBP, Jabid et al. [13] introduced Local Directional Pattern (LDP), which utilises edge response values instead of direct pixel intensity comparisons, making it more resistant to noise and illumination variation.

Recent studies have increasingly focused on underwater image enhancement and robust feature representation. Petrellis [14] demonstrated that morphological descriptors remain effective for fish monitoring in low-contrast underwater images. Deng et al. [15] and Abo El Rejal [16] proposed underwater image enhancement frameworks to improve image visibility and feature preservation under degraded underwater conditions. Yan et al. [17] introduced Contrastive Learning of Ignoring Background (CLIB) to reduce underwater background interference, while Wang et al. [18] proposed DyFish-DETR, a transformer-based underwater fish recognition framework. Siri et al. [19] further showed that integrating image enhancement with deep learning improves fish species recognition under natural underwater environments.

Feature fusion strategies have also shown promising retrieval performance. Zhao and Dong [20] proposed a Feature Fusion Visual Attention Graph Convolutional Network (FVAGCN) for underwater fish recognition. Zhang et al. [21] investigated modified Zernike Moment descriptors integrated with SURF-based feature extraction and reaffirmed the effectiveness of invariant shape descriptors. Mustaffa et al. [22] and Osman et al. [23, 24] demonstrated that combining moment-based descriptors with local directional and colour-based features substantially improves fish image retrieval performance in underwater environments. Poveda-Cuellar et al. [25] further demonstrated the effectiveness of deep learning-based visual feature representation for fish morphology analysis using field photography datasets.

Based on the literature reviewed, many existing fish image retrieval studies either rely heavily on controlled datasets or focus mainly on classification rather than retrieval tasks. Furthermore, several methods do not sufficiently address invariance to basic geometric transformations despite the fact that underwater fish images naturally vary in orientation, scale, and position. Therefore, invariant handcrafted descriptors remain attractive for fish-oriented CBIR applications due to their lower computational complexity and strong discriminative capability.

ZMInv and LDP provide complementary representations, where ZMInv captures invariant global shape features and LDP encodes local directional texture information. Motivated by these observations, this study extends the ZM-LDP fusion framework by incorporating wavelet-based shape encoding through 2D-DTCWT prior to LDP extraction, enabling the representation of invariant global shape characteristics together with enhanced local directional information for improved fish image retrieval.

3 Methods

ZMInvLDP is inspired from the works of [5]. Previous studies by Goyal and Walia [7, 8] demonstrated that ZM-LDP fusion achieved higher retrieval performance compared to several alternative descriptors. Nevertheless, this fusion is only invariant to rotation. The LDP descriptor is improved to be invariant to basic transformations by applying LDP on wavelet-based shape encoded images derived from momentgrams generated using ZMInv. Specifically, the momentgrams are first transformed using 2D-DTCWT to produce six directional sub-bands, which are subsequently superimposed prior to LDP computation. To extract features, the mask image is first resized to 64 × 64 pixels. ZMInv is later applied on the image, transforming the image to momentgram form and extracting the information of global shape features of the image. The LDP descriptor used in this study follows the modification proposed by Goyal and Walia [8]. Prior to LDP computation, 2D-DTCWT is applied to enhance local structural representation within the image. Its goal is to improve the image's edges. The extracted feature vectors are then indexed and stored in the database during offline processing.

During online retrieval, the query image is processed using the same feature extraction procedure, and the generated feature vectors are stored in the query feature database. Similarity matching is performed between the query representation and all database representations using Euclidean distance, where images with the smallest distance values are retrieved as the most relevant results. Lower distance values correspond to higher similarity between the images compared. Figure 1 shows the framework for ZMInvLDP.

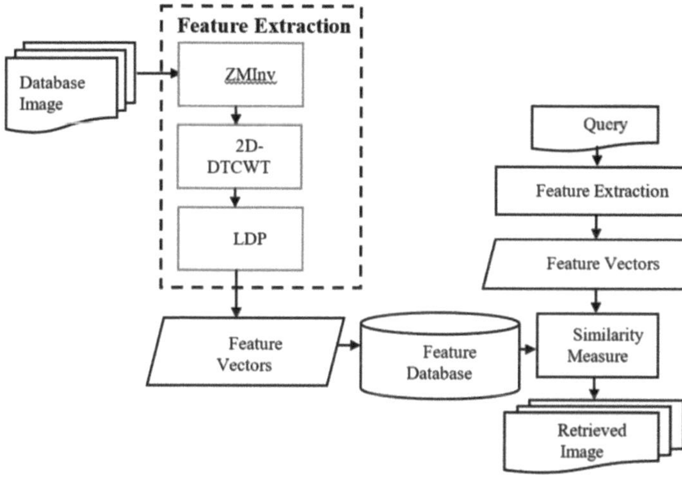

Fig. 1 CBIR framework for the ZMInvLDP

The term ZMInvLDP is retained to emphasise the fusion between the invariant moment representation and the final LDP descriptor, while the intermediate 2D-DTCWT stage serves as a wavelet-based enhancement mechanism within the feature extraction pipeline.

This study utilises the Fish4Knowledge mask image dataset containing 27,370 underwater fish images from multiple species categories. The original image sizes vary from 20 × 20 pixels to 200 × 200 pixels. Since different image sizes may influence the extracted features and retrieval accuracy, several commonly used image sizes, namely 32 × 32, 64 × 64, 128 × 128, and 256 × 256 pixels, were investigated. The experimental results showed that 64 × 64 pixels produced the best retrieval performance and was therefore selected for subsequent experiments. Images with original dimensions smaller than 64 × 64 pixels were resized to 64 × 64 pixels using interpolation to ensure a consistent input size for feature extraction. Images larger than 64 × 64 pixels were similarly resized to the selected resolution. Although enlarging low-resolution images may introduce interpolation artefacts and slight smoothing effects, the subsequent use of moment-based representation and directional feature extraction was found to preserve the discriminative characteristics required for retrieval. Since all images underwent the same preprocessing procedure, the influence of resizing was considered uniform across the dataset.

In this work, 2D Dual-Tree Complex Wavelet Transform (2D-DTCWT) is applied on the momentgram generated using Zernike Moment Invariants (ZMInv). The process produces six directional sub-bands, which are subsequently superimposed to form a wavelet-based shape encoded image following the method proposed in [8]. The generated histogram from the wavelet-based shape encoded image is illustrated in Fig. 2.

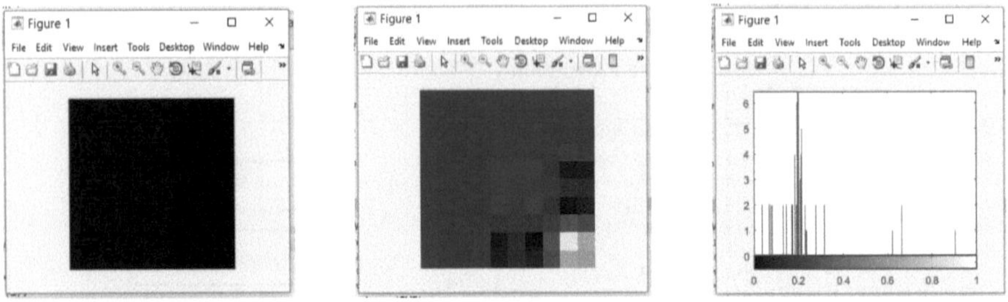

Momentgram

Wavelet-based Shape Encoded Image

Histogram of Wavelet-based Shape Encoded Image

Fig. 2 Generated histogram after 2D-DTCWT transformation on the momentgram

ZMInv is computed by assigning the order *n* and repetition *m*, where the value of *m*depends on *n*. Different ZMInv orders contribute differently to retrieval performance. Experimental analysis showed that order four achieved the highest retrieval accuracy. Before moment computation, the image is first represented using the Image Block Representation (IBR) method proposed in [26, 27]. The moment calculations follow the method described in [5]. Figure 3 illustrates the generated momentgram from the mask image.

Mask image

Momentgram

Fig. 3 Process of generating the momentgram from the mask image

Local Directional Pattern (LDP) is subsequently applied to the wavelet-based shape encoded image generated after performing 2D-DTCWT on the momentgram. Guided by the Kirsch Mask, LDP calculates relative edge responses in eight directions to highlight prominent corners, edges, and boundaries. The LDP computation is defined as:

$$LDPk(xc, yc) = \sum_{i=0}^{N-1} \int (m_i - m_k)2^i, \int (z) = \begin{cases} 1, z \geq 0 \\ 0, z < 0 \end{cases} \tag{1}$$

where N represents the number of neighbouring pixels, while m_i and m_k denote the directional edge responses generated using the Kirsch masks. In this work, $k = 3$ is selected because it is widely adopted in previous studies. Based on $k = 3$, the three most significant directional responses are assigned the value one, while the remaining responses are assigned zero, resulting in an 8-bit binary pattern. Since there are $C_8^3 = 56$ possible combinations, the descriptor produces 56 histogram bins as feature vectors. Figures 4 and 5 illustrate the LDP code generation process and the generated histogram, respectively.

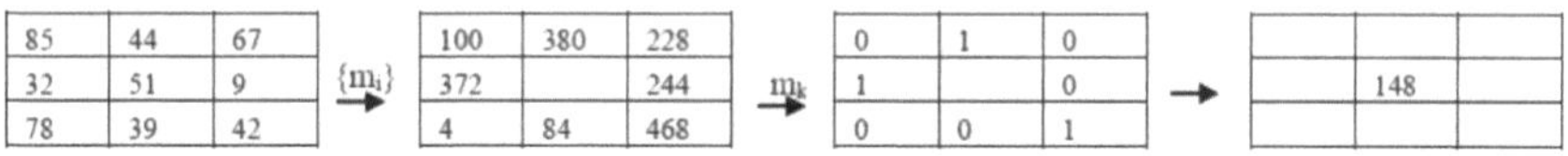

LDP binary pattern = 10010100, decimal representation = 148

Fig. 4 LDP pattern generation using a 3 × 3 neighbourhood

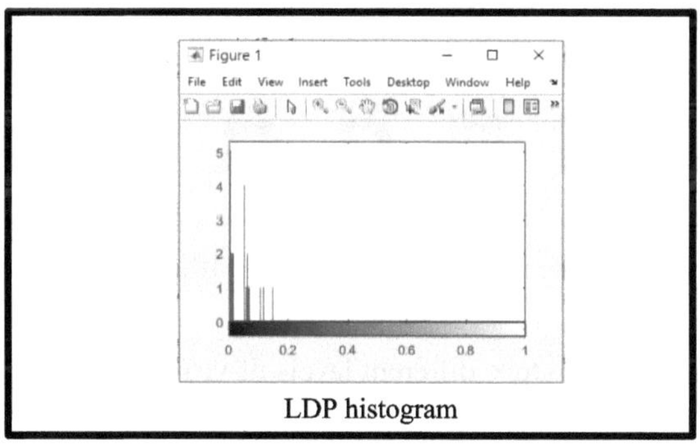

Fig. 5 Generated LDP histogram representation

Applying LDP on the generated momentgram yields 56 discriminative feature vectors for image representation. Compared to ZMInvLBP [5], which produces 256 feature vectors, the proposed descriptor achieves higher retrieval accuracy using fewer feature vectors.

Euclidean Distance is employed to evaluate similarity between the query and database descriptors. The distance calculation is defined as:

$$D(GDi, GQ) = \sqrt{\sum_{j=1}^{N} \left(G^{Dij} - G^{Qj}\right)^2} \tag{2}$$

where GD_i corresponds to the database descriptor, GQ represents the query descriptor, and N indicates the total number of extracted feature vectors.

4 Experimental Setup

Experiments were conducted using the Fish4Knowledge mask image dataset containing 27,370 fish images from 23 species categories. The images vary in terms of size, orientation, and position, representing natural underwater imaging conditions. Thirty percent of the dataset was randomly selected as query images to avoid selection bias. Fish detection and tracking were performed using software developed by [13], while the species categorisation was verified by marine biologists [14]. Figure 6 illustrates sample images from the dataset, while Table 1 summarises the fish species and number of detections.

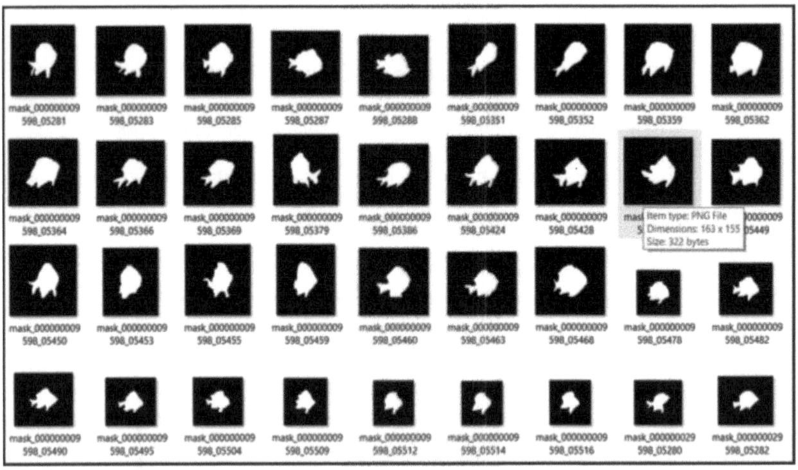

Fig. 6 Example fish images from the Fish4Knowledge dataset

Different image sizes produce different levels of visual information and may influence retrieval performance. Therefore, image sizes of 32 × 32, 64 × 64, 128 × 128, and 256 × 256 pixels were evaluated. Experimental results showed that 64 × 64 pixels achieved the best retrieval accuracy and was selected for subsequent experiments.

All experiments were implemented using MATLAB R2015b on a laptop equipped with an Intel® Core™ i7 processor and 8 GB RAM. Microsoft Excel was used for statistical analysis and graph generation.

Retrieval performance was evaluated using Mean Average Precision (MAP), Average 11-Point Precision-Recall, and Two-tailed paired t-test. Precision and recall are computed using Eqs. (3) and (4), respectively.

$$precision = \frac{retrieved\ relevant\ document}{retrieved\ documents} \tag{3}$$

$$recall = \frac{retrieved\ relevant\ document}{relevant\ documents} \tag{4}$$

Table 1 Fish ID, species name and number of detection

No.	Species ID	Number of Detection
01.	Dascyllus reticulatus	12112
02.	Plectroglyphidodon dickii	2683
03.	Chromis chrysura	3593
04.	Amphiprion clarkii	4049
05.	Chaetodon lunulatus	2534
06.	Chaetodon trifascialis	190
07.	Myripristis kuntee	450
08.	Acanthurus nigrofuscus	218
09.	Hemigymnus fasciatus	241
10.	Neoniphon sammara	299
11.	Abudefduf vaigiensis	98
12.	Canthigaster valentini	147
13.	Pomacentrus moluccensis	181
14.	Zebrasoma scopas	90
15.	Hemigymnus melapterus	42
16.	Lutjanus fulvus	206
17.	Scolopsis bilineata	49
18.	Scaridae	56
19.	Pempheris vanicolensis	29
20.	Zanclus cornutus	21
21.	Neoglyphidodon nigroris	16
22.	Balistapus undulatus	41
23.	Siganus fuscescens	25

The Average 11-Point Precision-Recall curve evaluates retrieval performance at recall levels between 0.0 and 1.0. The interpolated precision is defined as:

$$P_x = \max_{r' \geq r} precision\left(r'\right) \tag{5}$$

Average Precision (AP) and Mean Average Precision (MAP) are computed using Eqs. (6) and (7), respectively.

$$AP = \frac{\sum_{i=1}^{M} P_i}{M} \tag{6}$$

$$MAP = \frac{\sum_{i=1}^{N} AP_i}{N} \tag{7}$$

where M is the number of retrieved images and N is the total number of queries.

To determine statistical significance between the proposed and benchmark methods, a Two-tailed paired t-test with a significance level of 0.05 was performed. The mean, standard deviation, and t-value are calculated using Eqs. (8)–(10).

$$\mu = \frac{1}{N}\sum_{i=1}^{N} x_i \tag{8}$$

$$\sigma = \sqrt{\frac{1}{N}\sum_{i=1}^{N} (x_i - \mu)^2} \tag{9}$$

$$t = \frac{\overline{\mu_1} - \overline{\mu_2}}{\sqrt{\frac{\sigma_1^2}{N_1} + \frac{\sigma_2^2}{N_2}}} \tag{10}$$

A p-value of less than 0.05 indicates that the retrieval performances of the methods compared are significantly different.

5 Results and Discussion

Table 2 presents the MAP values obtained using different image sizes. The results show that image size has only a small influence on retrieval performance. Among the evaluated sizes, 64 × 64 pixels achieved the highest MAP value and were therefore selected for subsequent experiments.

Table 2 Performance comparison using different image sizes in terms of MAP

Image Size	MAP
32 × 32 pixels	48.29
64 × 64 pixels	49.04
128 × 128 pixels	48.9
256 × 256 pixels	48.97

Table 3 shows the retrieval performance for different ZMInv orders. The results indicate that order four achieved the highest MAP value. Although higher orders increased the number of feature vectors and processing time, no significant improvement in retrieval accuracy was observed beyond order four. Therefore, order four was selected for the proposed framework.

Figure 7 illustrates the Average 11-Point Precision-Recall curves of the proposed and benchmark methods, while Table 4 summarises the MAP values for all fish categories. The proposed ZMInvLDP descriptor achieved the highest overall MAP value of 91.3%, outperforming Fusion Method [20], ZMInvLBP [5], EZLH [21], and IZLH [22].

Table 3 Performance comparison using different orders of ZMInv in terms of MAP

Order of ZMInv	MAP
1	11.2
2	11.3
3	48.16
4	48.29
5	48.28
6	48.27
7	48.23
8	48.22
9	48.22
10	48.23
11	48.23
12	48.22

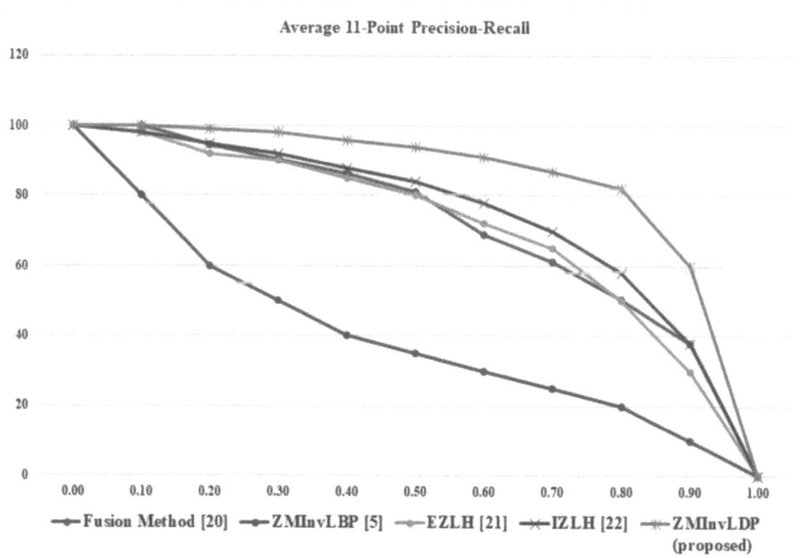

Fig. 7 Average 11 standard precision-recall graph of the proposed and benchmark methods

The experimental findings indicate that ZMInv successfully maintains invariant global shape characteristics, while LDP enhances local directional and texture representation. Compared to LBP, LDP produces more robust edge-based representations because the descriptor is generated using directional edge response magnitudes rather

Table 4 MAP value for each category of the proposed and benchmark methods

Species	Fusion Method [20]	ZMInvLBP [5]	EZLH [21]	IZLH [22]	ZMInvLDP (proposed)
1	94.54	64.33	76.49	76.79	95.1
2	87.51	77.49	69.09	85.12	95.4
3	91.61	67.67	65.98	84.45	86.09
4	92.46	73.31	84.61	85.64	87.08
5	93.76	59.86	75.29	76.55	85.42
6	68.21	66.07	68.64	90.35	80.62
7	84.13	62.26	82.7	84.58	86.49
8	73.47	71.45	73.72	74.83	90.26
9	52.34	82.7	83.93	98.8	86.15
10	59.48	71.96	77.96	97.63	89.24
11	89.88	64.86	57.24	91.88	90.38
12	100.00	64.87	73.13	79.14	85.55
13	73.48	57.93	71.73	72.29	89.87
14	55.76	84.44	63.86	72.93	88.46
15	59.75	77.62	89.89	90.06	85
16	56.43	94.6	72.18	77.8	95.83
17	50.00	63.89	86.38	87.61	95.9
18	70.43	77.7	82.97	91.65	93
19	29.17	69.3	79.82	80.15	100
20	36.17	66.62	70.16	84.06	100
21	35.29	73.89	80.26	88.33	100
22	75.50	81.9	79.43	85.51	84.03
23	27.75	69.45	73.87	79.75	87.08
MAP	**67.70**	**71.48**	**75.62**	**84.17**	**91.3**

than neighbouring pixel intensities. This property improves robustness against background noise and underwater image variations. Consequently, the integration of ZMInv and LDP provides more discriminative and invariant feature representation for fish image retrieval.

To evaluate statistical significance, a two-tailed paired t-test was performed between the proposed and benchmark methods. Table 5 shows that all *p*-values are below the significance level of 0.05, indicating that the retrieval performance improvements achieved by ZMInvLDP are statistically significant. Among the compared methods, IZLH [22] achieved the closest performance to the proposed descriptor, while ZMInvLBP [5] showed the largest performance difference.

Table 5 Paired t-test at a significant level of 0.05. P-values <0.05 indicate significantly different results

Method	*t*-value	*p*-value	Null Hypothesis (H_0)	Remark
Fusion Method [20] vs. ZMInvLDP (proposed)	8.2143	2.1845E-10	Reject	ZMInvLDP (proposed) > Fusion Method
ZMInvLBP [5] vs. ZMInvLDP (proposed)	9.5328	4.7213E-12	Reject	ZMInvLDP (proposed) > ZMInvLBP
EZLH [21] vs. ZMInvLDP (proposed)	5.8731	3.1146E-07	Reject	ZMInvLDP (proposed) > EZLH
IZLH [22] vs. ZMInvLDP (proposed)	3.9625	2.4872E-04	Reject	ZMInvLDP (proposed) > IZLH

6 Conclusion

This study proposed an invariant fish-oriented CBIR descriptor known as ZMInvLDP, which integrates Zernike Moment Invariants (ZMInv) and Local Directional Pattern (LDP) for fish species image retrieval. The proposed framework effectively combines invariant global shape representation with local directional texture extraction for robust fish image retrieval. Experimental results using the Fish4Knowledge dataset demonstrated that ZMInvLDP achieved superior retrieval performance compared to the benchmark methods, obtaining the highest MAP value of 91.3%.

The results also showed that applying LDP on the generated momentgram improves the discriminative capability of the descriptor and enhances robustness against underwater image variations and background noise. The statistical evaluation based on the two-tailed paired t-test also verified the superiority of the proposed descriptor over the benchmark methods.

For future work, background elimination techniques may be incorporated to improve feature extraction by focusing only on the fish object. In addition, the integration of classification methods and alternative similarity measures such as Manhattan Distance and Jaccard Similarity may further enhance retrieval performance.

References

1. Zheng, X., Zhong, J., Zhang, Y.: A SOA-based fish recognition system prototype. In: International Conference on Intelligent Human-Machine Systems and Cybernetics (IHMSC), pp. 66–69 (2011). https://doi.org/10.1109/IHMSC.2011.86

2. Hu, J., Li, D., Duan, Q., Han, Y., Chen, G., Si, X.: Fish species classification by color, texture and multi-class support vector machine using computer vision. Comput. Electron. Agric. **88**, 133–140 (2012). https://doi.org/10.1016/j.compag.2012.07.008
3. Yao, H., Duan, Q., Li, D., Wang, J.: An improved k-means clustering algorithm for fish image segmentation. Math. Comput. Model. **58**(3–4), 790–798 (2013). https://doi.org/10.1016/j.mcm.2012.12.025
4. Singh, C., Sharma, P.: Performance analysis of various local and global shape descriptors for image retrieval. Multimed. Syst. **19**(4), 339–357 (2013). https://doi.org/10.1007/s00530-012-0288-7
5. Papakostas, G.A., Macedonia, E.: Moment-based local binary patterns: a novel descriptor for invariant pattern recognition applications. Neurocomputing. **99**, 358–371 (2013). https://doi.org/10.1016/j.neucom.2012.06.031
6. Singh, C.: Improving image retrieval using combined features of Hough transform and Zernike moments. Opt. Lasers Eng. **49**(12), 1384–1396 (2011). https://doi.org/10.1016/j.optlaseng.2011.07.009
7. Goyal, A., Walia, E.: An analysis of shape based image retrieval using variants of Zernike moments as features. Int. J. Imag. Robot. **7**(S12), 44–69 (2012)
8. Goyal, A., Walia, E.: Variants of dense descriptors and Zernike moments as features for accurate shape-based image retrieval. SIViP. **8**(7), 1273–1289 (2014). https://doi.org/10.1007/s11760-012-0353-x
9. Farokhi, S., Sheikh, U.U., Flusser, J., Yang, B.: Near infrared face recognition using Zernike moments and Hermite kernels. Inf. Sci. **316**, 234–245 (2015). https://doi.org/10.1016/j.ins.2015.04.030
10. Haralick, R.M., Shanmugam, K.: Textural features for image classification. IEEE Trans. Syst. Man Cybern. **SMC-3**(6), 610–621 (1973). https://doi.org/10.1109/TSMC.1973.4309314
11. He, D.C., Wang, L.: Texture unit, texture spectrum, and texture analysis. IEEE Trans. Geosci. Remote Sens. **28**(4), 509–512 (1990). https://doi.org/10.1109/TGRS.1990.572934
12. Ojala, T., Pietikäinen, M., Harwood, D.: A comparative study of texture measures with classification based on featured distributions. Pattern Recogn. **29**(1), 51–59 (1996). https://doi.org/10.1016/0031-3203(95)00067-4
13. Jabid, T., Kabir, M.H., Chae, O.: Facial expression recognition using local directional pattern (LDP). In: 17th IEEE International Conference on Image Processing (ICIP), pp. 1605–1608 (2010). https://doi.org/10.1109/ICIP.2010.5652374
14. Petrellis, N.: Fish monitoring from low-contrast underwater images. Electronics. **12**(15), 3338 (2023). https://doi.org/10.3390/electronics12153338
15. Deng, X., et al.: Underwater image enhancement model for domain adaptation. Front. Mar. Sci. **10** (2023). https://doi.org/10.3389/fmars.2023.1138013
16. Abo El Rejal, A.H.: An end-to-end CNN approach for enhancing underwater images using spatial and frequency domain techniques (2023)
17. Yan, Q., et al.: CLIB: contrastive learning of ignoring background for underwater fish image classification. Front. Neurorobot. **18** (2024). https://doi.org/10.3389/fnbot.2024.1423848
18. Wang, Z., et al.: DyFish-DETR: underwater fish image recognition based on transformers. J. Mar. Sci. Eng. **12**(6), 864 (2024). https://doi.org/10.3390/jmse12060864
19. Siri, D., et al.: Enhanced deep learning models for automatic fish species identification under underwater environments. Heliyon. **10**(12) (2024). https://doi.org/10.1016/j.heliyon.2024.e35217
20. Zhao, J., Dong, X.: Underwater fish species recognition based on feature fusion visual attention graph convolutional networks. Intell. Mar. Technol. Syst. (2025). https://doi.org/10.1007/s44295-025-00064-5

21. Zhang, L., et al.: An improved SURF and modified Zernike moments descriptor for invariant object recognition. Electronics. **14**(5), 1025 (2025). https://doi.org/10.3390/electronics14051025
22. Mustaffa, M.R., Norowi, N.M., Yee, S.M.: Content-based feature fusion representation for marine invertebrates. Malaysian J. Comput. Sci. **33**(3), 170–187 (2020). https://doi.org/10.22452/mjcs.vol33no3.1
23. Osman, N.S., Mustaffa, M.R., Doraisamy, S.C., Madzin, H.: Content-based image retrieval for fish based on extended Zernike moments-local directional pattern-hue colour space. Int. J. Innov. Technol. Explor. Eng. (IJITEE). **8**, 173–183 (2019)
24. Osman, N.S., Mustaffa, M.R., Doraisamy, S.C., Madzin, H.: Fish image analysis: fusion moment-based and directional features colour space. Malays. J. Comput. Sci. **38**, 111–129 (2025)
25. Poveda-Cuellar, J.L., Morantes Duarte, D., Martínez-Carrillo, F., García-Melo, J.E., Marchant, S., Reu, B.: Deep learning on field photography reveals the morphometric diversity of Colombian freshwater fish. Zoomorphology. **144**(3), 1–16 (2025). https://doi.org/10.1007/s00435-025-00747-x
26. Spiliotis, I.M., Mertzios, B.G.: Real-time computation of two-dimensional moments on binary images using image block representation. IEEE Trans. Image Process. **7**(11), 1609–1615 (1998). https://doi.org/10.1109/83.725368
27. Shu, H., Zhang, H., Chen, B., Haigron, P., Luo, L.: Fast computation of Tchebichef moments for binary and grayscale images. IEEE Trans. Image Process. **19**(12), 3171–3180 (2010) International Journal of Innovative Technology and Exploring Engineering (IJITEE) 8, 173–183 (2019)). https://doi.org/10.1109/TIP.2010.2052276

Research on Islamic Studies in the ICT Field: A Bibliometric Review in the Scopus Database

Akram M. Zeki[1](✉), Shafiullah Wasiq[1], Zainab S. Attarbashi[2], and Manar Mustafa[2]

[1] Department of Information Systems, International Islamic University Malaysia, Kuala Lumpur, Malaysia
akramzeki@iium.edu.my

[2] Department of Computer Science, International Islamic University Malaysia, Kuala Lumpur, Malaysia

Abstract. The integration of Information and Communication Technology (ICT) into Islamic Studies has created significant opportunities for research, education, and knowledge sharing. However, the scholarly landscape of this interdisciplinary field remains underexplored. This study provides a comprehensive bibliometric analysis of research trends in Islamic Studies within the ICT domain, focusing on publications from 1990 to 2025. Using Scopus, the analysis includes 4,190 English-language journal articles and conference papers in computer science and engineering. Bibliometrix, VOSviewer, and Excel were used to examine publication trends, keywords, citations, and collaborative networks. Results show a steady rise in publications from 1990 to the late 2000s, followed by substantial growth after 2009, especially after 2015. Research productivity peaked in 2019. The dataset covers 762 sources and 10,379 authors, reflecting a broad and collaborative research community. Geographical analysis highlights contributions from 99 countries, with Malaysia, Indonesia, and Saudi Arabia as the most productive. Citation analysis indicates that Malaysia leads in total citations, while China and Canada have higher citation impact per publication. Keyword and network analyses reveal major research themes, including Qur'anic studies, Islamic finance, computational text analysis, and digital Islamic education, underscoring the multidisciplinary nature of research at the intersection of Islamic Studies and ICT. This study offers valuable insights for researchers, policymakers, and practitioners, and lays the groundwork for future analyses in emerging technologies, including the Internet of Things (IoT), artificial intelligence (AI), and Islamic Studies.

Keywords: Islamic Studies and ICT · Bibliometric Analysis · Research Trends · Scopus Database

1 Introduction

Information and Communication Technology (ICT) play a significant role in daily life and has transformed many aspects of education, including Islamic Studies. Technology has expanded access to learning and made lessons more engaging and accessible. In Islamic Studies, traditional teaching methods have evolved, allowing students to benefit

N. Mohd Sharef et al. (Eds.): CAMP 2026, CCIS 3093, pp. 278–292, 2027.
https://doi.org/10.1007/978-981-92-4302-0_20

from interactive and flexible learning experiences. Online platforms, mobile applications, and virtual classrooms now enable students to study Islam in ways that reflect current technological trends (Alhassan et al., 2023). Teachers and students of Islamic Studies have a lot to benefit from ICT to boost their careers and develop their field of studies and area of specialization (Alhassan et al., 2023).

Since the early 2000s, interdisciplinary research at the intersection of Islamic Studies and Information and Communication Technology (ICT) has expanded considerably, reflecting both the urgency to digitize classical sources and the broader imperative to reimagine religious education for a digital age. This emerging field, often described as 'Digital Islam' or 'Digital Humanities in Islamic Studies,' brings together established theological, legal, and historical scholarship with contemporary technological approaches such as big data analysis, artificial intelligence, and social media analytics (Wahid, 2024). Rather than simply eroding traditional religious authority, ICT has created a dynamic space where classically trained ulama now interact and sometimes compete with millennial Muslims and a new generation of Da'i. The proliferation of the Internet and ICT has also transformed religious practice, making it possible for diverse audiences to access, produce, and engage with Islamic knowledge in ways that were previously unimaginable (Nurdin, 2022).

Despite, ICT has undoubtedly expanded access to knowledge, enhanced research capabilities, and facilitated scholarly networking within Islamic studies, the field still lacks a comprehensive bibliometric mapping that reflects its interdisciplinary breadth. Most existing analyses remain fragmented, often concentrating on specific areas such as Islamic Finance, E-Learning in Islamic Studies, Islamization of Technology and Ethics, Digital Islam, or the intersection of AI and modern technology with Islamic contexts. This focus on isolated niches has limited our understanding of how ICT is integrated across the wider landscape of Islamic studies. Furthermore, bibliometric studies using the Scopus database tend to emphasize theoretical frameworks, regulatory issues, or particular applications, leaving the broader evolution, intellectual structure, and collaborative networks within this domain insufficiently explored. To address these gaps, the present study undertakes a systematic bibliometric analysis of Islamic studies in the context of ICT, drawing on data from the Scopus database.

This study systematically maps the academic landscape of Islamic studies in information and communication technology (ICT) by identifying key journals, influential authors, significant publications, the current state of the field, emerging themes, and future research directions. The findings aim to benefit researchers and recognize academics who have made substantial contributions to this area. Furthermore, the study incorporates a geographic perspective to examine regional differences in the effectiveness, perception, and implementation of ICT within Islamic studies globally. This comprehensive approach seeks to deepen understanding of ICT development in Islamic studies and to provide actionable insights for researchers, policymakers, and scholars engaged in advancing ICT-driven initiatives across diverse Islamic contexts.

2 Literature Review

In recent years, bibliometric studies have become more common for identifying research trends (Ahmi & Mohamad, 2019). This growth is linked to easier access to bibliometric software and databases, and the method's use in many fields. Bibliometric analysis is especially valued for handling large amounts of scientific data and making a strong impact on research (Donthu et al., 2021).

ICT plays an important role in Islamic studies. Using technology in Islamic education has changed traditional teaching methods, making learning more interactive, engaging, and accessible. Digital tools like e-learning platforms, mobile apps, and virtual classrooms give students new ways to connect with Islamic teachings that fit the digital age. For example, platforms like Google Classroom and Zoom have helped teachers in schools and madrasas teach remotely, keeping education going during disruptions like the COVID-19 pandemic (Muhsinin et al., 2024; Sartika et al., 2020).

Many studies have looked at how digital technology and Islam interact, covering topics like Islamic ethics in technology (Mujahidul Islam, 2021), changes in Islamic studies in the digital age, and how digital tools, social media, and artificial intelligence affect traditional Islamic practices and beliefs (Abdelnour, 2025; Mohamad Kamal et al., 2025; Prima et al., 2025). Shafira Fatimah Azzahra and Lubis (2024) examined the role of ICT in Islamic education, highlighting both its challenges and opportunities. They found that ICT helps increase access to Islamic knowledge, improves teaching, and encourages active learning. Although combining ICT and Islamic studies is important, research in these areas remains limited. While there is more literature on technology in Islamic finance, the overall understanding is still fragmented and lacks a clear overview. Most previous studies focus on specific topics, such as consumer behavior in Islamic banking or fintech (Zafar et al., 2026).

Previous bibliometric and review studies have explored topics such as Digital Islam, Islamic broadcasting, Islamic management, Islamic finance, and technology-enhanced Islamic education (Idris et al., 2019; Mujahidul Islam, 2021; Shafira Fatimah Azzahra & Lubis, 2024; Alit et al., 2025; Zafar et al., 2026). For example, Alit et al. (2025) utilized the Scopus database to analyze Islamic broadcasting in Muslim countries, focusing on co-citation analysis, international collaborations, and keyword patterns related to Islam and media. Similarly, Idris et al. (2019) applied bibliometric methods with Scopus to examine global research on Islamic management. Although these studies have yielded valuable insights into specific intersections of Islam and technology, most focus on particular themes rather than providing a comprehensive overview of Islamic Studies within the broader information and communication technology (ICT) domain. The present study aims to address this gap by systematically analyzing research trends, identifying key contributors, and evaluating the impact of scholarship in this field. In doing so, this research offers a more comprehensive bibliometric analysis of trends in Islamic Studies within the ICT domain.

3 Methodology

The present study adopted a bibliometric analysis, which involves systematically collecting and analyzing relevant literature. Bibliometrics is a statistical method for measuring and analysing the number of publications and their citations within a specific study domain; it enables scholars to identify the theoretical structure, publication trends, and future direction of study (Garfield, 1979). Pritchard (1969) initially used the word "bibliometrics," substituted for the earlier phrase "bibliography of statistics. He characterised it as "the application of mathematical techniques in relation to books and other means of communication. Additionally, bibliometrics analysis is used the extraction of measurable data from published studies through statistical analysis and the evaluation of how the information in a publication is used (Sogi, 2023). Since its introduction, bibliometric analysis has grown quickly. Researchers now use it to explore the structure of a field, identify research collaborations, and spot new trends in published work. It also helps generate new research ideas, find gaps in a discipline, and contribute directly to the field (Donthu et al., 2021).

3.1 Database and Search Strategy

This study used only the Scopus database as its data source. Scopus was selected because it is widely used for bibliometric analysis, covers many disciplines, and offers broad coverage and powerful search capabilities (Effendi et al., 2021; Baas et al., 2020). It is efficient at covering several research areas and delivering precise outcomes (Singh et al., 2021). Following an initial query in the Scopus database, we identified terms related to ICT in the context of Islamic Studies. Sweileh (2018) notes that choosing the right keywords is essential for a high-quality bibliometric analysis, as they directly affect the quality and relevance of the results. After reviewing the literature on ICT and Islam, we selected 9 keywords and their various spellings.

The data collection followed the recommended reporting items for Systematic Reviews and Meta-Analyses (PRISMA) standards, as shown in Fig. 1, to ensure the quality of the review. To collect data, we entered the keywords "Quran", "Hadith", "Sunnah", "Surah", "Islam", "Islamic", "Sharia", "Makasid", and "Muslim" into the Scopus search field. We included different spellings of these terms because there is no single standard spelling, so authors often choose their own versions.

To select accurate and common spelling variants for Arabic keywords, we consulted sources such as the Encyclopaedia Britannica and used the New York University transliteration tool (ALA-LC, 2021). This helped ensure the precise Romanization of the keywords. For example, Qur'an appeared as 'Qur'ān', 'Quran', 'Koran', and sometimes as 'Holy Quran'.

We then limited the subject areas to "Computer Science" and "Mathematics," included only English-language literature, and retrieved relevant literature from Scopus from 1990 to 2025, excluding book series, book chapters, and books. This resulted in 4,190 journal articles and conference proceedings. The data were exported to a Microsoft Excel file, which included author, year, title, source title, citations, affiliations, abstract, and keywords. Several limitations should be noted in the interpretation of these findings. First, the data set was only derived from the Scopus database. While Scopus provides thorough coverage, it does not encompass the entire body of academic research in the topic. Secondly, the absence of non-English publications is likely to lead to an absence of research from Arabic-speaking countries and other regions of the world. Third, the use of specified keywords and subject-area filters may have led to the exclusion of relevant research that used different language. Future research should seek to utilise more databases, including non-English sources, and employ broader search techniques to obtain a broader view of the literature in order to overcome these limitations.

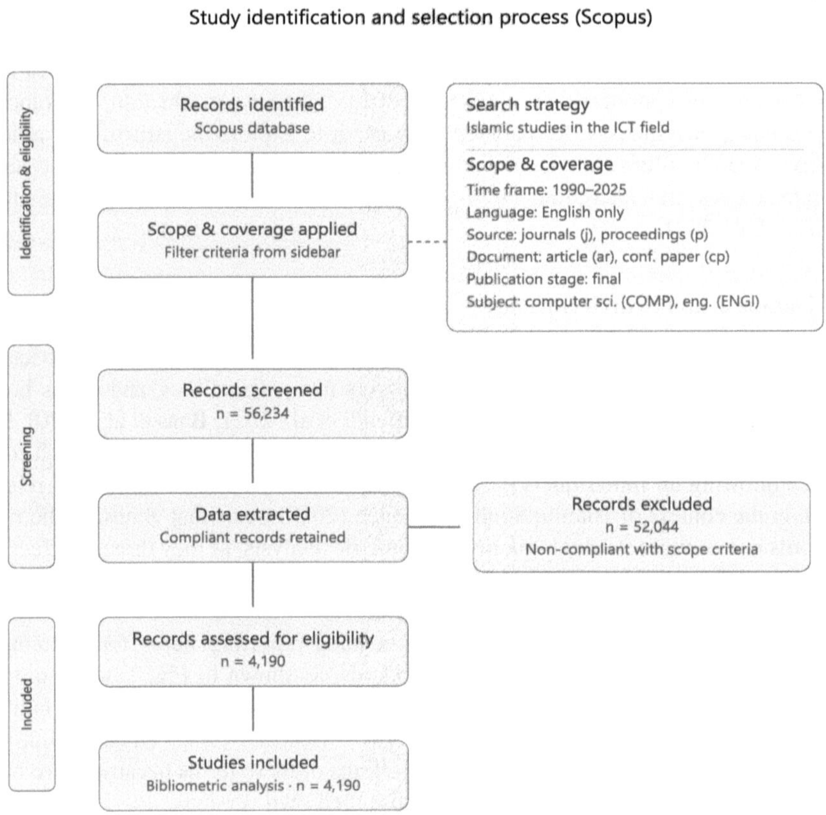

Fig. 1. Prisma flow diagram

3.2 Data Analysis

In this study, Bibliophagy software in R-Studio is used to conduct a bibliometric analysis and map key developments, trends, and dynamics in the field of Islamic studies in ICT. Bibliometric is an open-source software that enables comprehensive science mapping analysis and can create and map visual representations of information in an understandable way based on citation data, keywords, and other bibliographic metadata (Tomaszewski, 2023). In addition, the VOS Viewer was used in this study. This freely available software package is used for visualizing bibliometric maps and for displaying bibliometric relationships across a variety of variables (Kirby, 2023). The software is ideal for analyzing, mapping, and visualizing data (Simandan, 2023). After importing the Excel sheets into the VOS viewer, several operations were run, highlighting important relationships between variables (see results and discussion). The analyses included keyword co-occurrence and examined co-authorship among researchers, institutions, and countries.

4 Result and Discussion

4.1 Descriptive Overview of the Bibliometric Dataset (1990–2025)

The bibliometric analysis identified 4,190 publications on Islamic Studies within the ICT field indexed in Scopus between 1990 and 2025, disseminated across 762 distinct sources, indicating a highly dispersed publication structure (Table 1). The findings show a strong growth in Islamic studies and ICT research. With annual growth rate of 14.84%. This trend reflects the rapid integration ICT and Islamic studies by adopting digital learning systems online religious resources and AI applications.

The citation results indicate that research on Islamic studies and ICT is continuing to gain academic attention. While average count of 8.165 per publication reflects a moderate level of influence, the extensive reference based on 32,423 citation shows that researchers are building upon established body of knowledge.

Most of the research at the nexus of ICT and Islamic Studies is carried out collectively. The significance of collaboration in this subject is demonstrated by the participation of almost 10,000 authors in 4,190 publications, with an average of 3.37 co-authors for every article. The need for expertise from several areas is reflected in the low percentage of single-authored research. The topic is advancing in terms of both academic knowledge and technical applications, as evidenced by the high representation of both journal publications and conference papers.

Table 1. Descriptive overview of the bibliometric dataset (1990–2025)

Metric	Value	Metric	Value
Timespan	1990–2025	Authors	10,379
Sources	762	Single-authored docs	658
Documents	4,190	Co-authors per doc	3.37
Growth Rate (%)	14.84	Int. Co-authorship (%)	17.92
Avg. Age	7.82	Keywords Plus	13,802
Avg. Citations	8.17	Author Keywords	11,129
References	32,423		

4.2 Publication Trends Over Time (1990–2025)

In this study before looking at the performance of publications, collaborations and co-authorships between countries in the Scopus database. The annual distribution of publication on Islamic studies and ICT index in Scopus from 1990 to 2025 has been revealed in three phases.

The emerging stage started from 1990–2005 with low but steady growth. The publication in these years remain in the single digits to low tens. It indicate scholarly interest in integrating islamic studies with ICT was very limited. The second phase was the the growth stage started from 2006–2015 saw a gradual increase in the number of publications. In the year 2008 the number of publications has increase from 60 documents to 97 in 2011 and 183 by 2015. This year has witnessed development in digital infrastructure in the world which contribute to the growth.

The third phase, the acceleration stage started form 2016–2025 marked a rapid growth in authors productivity. From 2016 onward the number publication rose sharply, peaking at 414 documents in 2019 with slight decrease in the subsequent years. The document published for the year 2023 with 344 and 2024 with 373 shows that the field has maintained its momentum while the lower count for the year 2025 likely reflect incomplete indexing rather than decline in the research activity. The overall distribution of articles output shown in (Fig. 2).

Based on the result there could be several growth factors in publication on ICT and Islam including technological advancements, interest in leveraging technology for religious studies and the increasing participation of Arabic-speaking scholars in the field.

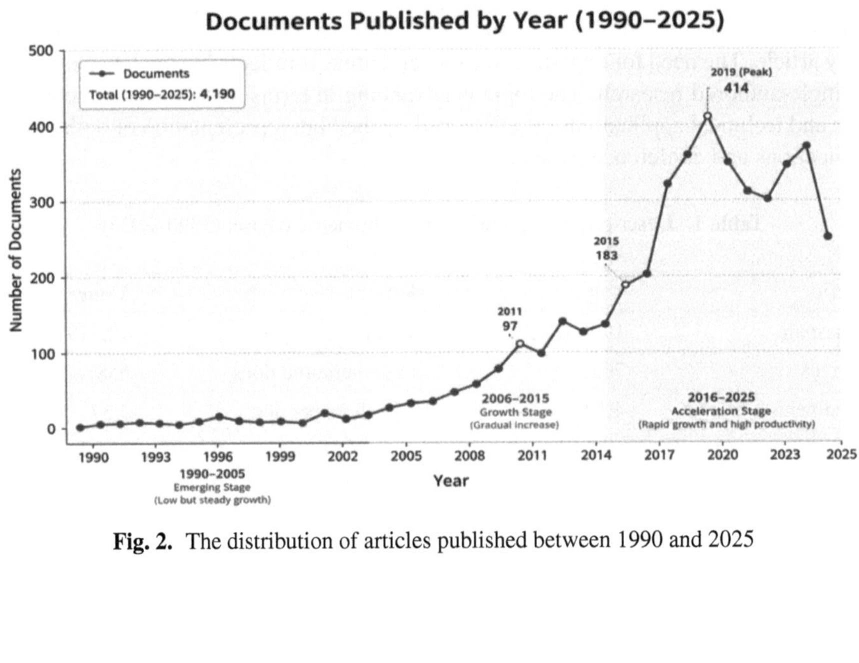

Fig. 2. The distribution of articles published between 1990 and 2025

4.3 Geographical Distribution of Research

Figure 3 displays the top 15 most productive countries in Islamic Studies and ICT research. Malaysia ranks first with 1,169 publications, followed by Indonesia with 991 and Saudi Arabia with 387. These findings suggest that Southeast Asia and the Middle East are the primary contributors to research in this field, although the United States and the United Kingdom also exhibit significant research activity.

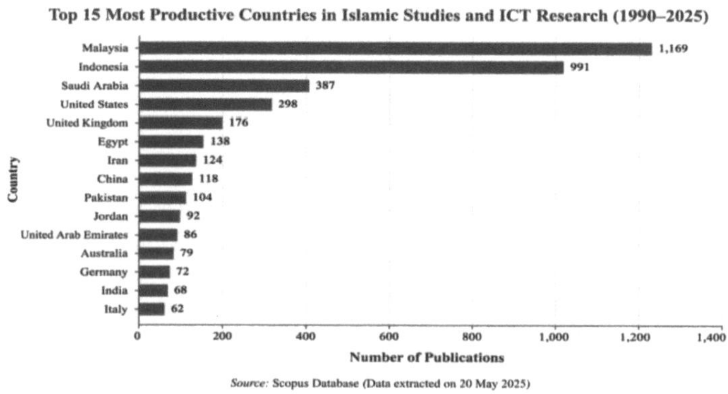

Fig. 3. Number of documents published by country

4.4 Most Cited Countries

Countries most cited are analyzed to understand the country's performance based on the number of citations. Malaysia is the country most frequently cited based on the research period in the Scopus database, namely 3,937 times, followed by China 2,675 times, United States (2,195), with Indonesia (1,695) and Saudi Arabia (1,643) had significant citation rates. Despite publishing fewer articles, Canada (33.00) and China (30.40) had the highest citation impact, indicating that their publications were cited more frequently. This result indicate that research productivity and research influence do not necessarily correlate because nations with a small number of publications may produce studies with greater academic impact (Table 2).

Table 2. Top 10 most cited countries

Country	TC	Avg. Citations
Malaysia	3937	8.60
China	2675	30.40
USA	2195	18.80
Indonesia	1695	4.10
Saudi Arabia	1643	9.60

(*continued*)

Table 2. (*continued*)

Country	TC	Avg. Citations
United Kingdom	1057	15.30
Canada	693	33.00
Iran	589	8.10
Australia	577	14.80
Jordan	480	11.20

4.5 Most Cited Authors

Table 3 presents the top ten most cited authors in the field. Among the 10,379 authors identified in the dataset,Bijaksana Moch Arif and Mujani, Wan Kamal are the most prolific, having published 19 articles. This is Followed by Akram, M. Zeki with 16 publications, and Huda, Arief Fatchul with 15 publications. A number of authors, including Agustina, Maulidar, Majid, M. Shabri Abd, and Saad, Saidah, who each contributed 14 papers, also showed impressive results. A relatively focused contribution among the top scholars in this field is demonstrated by the remaining authors in the top fifteen producing between 10 and 12 documents.

Table 3. Top 10 most cited authors

Author	Documents	Percentage (%)
Bijaksana, Moch Arif	19	0.45
Mujani, Wan Kamal	19	0.45
Akram, M. Zeki	16	0.38
Huda, Arief Fatchul	15	0.36
Agustina, Maulidar	14	0.33
Majid, M. Shabri Abd	14	0.33
Saad, Saidah	14	0.33
Jamsari, Ezad Azraai	13	0.31
Muslim Lhaksmana, Kemas Muslim	13	0.31
Mahmud, Murni Bt	13	0.31
Sukmana, Husni Teja	12	0.29
Susanty, Aries	12	0.29
Mustapha, Aida Binti	11	0.26
Abd Rahman, Nurazzah Abd	11	0.26
Adiwijaya, Adiwijaya	10	0.24

4.6 Collaborations and Co-authorships Between Countries

Figure 4 presents the country co-authorship network in Islamic Studies and ICT research. The analysis reveals that the most frequent collaborations occur between countries with relevant research output in these areas. The network analysis shows that Malaysia is a key center in the collaboration network. Malaysia has strong and close connections with Indonesia, Saudi Arabia, the United Kingdom, Nigeria, and Palestine Malaysia and Indonesia have one of the strongest partnerships in the data, indicating that proximity and common academic interests are important factors in how nations collaborate. Another notable partner is Indonesia, which maintains close ties not only with Malaysia but also with the US, Taiwan, Thailand, and Hungary. This demonstrates Indonesia's increasing participation in regional and global research networks. By establishing close ties with Saudi Arabia, India, Germany, and Bangladesh, the United Kingdom serves as a link between Middle Eastern and European nations. This demonstrates its function as a global link that facilitates collaboration between many places.

Similarly, the United States forms a strong partnership with South Korea, Turkey, Brazil, and Italy, demonstrating the strength of cross-continental ties. European nations that collaborate closely inside Europe, such as France, Germany, Portugal, and Slovenia, frequently establish smaller, related groups. In general, the co-authorship network demonstrates that while there are numerous organisations, they are all globally interconnected. While the United States and the United Kingdom act as global bridges connecting various regions of the world, Malaysia and Indonesia stand out as prominent nations in their respective regions. This trend demonstrates both global scientific cooperation and significant local ties, such as in Southeast Asia and the Middle East.

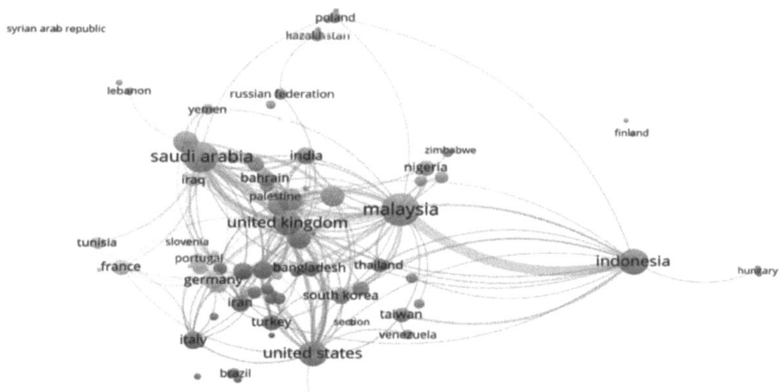

Fig. 4. Country co-authorship network visualization

Figure 5 illustrates the sequential evolution of international research collaborations. The color gradient represents the average publication year, ranging from earlier collaborations (blue) to more recent collaborations (yellow). The results indicate that Malaysia and Indonesia continue to play leading roles in recent collaborative research activities. Several countries, including Indonesia and Saudi Arabia, exhibit relatively recent patterns of collaboration, suggesting growing international engagement in Islamic Studies and ICT research. Overall, the overlay visualization demonstrates the expansion and diversification of global research networks over time.

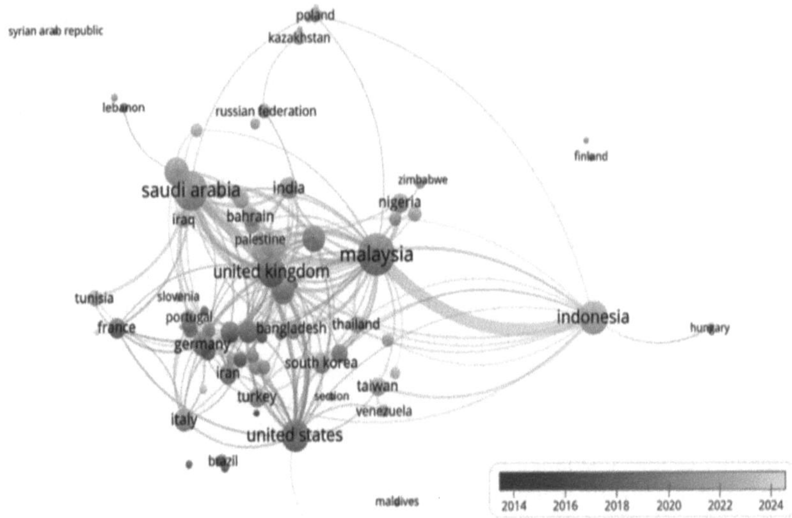

Fig. 5. Overlay visualization of country co-authorship network

4.7 Most Influential Academic Institutions

Figure 6 presents the leading academic institutions in Islamic Studies and ICT, ranked by publications. Universiti Kebangsaan Malaysia produced the highest number of publications (202), followed by International Islamic University Malaysia (191) and Universiti Teknologi MARA (188). Additional notable institutions include Universiti Utara Malaysia (89 publications), Umm Al-Qura University (85 publications), and Universiti Malaya (81 publications). These findings demonstrate that Malaysian universities are the primary contributors to research output in Islamic Studies and ICT, underscoring Malaysia's prominent role in advancing interdisciplinary research at this intersection.

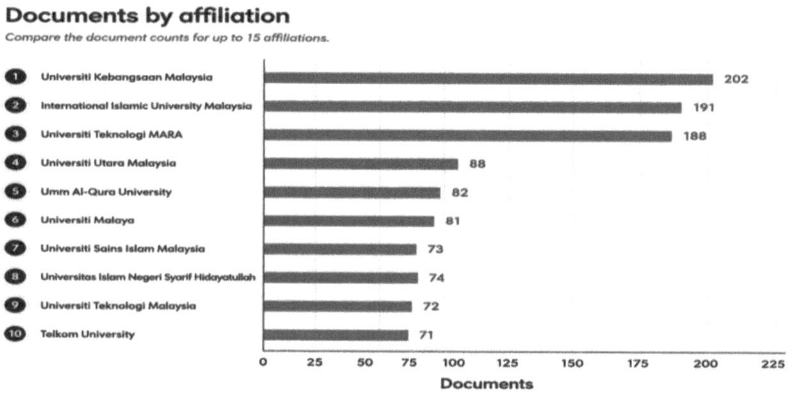

Fig. 6. Most influential academic institutions

4.8 Keyword Analaysis

The network visualization illustrates the relationships and co-occurrence patterns among keywords within the selected publications. The key research themes in the discipline can be found in several significant clusters. One notable cluster focuses on the "Quran," "hadith," and "science," emphasising research on Islamic writings and their relation to Islamic studies and scientific understanding. Keywords like "Islamic bank," "risk," "conventional bank," and "evidence" appear to be highly connected in another cluster that focuses on Islamic finance and banking, suggesting substantial research activity in the financial and economic facets of Islamic studies as shown in Fig. 7.

Another group of studies focuses on technology and analysis, using keywords like "classification," "identification," "modeling," and "activity." This points to the use of computational methods, data analysis, and artificial intelligence in Islamic-related research. Geographic keywords such as "Saudi Arabia," "Makkah," "Iran," and "Pakistan" are also common, showing that many studies focus on these regions. Other main themes include social and cultural topics, with terms like "religion," "woman," "identity," and "culture."

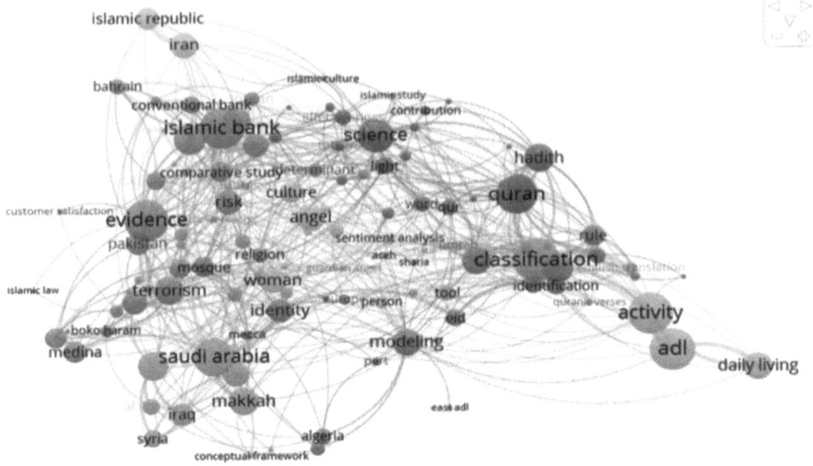

Fig. 7. Keyword co-occurrence network visualization

5 Conclusion

The way people access, study, and share Islamic knowledge is changing because of information and communication technology (ICT). As digital tools such as computers, the internet, and online repositories become more widely used, Islamic resources are becoming easier to find and use. These digital sources might soon be more important than traditional printed materials. Scholars in Islamic Studies can use ICT for teaching, learning, and research, but developing advanced technologies often requires collaboration with experts in computer science and technology.

The study examined Islamic Studies and ICT across 4,190 papers indexed in Scopus between 1990 and 2025. The study found an annual growth rate of 14.84% in publications. There is a difference between production and research influence; geographically, Malaysia, Indonesia, and Saudi Arabia produce the most research, yet Canada and China have a greater citation effect per article. The field's wide multidisciplinary scope is further demonstrated by keyword and network analyses, which show that Qur'anic and Hadith studies, Islamic finance, computational text analysis, and digital Islamic education are the main research subjects.

The finding in this study highlights how research on Islamic Studies and ICT has developed over time. It also indicates the connection between Islamic knowledge and digital technology. In addition, the study found that most of the research comes from just a few countries which indicate a necessity for global participation as the field is expanding. Malaysia emerged as the dominant contributor to the field, while Indonesia and Saudi Arabia also demonstrated substantial research activity, indicating a strong engagement of Southeast Asian and Middle Eastern institutions.

The finding gives a clear picture to the future study on key areas to be investigated. It also contributes to the educators and policymakers who can use these findings effectively in helping people better understand Islamic beliefs and practices using ICT.

References

Abdelnour, M.G.: Artificial intelligence and the Islamic theology of technology: from "Means" to "Meanings" and from "Minds" to "Hearts". Religions. **16**(6) (2025). https://doi.org/10.3390/rel16060796

Ahmi, A., Mohamad, R.. Bibliometric Analysis of Global Scientific Literature on Web Accessibility (2019)

Alhassan, A.S., Inuwa, H., Haruna, S.: Islamic studies in the age of ICT. Prob. Prospect. (2023) www.hummingbirdjournals.com

Alit, N.A., Sakri, M.L.A., Mohd Azmi, M.H.A.: Research trend on Islamic broadcasting in Muslim nations: a bibliometric analysis. Int. J. Educ. Psychol. Counsel. **10**(58), 527–540 (2025). https://doi.org/10.35631/ijepc.1058035

Donthu, N., Kumar, S., Mukherjee, D., Pandey, N., Lim, W.M.: How to conduct a bibliometric analysis: an overview and guidelines. J. Bus. Res. **133**, 285–296 (2021). https://doi.org/10.1016/J.JBUSRES.2021.04.070

Effendi, D.N. et al.: Bibliometric analysis of scientific literacy using VOS viewer: analysis of science education. J. Phys. Conf. Ser. **1796**(1) (2021). https://doi.org/10.1088/1742-6596/1796/1/012096

Garfield, E.: Is citation analysis a legitimate evaluation tool? Scientometrics. **1**(4), 359–375 (1979). https://doi.org/10.1007/bf02019306

Idris, M., Zakaria, W.F.A.W., Long, A.S., Salleh, N.: Work quality in organization: an Islamic management perspective. Int. J. Islam. Thought. **15**, 60–70 (2019). https://doi.org/10.24035/ijit.15.2019.006

Kirby, A.: Exploratory Bibliometrics: using VOSviewer as a preliminary research tool. Publications. **11**(1) (2023). https://doi.org/10.3390/publications11010010

Mohamad Kamal, R., Widad Roslan, N.: The impact of digital items on Islamic education among Muslim youths. Int. J. Res. Innov. Soc. Sci. (IJRISS). (2025). https://doi.org/10.47772/IJRISS

Muhsinin, U., Al-furqan Makassar, S., Al-Biruni Babakan Ciwaringin, S., Sulthan Thaha Saifuddin Jambi, U.: The role of technology in improving the quality of Islamic religious education in schools and madrasas in the digital era. Join J. Soc. Sci. **1**(1) (2024) https://ejournal.mellbaou.com/index.php/join/index

Mujahidul Islam: Islamic Ethics and The Rise of Digital Technology - Maydan (2021). https://themaydan.com/2021/07/islamic-ethics-and-the-rise-of-digital-technology/

Nurdin, N.: Proceeding of International Conference on Islamic and Interdisciplinary Studies (ICIIS), 2022 Impact of Internet Development on Muslim Interaction with Islam (2022). https://jurnal.uindatokarama.ac.id/index.php/iciis/issue/archive

Prima, A., Julhadi, J., Mursal, M.: Transformation of Islamic studies in the digital era: opportunities and challenges. J. Neosantara Hybrid Learn. **2**(3), 505–514 (2025). https://doi.org/10.70177/jnhl.v2i3.1938

Sartika, F., Ritonga, M., Rasyid, A.: Implementation of Islamic religious education in madrasah Ibtidaiyah during Covid-19 pandemic. Khalifa J. Islam. Educ. **4**(2), 97 (2020)

Azzahra, S.F., Lubis, M.: Role of information and communication technology in Islamic Education: challenges and opportunities. Accelerat. Quant. Inform. Technol. Algorithm J. **1**(1), 23–27 (2024). https://doi.org/10.62123/aqila.v1i1.26

Simandan, R.: Electronic commerce research during the COVID-19 pandemic: a bibliometric analysis. J. Int. Comm. Econ. Policy. **14**(2), 2350010 (2023). https://doi.org/10.1142/S1793993323500102

Sogi, G.: Bibliometrics - need to look beyond numbers? Contemp. Clin. Dentist. **14**(1), 1–2 (2023). https://doi.org/10.4103/ccd.ccd_99_23

Sweileh, W.M.: Research trends on human trafficking: A bibliometric analysis using Scopus database. Glob. Health. **14**(1) (2018). https://doi.org/10.1186/s12992-018-0427-9

Tanudjaja, I., Kow, G.Y.: Exploring Bibliometric Mapping in NUS Using BibExcel and VOSviewer (2018)

Tomaszewski, R.: Visibility, impact, and applications of bibliometric software tools through citation analysis. Scientometrics. **128**(7), 4007–4028 (2023). https://doi.org/10.1007/s11192-023-04725-2

Visser, M., van Eck, N.J., Waltman, L.: Large-scale comparison of bibliographic data sources: scopus, web of science, dimensions, crossref, and microsoft academic. Quantitative Science Studies. **2**(1), 20–41 (2021). https://doi.org/10.1162/qss_a_00112

Wahid, S.H.: Exploring the intersection of Islam and digital technology: a bibliometric analysis. Soc. Sci. Human. Open. **10** (2024). https://doi.org/10.1016/j.ssaho.2024.101085

Zafar, M.B., Yasin, T., Ali, H.: Emerging technologies and Islamic finance: a bibliometric review. J. Islam. Market., 1–20 (2026). https://doi.org/10.1108/JIMA-05-2025-034

Retrieval Augmented Generation (RAG)-Based Information Bot for Higher Education Institutions

Cai Ni Lim[1(✉)], Justtina Anantha Jothi A. P. C. M. E. John[1], J. Joshua Thomas[1], Gik Hong Yeap[2], Joel Yeo[3], Neam Heng Lee[3], and Eu Gene Ooi[3]

[1] Department of Computing, UOW Malaysia KDU Penang University College, Pulau Pinang George Town, Malaysia
0206074@student.uow.edu.my
[2] University of Wollongong Malaysia, Shah Alam, Selangor, Malaysia
[3] Intel Technology Sdn. Bhd, Penang, Malaysia

Abstract. Higher education information is often strewn throughout university websites and pdf brochures, making it hard for prospective students and staff to find reliable and consistent information. Program formats, enrollment requirements, fees, intakes, and subsidies are often scattered. Current academic information retrieval approaches are examined, and a retrieval-augmented generating assistant can be hosted locally to update trustworthy institutional material for Penang higher education institutions. Vector-based similarity search retrieves official university materials from our analysis and database, allowing users to access relevant information using a conversational interface. To provide transparency and reliability, a local large language model constructs grounded replies from retrieved content. Common student information inquiries were used to design and test the suggested system. A cost-effective and reliable local retrieval-augmented assistant improves academic information access and manual browsing.

Keywords: Retrieval-Augmented Generation · Academic Information Retrieval · Vector Database · Large Language Models · Information Retrieval · Local AI Deployment

1 Introduction

Artificial intelligence is transforming how we locate and share knowledge in many areas, including higher education. In Malaysia, especially in locations with a large number of private universities like Klang Valley and Penang, prospective students are typically met with material that is fragmented and inconsistent throughout the many university websites and downloadable PDF documents. Users are thus left with manual browsing and keyword searches that are ineffective for information-heavy judgment [1]. AI-powered conversational interfaces provide a viable alternative where users can ask queries in natural language and get targeted responses. However, conventional rule-based chatbots are constrained by inflexible, predefined scripts and static decision trees, and are inherently unable to adapt to new user inputs or context-sensitive interactions [2].

N. Mohd Sharef et al. (Eds.): CAMP 2026, CCIS 3093, pp. 293–303, 2027.
https://doi.org/10.1007/978-981-92-4302-0_21

Retrieval-Augmented creation (RAG) solves this restriction by adding retrieval of up-to-date source material, together with answer creation based on the retrieved documents. But a key missing piece is the operationalization of RAG designs in resource-constrained institutional settings. This dependency poses two major risks: (1) high data governance risks, with proprietary institutional data and student queries being sent externally through unmanaged third-party APIs, and (2) exorbitant, unpredictable operational costs based on tokens that scale linearly with user count, which are untenable for localized or medium-sized institutions [6]. There is a clear gap in the study on how to easily connect open-source, quantized local LLMs with tailored retrieval pipelines to achieve high-faithfulness grounding without overloading local infrastructure or subscription overheads.

This research provides a low-cost locally-hosted RAG-based assistant with higher education information for Penang universities. A searchable knowledge base of official university web pages and documents is accessed by a web-based conversational interface. A backend service comprehends user queries and fetches relevant passages, producing grounded responses with a local language model and explicit references to the original sources. The technology can keep recurrent costs low by running exclusively on local infrastructure and gives organizations more control over data governance.

The main contributions of this work are as follows:

- Designing a local RAG system to manage a fragmented higher education information in Penang.
- A structured ingesting and indexing pipeline for populating a searchable knowledge base with official institutional web sites and documents.
- An intent-aware retrieval and production pipeline to create grounded responses with explicit source attribution.
- An end-to-end evaluation with realistic student information queries to show reduced search effort and enhanced accessibility.

2 Literature Review

2.1 Overview

The history of chatbots has seen evolution from rule-based and retrieval-based systems to neural LLMs, but a persistent limitation has been static knowledge which is difficult to keep up with fast-changing educational details [8–11]. RAG overcomes this shortcoming by combining focused retrieval from official web pages and documents with controlled language generation, to produce brief and context-aware answers with sources for transparency [12–14]. Operational factors (cost, governance) also motivate deployments that favor controllable data pipelines and clear provenance [6, 14, 15].

2.2 Retrieval-Augmented Generation: Concepts and Design Choices

RAG is a framework that combines a retriever and a generator, where solutions are constructed using evidence retrieved at inference time from an external corpus [12–14]. The corpus for higher education comprises of official university web pages and PDF documents which are cleaned, divided into meaningful chunks with metadata, placed in a vector store, and retrieved using similarity search as shown in Fig. 1 [13, 14].

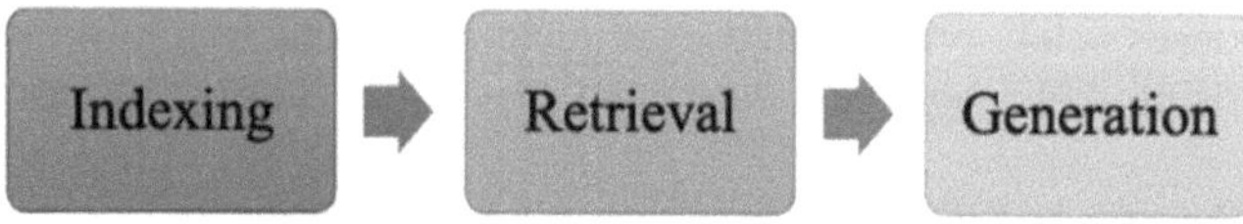

Fig. 1 Three-step RAG pipeline (ingestion, retrieval, generation) [16].

2.2.1 Chunking, Metadata and Reranking

Chunk size impacts retrieval precision and latency. Chunks of sentence or short paragraph size are balanced in coherence and significance [17]. Section labels (e.g. Programme, Entry Requirements, Fees) as metadata allow scoped retrieval and reduce off-topic results [13, 14]. Reranking boosts precision by pushing the most important passages up, and in the case of conditional reranking, accuracy vs. latency tradeoffs are made [13, 18].

2.2.2 Faithfulness and Evaluation

RAG systems ensure grounding by requiring citation in prompts and abstention when evidence is lacking. Evaluation decouples the quality of retrieval (recall, precision, ranking metrics) from faithfulness of generation [13]. Benchmarks such as RAGTruth allow systematic research of hallucinations in retrieval-augmented environments and are used to examine faithfulness to the recovered evidence [19].

2.3 RAG in Higher-Education Information Systems

RAG performs effectively in this setting as it bridges the gap between formal institutional databases and unstructured textual material (e.g. promotional PDFs and handbooks). These frameworks have specialized versions in the recent literature. For example, Chen et al. [4] built a RAG-powered university admission helper specifically for parsing complicated policy papers and demonstrated considerable improvements in backlogs of human inquiries. Similarly, the deployment of JayBot highlighted the efficacy of large language models in guiding potential candidates via multi-criteria program suggestions by matching user profiles with institutional offering [5]. Beyond pure informative tasks, integrated systems assessed on industry benchmarks underline that document-grounded text creation directly improves user trust compared to ungrounded parametric models [20].

In the current educational assistants, however, there is a different architectural pattern: heavy reliance on commercial, external cloud loops, such as GPT API layers or cloud-managed vector indices. This dependence forms a key gap in literature towards the performance of local, quantized open-source pipelines when limited to local campus servers. This constraint is a contextual demand that is very important in the fragmented higher education environment of Malaysia and Penang [21, 22].

As systematized in Table 1, traditional information systems have challenging tradeoffs between update overhead and contextual openness. Cloud-hosted models do hold information in the cloud but bring in unknown subscription costs and pass off control of sensitive data to outside platforms. The proposed local RAG helper addresses these difficulties by managing data processing on private infrastructure. Therefore, institutions can base responses solely on validated local materials, protecting information privacy and avoiding rising API usage fees.

Table 1 Comparison of Higher-Education Information Access Approaches

System type	Freshness	Transparency	Maintenance burden	Typical failure mode
Static FAQ pages and site search	Manual updates only; highly prone to information staleness [7].	Low; pages rarely cite provenance beyond the page itself.	High; decentralized content authors must manually track changing data.	Outdated or generic answers when academic structures or fees change.
Traditional non-retrieval chatbot	Dependent on pre-authored intents and answers; updates require content edits [2].	Low to moderate; lack of granular, automated inline citation mechanisms [9].	Continuous structural authoring of intent trees, slot-filling rules, and scripts.	Brittle outside scripted paths; stale text returned from old knowledge bases [3].
Retrieval-Augmented Generation	Automated index refresh updates knowledge at inference time [12, 14].	High; generation is explicitly bounded by extracted document context [13].	Curate and refresh corpus; maintain retriever and evaluation.	Latency and quality depend on corpus coverage and chunking [6].

2.4 Cost, Governance and Transparency Gaps in Existing Approaches

As an alternative, the literature suggests local or institution-controlled RAG deployments [13, 14]. Recent academic implementations rely on state-of-the-art generative models and are largely biased towards public cloud environments like as OpenAI's GPT architectures or centralized cloud-based RAG pipelines [4, 5, 20]. These cloud systems are linguistically fluent yet have important operational limits. The gap between current chatbot solutions and the requirement for cost-aware, transparent and updateable systems is the motivation behind the strategy adopted in this project [13, 14].

3 System Architecture and Design

3.1 System Architecture

The suggested system architecture is based on a three-tier approach, which divides the web interface, the application logic, and the data and knowledge sources. The user interacts with the Presentation Tier (Web UI) via a browser-based chat interface built with standard HTML, JavaScript and Bootstrap. This frontend has a simple text box for queries, streamed answers, and source links backing them. The backend API is called by the browser via HTTP requests when a user makes a query or feedback. Figure 2 shows an overview of the key components of the system and their data flow.

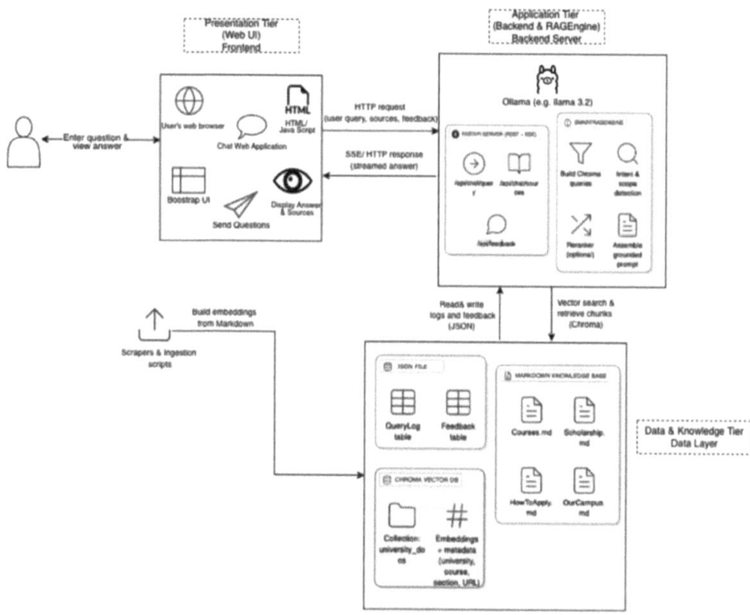

Fig. 2 System architecture of the locally hosted RAG-based higher education information assistant

Application Tier: It is built with FastAPI. FastAPI exposes essential endpoints like /api/chat/query for user questions, /api/chat/sources to get documents and /api/chat/feedback for user reactions. The SmartRAGEngine component has the basic orchestration logic. The module identifies the user intent and scope, produces metadata-aware queries to the vector database, optionally re-ranks the resulting chunks, and builds a grounded prompt. The final output is provided by Ollama which is hosting a local language model such as llama3.2.

Figure 3 summarises the end-to-end control flow for retrieval, reranking, generation, and citation display.

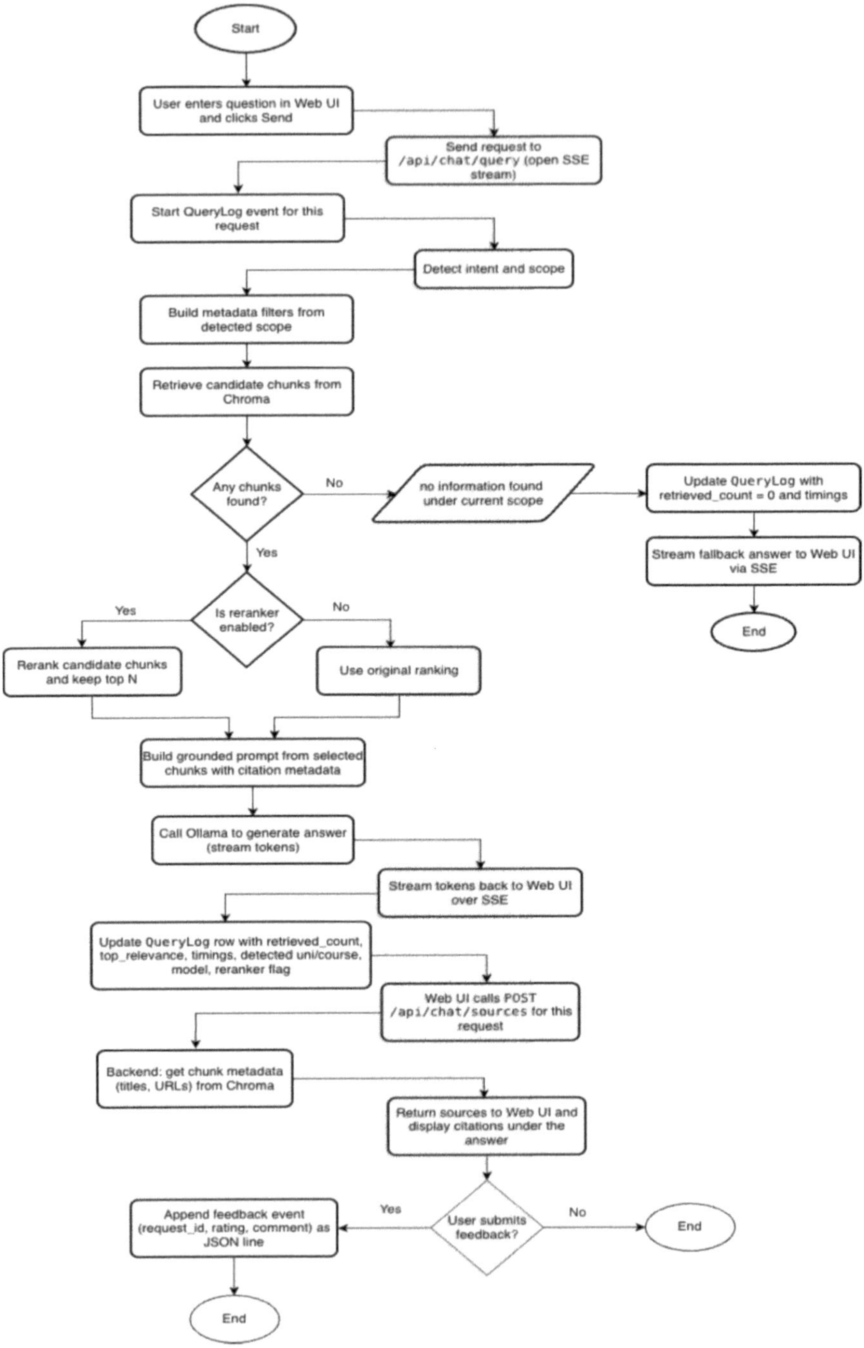

Fig. 3 End-to-end flow for retrieval and answer generation in the RAG-based chatbot

3.2 Data Collection, Preprocessing, and Knowledge Base

The Data and Knowledge Tier merges a structured logging store with a domain-specific knowledge base. The factual knowledge for answering questions is stored in two kinds. For each institution participating in the program, a curated markdown knowledge base exists, containing documents like Courses.md, Scholarship.md, HowToApply.md, and OurCampus.md. Secondly these markdown files are analyzed by special scrapers and ingestion programs, extracting key sections and building dense vector representations. The generated chunks, along with extensive metadata like university, course, section and URL are saved in a Chroma vector database under the university_docs collection.

3.3 Runtime Query Flow, Retrieval, and Generation

A user types an inquiry into the web interface. The Web UI transmits it to the backend at /api/chat/query, and then opens a server-sent events stream so the reply can come back token by token. Backend generates a new request_id for this query and passes it to the SmartRAGEngine. The SmartRAGEngine identifies the intent and scope of the query, e.g. whether it targets simple documents or course sections, and if possible infers the university, course type and section. Using these hints, the engine queries Chroma with metadata filters and over-fetches the most relevant chunks. If the reranker is enabled, the engine sends the retrieved candidate chunks to the cross-encoder and keeps the re-ordered top results before building a grounded prompt from the selected chunks. When no chunks are found, the system streams a fallback answer back to the web interface via SSE.

From the detected scope the engine builds metadata filters (institution, document type, section) and uses them to retrieve candidate chunks from Chroma. When no chunks are found, the system marks that no information was found under the current scope, updates the QueryLog entry with retrieved_count = 0 and timing information, and streams a fallback answer back to the web interface via SSE.

Next, the engine calls Ollama to generate the answer. Tokens stream back to the backend and are immediately forwarded to the browser over SSE. After the answer is shown, the Web UI calls /api/chat/sources for the same request. The backend retrieves the chunk metadata, including stored URLs, and returns these sources to the UI for display.

4 Evaluation and Results

The evaluation scenarios and outcomes for the information bot based on HEIs in Penang are presented in this part. Testing phase consists of white-box and black-box methods.

4.1 Evaluation Scenarios (Student Queries)

Evaluation was organised around realistic student questions that reflect the core functions of the chatbot:

1. listing universities in Penang,

2. returning scholarship information for a selected institution,
3. listing programmes for one university grouped by level,
4. listing universities offering a specific programme at a given level,
5. returning programme structure (subjects) for a selected course,
6. returning top-3 cheapest fees for a programme based on RM amounts,
7. handling out-of-scope questions with a fixed fallback message, and
8. logging thumbs up/down feedback with optional comment and sources.

4.2 White-Box Evaluation (Unit Tests)

White-box unit tests tested the major backend conditions and branches. For each unit test, the input query, preconditions (e.g., Chroma DB contains relevant documents), expected output format (title, bullet structure, and sources), and code conditions (success vs. fallback branches) were defined and executed. All defined unit tests passed.

4.3 Black-Box Evaluation (User Interface Tests)

The same situations were tested from the end user side using the web interface in black-box testing. Each test scenario started with a genuine user action in the chatbot interface, and the actual answer was compared with the required functional criteria (including answer structure, content accuracy, availability of source URLs and behaviour when information is not available). All defined black box test scenarios passed.

4.4 Accuracy, Relevance, and Qualitative Observations

During the testing period, the chatbot always gave nicely formatted replies, with clear headings and bulleted lists. It also always provided direct connections to sources when the information was present in the knowledge base. Furthermore, the system handled the transition well when a scoped search returned no matching material, by simply providing a fallback message that no records were found. Likewise, for completely out-of-scope questions, the interface prompted for a fixed default answer, thus preventing the generation of unnecessary program, tuition or financial assistance information. Overall, these results demonstrate that the platform performs well in satisfying its core functional criteria for typical higher education information requests.

4.5 Benchmark Results (Accuracy and Latency)

Together with the qualitative evaluation suite, we isolated a targeted exploratory benchmark set ($n = 4$) of representative student queries to enable extensive end-to-end performance and empirical profiling. This small sample size allows a first feasibility baseline in a resource-constrained local infrastructure. For the purpose of objectivity, the answer correctness and source-grounding alignment were checked by an internal expert panel, with the principal system developer and the academic research supervisor as lead domain referee. This human in the loop verification guarantees 100% faithfulness to official curriculum data with a clear correctness rate of 100% (4/4) for the target profiles under review. The end-to-end latency observed for each query complexity was different: How to apply INTI? (99.198 s), Where is ATC campus located? (47.169 s), List scholarships for UOW (132.240 s), and What is the fees of MSU bachelor-information-technology-mobile-wireless? (10.367 s). The results of the baseline show that the assistant is able to offer completely precise and contextual replies on the local infrastructure.

4.6 Performance Observations (Scalability)

SSE gradually provides responses to the web interface, enhancing the responsiveness user experience. We also did some basic concurrency tests to see the latency under concurrent access. The recorded latency statistics with the concurrency of 8 were: average 53.308 s, 95th percentile 60.664 s, minimum 19.624 s, maximum 100.157 s. In addition, the system provides monitoring outputs for per query behaviour and scalability (concurrency versus latency), which allow practical evaluation of response latency under typical usage.

5 Discussion

This section analyzes the strengths, limits and practical consequences of the information bot for the HEIs. The system combines online scraping, with a structured knowledge base, a retrieval-augmented generation (RAG) pipeline and a local LLM to deliver condensed programme information to the user.

5.1 Strengths of the Approach

One of the great benefits of this strategy is that it provides improved access to informa tion on higher education. Such a coherent synthesis is especially valuable for students actively comparing academic options at multiple colleges. Another key strength is the underlying technical architecture. A retrieval-augmented generation pipeline (Chroma) from a localized large language model (Ollama) is utilized to guarantee precise document retrieval and seamless natural language generation. Finally, the source Markdown texts are chunked section-wise so that the retrieval engine can extract the relevant modules such as financial structures or admittance criteria without the context being polluted by unrelated data. This is a unique compliance and economic benefit for institutional use, as it provides increased data protection and reduced operational reliance on external cloud services by running the language model only on local infrastructure.

5.2 Limitations

However, several essential limitations of the present prototype have to be acknowledged. The knowledge base of the system is limited in scope and coverage to a target cohort of regional higher education institutions and specified academic programs, and hence cannot provide complete answers for universities not indexed. In addition, retrieval fidelity is still constrained by the baseline quality and structural heterogeneity of public institutional websites. Deeply nested subpages, irregular HTML layouts, and non-text assets, for example, fee structures embedded in rasterized images significantly reduce the effectiveness of automated web-scraping extraction. The other limitation is lag in operations when users search for something complicated or wide in scope. Narrow yet highly filtered questions are answered within acceptable response times but broad questions requiring large-scale multi-document retrieval and cross-encoder reranking increase the processing overhead and lengthen the token generation times. This variation in latency is a common performance trade-off for localized, resource-constrained device deployments.

6 Conclusions and Future Work

The prototype can be extended in many directions for future work. Firstly the information base can be extended to other institutions, programmes and scholarship schemes other than the current Penang-centric set. Websites change and material may be scraped and validated periodically to keep it up to date. Secondly, extraction can be improved for sites with crucial information, such fees in images, e.g. by adding OCR-based extraction. Third, query comprehension can be enhanced by establishing a small domain specific dataset of authentic user inquiries and their verified responses and using the dataset to enhance the intent categorization and routing, eliminating the requirement on heuristic rules for imprecise queries. Fourth, retrieval quality can be enhanced by experimenting with embedding models and rerankers fitted to educational content. Finally, if the system is opened to a larger user base, the deployment pipeline can be further expanded to more robust systems, and the performance and scalability can be improved via GPU-enabled deployment and inference optimization frameworks (e.g., TensorRT or OpenVINO). A modest user research with students and counselors would also provide more systematic evidence on usability and impact.

References

1. Nagel, M.T., et al.: How do university students' web search behavior, website characteristics, and the interaction of both influence students' critical online reasoning? Front. Educ. **5**, 1–14 (2020)
2. Ezeife, C., Adepoju, O., Planas, M.: Machine learning-powered chatbots for real-time business intelligence and analytics. Int. J. Manag. Organ. Res. **2**(1), 43–55 (2021)
3. Izadi, S., Forouzanfar, M.: Error correction and adaptation in conversational AI: a review of techniques and applications in chatbots. AI. **5**(2), 803–841 (2024)
4. Chen, Z., et al.: Facilitating university admission using a chatbot based on large language models with retrieval-augmented generation (2024)
5. Odede, J., Frommholz, I.: Jaybot – aiding university students and admission with an LLM-based chatbot, pp. 391–395 (2024)
6. European Data Protection Board: Opinion 28/2024 on certain data protection aspects related to the processing of personal data in the context of AI models (2024). [Online]. Available: https://www.edpb.europa.eu/system/files/documents/2024-12/edpb_opinion_202428_ai-models_en.pdf
7. Labadze, L., Grigolia, M., Machaidze, L.: Role of AI chatbots in education: systematic literature review. Int. J. Educ. Technol. High. Educ. **20**(56) (2023)
8. Cerny, M.: The history of chatbots: the journey from psychological experiment to educational object. J. Appl. Tech. Educ. Sci. **12**(3) (2022)

9. Caldarini, G., Jaf, S., McGarry, K.: A literature survey of recent advances in chatbots. Information. **13**(1) (2022)
10. Pandey, S., Sharma, S.: A comparative study of retrieval-based and generative-based chatbots using deep learning and machine learning. Healthcare Analy. **3**, 100198 (2023)
11. Huang, L., et al.: A survey on hallucination in large language models: principles, taxonomy, challenges, and open questions. ACM Trans. Inform. Syst. **43**(2) (2025) [Online]. Available: https://dl.acm.org/doi/10.1145/3703155
12. Lewis, P., et al.: Retrieval-augmented generation for knowledge-intensive NLP tasks. In: Proceedings of the 34th International Conference on Neural Information Processing Systems, ser. NIPS '20. Curran Associates Inc., Red Hook (2020)
13. Fan, W., et al.: A survey on RAG meeting LLMs: towards retrieval-augmented large language models (2024)
14. Gao, Y., et al.: Retrieval-augmented generation for large language models: a survey. ArXiv. **abs/2312.10997** (2023) [Online]. Available: https://api.semanticscholar.org/CorpusID:266359151
15. National Institute of Standards and Technology (NIST): AI risk management framework (AI RMF 1.0), NIST, Tech. Rep. (2023)
16. Bora, A.: Comparative analysis of retrieval-augmented generation (RAG) based large language models (LLM) for medical chatbot applications (2024)
17. Chroma Core Research Team: Evaluating chunking strategies for retrieval-augmented generation. Chroma AI Research Division, Technical Report (2024). [Online]. Available: https://www.trychroma.com/research/evaluating-chunking
18. Santhanam, K., et al.: ColBERTv2: Effective and efficient retrieval via lightweight late interaction. In: Proceedings of the 2022 Conference of the North American Chapter of the Association for Computational Linguistics: Human Language Technologies (NAACL-HLT), pp. 3715–3734 (2022)
19. Niu, C., et al.: RAGTruth: a hallucination corpus for developing trustworthy retrieval-augmented language models. In: Proceedings of the 62nd Annual Meeting of the Association for Computational Linguistics (Volume 1: Long Papers), pp. 10862–10878. Association for Computational Linguistics, Bangkok (2024)
20. Garcia, I.G., et al.: Design and evaluation of an educational virtual agent based on retrieval-augmented generation for the entrepreneurship culture course in higher education. IEEE Revista Iberoamericana de Tecnologias del Aprendizaje, 1–1 (2026)
21. Ministry of Higher Education (Malaysia): Public universities (2021). [Online]. Available: https://jpt.mohe.gov.my/index.php/en/hei/public-hei/public-universities
22. Penang CAT Centre: Institutes of higher learning in Penang (2018). [Online]. Available: https://www.penangcatcentre.my/wp-content/uploads/2018/07/List-of-Institutes-Of-Higher-Learning-In-Penang.pdf; (penangcatcentre.my in Bing) Accessed 28 Nov 2025

The Role of Information Retrieval in the Development of Science: A Case Study of Biotechnology

Mahyuddin K. M. Nasution[1(✉)], Raditya Macy Widyatamaka Nasution[2], and Tutut Herawan[3]

[1] Data Science and Artificial Intelligence Study Program, Fakultas Ilmu Komputer dan Teknologi Informasi (Fasilkom-TI), Medan 20155, Sumatera Utara, Indonesia
mahyuddin@usu.ac.id
[2] Biotechnology Program, Universiti Malaya, Kuala Lumpur 50603, Malaysia
[3] Faculty of Computer Science and Information Technology, Universiti Malaya, Kuala Lumpur 50603, Malaysia
tutut@um.edu.my

Abstract. The development of science requires procedures for disclosing appropriate information. This information is not only contained within an information space but also requires the role of information retrieval, where metrics challenge measurement engagement, which requires a case study, a development in Biotechnology science. Based on the Scopus indexed database, the role of IR is revealed following data trends, which divide it into three datasets. However, scientific development continues to thrive due to the interaction between science and science, which is mapped with strength relations, and therefore requires categories to strengthen that mapping. Computation notes that the involvement of categories seems better than direct measurement through trends.

Keywords: average · standard deviation · category · data · behavior

1 Introduction

The development of science through the activities of the Tri Dharma (three pillars) of Higher Education produces various outcomes as evidence of each activity [1,2]. In addition to graduates as alumni, throughout their journey, every academic - as students or lecturers - carries out activities and consolidates their knowledge by writing and publishing their scientific work as articles, papers, book-(chapters), or registering them as intellectual property rights or other forms of scientific publication [3–6].

Sciences grow and accumulate, giving rise to technology as a milestone in their development [7]. Together with the results of studies - scientific publications [8] - administratively [9], this journey leads to the establishment of study programs [10], such as Biotechnology [11], the case study in this paper, where

N. Mohd Sharef et al. (Eds.): CAMP 2026, CCIS 3093, pp. 304–318, 2027.
https://doi.org/10.1007/978-981-92-4302-0_22

Information Retrieval (IR) plays a role in demonstrating it as a case of science development. The role of IR in this paper is to find and present relevant information from large, semi-structured data sets, Scopus, to meet user needs quickly and accurately, through the use of queries as data feasibility management to demonstrate the development of science [12,13], Biotechnology [14,15].

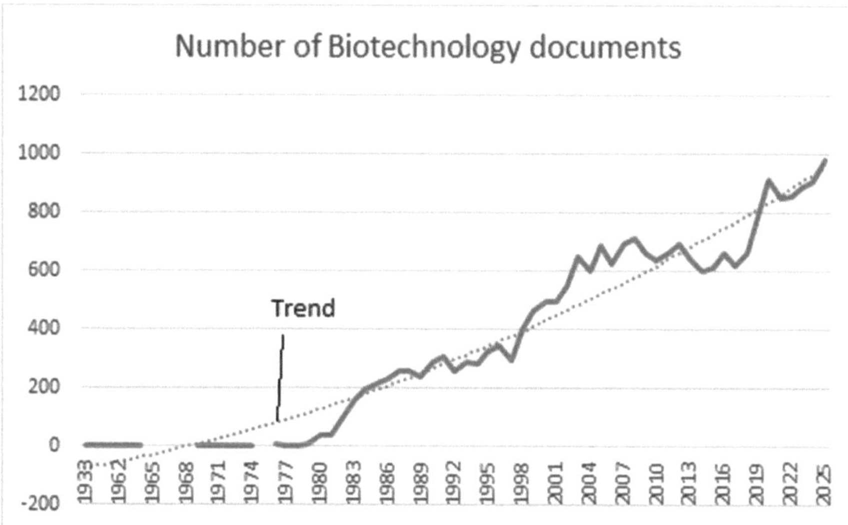

Fig. 1. Number of Biotechnology documents in Scopus database 1933–2025.

2 Methodology

A science, such as biotechnology [16], in its development is based on an initial concept, namely a definition [D1] which states that "biotechnology is a discipline combining biology and technology" [17]. With a strategy for sustainability in the world of education [18], the development continues to strengthen as expressed in Fig. 1, based on the process of searching and filtering documents by IR to obtain managed data [19]. With the role of IR in connecting researchers to various global repositories [20], in this case, the Scopus reputable database is used to lift appropriate data through the query $q(\text{Article title}) =$ "Biotechnology", or filtering data on the number of biotechnology documents through queries [21]. The number of documents on biotechnology per year shows a linear upward trend.

The trend serves as an indication of the demand for ideas on biotechnology studies based on what has been developed. In other words, the evidence of the earlier period becomes the basis for the evidence of the later period (present), but reveals further (future) evidence.

Table 1. Datasets for number of biotechnology documents

	Subject area	Year		
n	X	1933–2023	2024	2025
1	Agricultural and Biological Sciences	3982	313	324
2	Arts and Humanities	656	12	21
3	Biochemistry, Genetics and Molecular Biology	9620	495	525
4	Business, Management and Accounting	1725	89	51
5	Chemical Engineering	4647	199	246
6	Chemistry	1389	76	66
7	Computer Sciences	513	46	50
8	Decision Sciences	230	10	11
9	Dentistry	17	1	1
10	Earth and Planetary Sciences	404	25	37
11	Economics, Econometrics and Finance	1096	54	27
12	Energy	452	42	43
13	Engineering	3875	265	267
14	Environmental Sciences	1941	124	155
15	Health Professions	150	16	7
16	Immunology and Microbiology	4176	207	211
17	Materials Science	658	41	38
18	Mathematics	144	3	5
19	Medicine	3025	154	170
20	Multidisciplinary	668	15	14
21	Neuroscience	81		4
22	Nursing	305	5	7
23	Pharmacology, Toxicology and Pharmaceutics	1308	46	55
24	Physics and Astronomy	379	32	21
25	Psychology	129	4	6
26	Social Sciences	2108	78	83
27	Veterinary	256	17	10
	Total	43934	2369	2455
	Average, $\mu =$	1627	88	91
	standard deviation, $\sigma =$	8142	439	455

2.1 An Approach to Data Behavior

In its development, science does not no-grow linearly, science increasingly influences human life, and distributes itself to other scientific fields [22]. When the subject area states taxonomically scientific clusters, as in indexing database Scopus, suppose $X = \{x_i : i = 1, \ldots, n = 27\}$ is the number of documents per subject area [23], with the initial behavior found in the following measurements [24,25]: average

$$\mu = \frac{\sum_{i=1}^{n} x_i}{n} \tag{1}$$

and standard deviation

$$\sigma = \sqrt{\frac{\sum_{i=1}^{n} (x_i - \mu)^2}{n}}. \tag{2}$$

Then, by using Eq. (1), the subject areas are grouped into two groups: $x_i < \mu$ and $x_i \geq \mu$, and with the treatmen of Eq. (2) reveal the behavior of the data,

$$z_i = \frac{x_i - \mu}{\sigma} \tag{3}$$

There is a distribution of scientific documents with titles containing the word Biotechnology, from 1933 to 2023, followed by 2024 and 2025, into 27 subject areas, Table 1, each dataset has its own behavior. Each subject area for each year meets the considerations of Eq. (3), where $z_{-\alpha=-0.025} \leq z_i \leq z_{\alpha=0.025}$, and overall the data are in a normal distribution, or random even though $n < 30$ [26].

X	1	2	3	4	5	6	7	8	9	10	11	12	13	14	15	16	17	18	19	20	21	22	23	24	25	26	27
1	3982	68	1854	60	121	160	14	9		57	243	31	579	527	37	370	21	6	285	6	11	63	52	18	1	191	148
2	69	656	61	40	1	1	30	1		7	43	3	62	79	4	2	6	1	72	7	30	31	1	1	40	348	
3	1854	61	9620	286	3502	501	82		1	22	351	143	2000	486	62	3669	78	21	1017	9	58	41	536	40	1	114	54
4	60	40	286	1725	168	19	65	185		11	648	23	271	61	14	4	46	4	51		1	23	11	3	39	272	
5	121	1	3502	168	4647	561	76			13	2	178	1832	237	4	2188	215	6	131				68	58		21	
6	160	1	501	19	561	1389	34			17		105	248	170	14	53	200	12	86	6			108	93		9	
7	14	30	82	65	76	34	513	36			21	34	180	23	3	3	56	85	36	2	3		1	58	2	160	
8	9	1		185			36	230			4	2	34	2				14	1							15	
9			1						17						1				6								
10	57	7	22	11	13	17				404	3	40	78	260		4	16	1	4			5	3	5		32	
11	243	43	351	684	2	170	21	4		3	1096	7	23	74	12		2	2	19				5	2	4	204	1
12	31	3	143	23	178		34	2		40	7	452	72	295		45	29	8	10					6		53	
13	579	62	2000	271	1832		180	34		78	23	72	3875	302	17	938	409	59	499	4	2	3	253	145		179	13
14	527	79	486	61	237		23	2		260	74	295	302	1941	5	121	52	4	105		2	2	39	7		340	9
15	37	79	62	14	4	14	3		1		12		17	5	150	22	1		114		1	4	35	2	1	18	8
16	370	2	3669	4	2188	53	3			4		45	938	121	22	4176	11	1	375		12		106	3		18	17
17	21	6	78	46	215	200	56			16	2	29	409	52	1	11	658	34	26		1		11	157		1	
18	6	1	21	4	6	12	85	14		1	2	8	59	4		1	34	144	2		2		2	41		19	
19	285	72	1017	51	131	86	36	1	6	4	19	10	499	105	114	375	26	2	3025	3	16	218	509	14	16	193	19
20	6	7	9			6	2						4						3	668				7		4	
21	11	30	58	1			3						2	2	1	12	1	2	16		81	2	14		5	2	
22	63	31	41	23						5			3	2	4				218		2	305	10		3	97	
23	52	1	536	11	68	108	1			3	5		253	39	35	106	11	2	509		14	10	1308			13	3
24	18	1	40	3	58	93	58			5	2	6	145	7	2	3	157	41	14	7				379		3	
25	1	40	1	39			2				4				1				16		5	3			129	67	
26	191	348	114	272	21	9	160	15		32	204	53	179	340	18	18	1	19	193	4	2	97	13	3	67	2108	
27	148		54								1		13	9		17			19				3				256

Fig. 2. Interaction between subject areas in the number of Biotechnology documents in the Scopus database, 1933–2023.

Furthermore, the development of science is driven by the interaction between 27 subject areas where there are $n(n-1) = 702$ data, as in Fig. 2, which maps the development behavior, where IR has the potential to show the acceleration of the growth of scientific literature, related to the case raised, and requires some treatment [27,28].

2.2 An Approach for Treatment

The growth of documents on biotechnology shows an upward trend, but confirmation is needed to state the continuity of development by linking one set of information with another set of information [29]. For this reason, separating data from the beginning to 2023 as the baseline dataset (X_1) that has determined biotechnology as a current study program, and data in 2024 (X_2) and data in 2025 (X_3) respectively as comparison datasets [30].

Suppose it stated that $tp = |X_i \cap X_j = \{x_i \cap x_j : x_i \geq \mu \wedge x_j \geq \mu\}|$, $fp = |X_i = \{x_i < \mu\}|$ and $fn = |X_j = \{x_j < \mu\}|$, then precision ($pr$) answers about "out of the all the instances the model predicted as positive", or

$$pr = \frac{tp}{tp + fp}, \tag{4}$$

and recall (re) answers about "out of all the actual positive instances", or

$$re = \frac{tp}{tp + fn}, \tag{5}$$

where tp (true positive) is correctly predicted positive, fp (false positive) is incorrectly predicted positive, and fn (false negative): incorrectly predicted. The metrics for the treatment of IR used are pr and re. Next, $tn = |X_i \cap X_j = \{x_i \cap x_j : x_i < \mu \wedge x_j < \mu\}|$ as a complementary metric [31].

Table 2. Precision and recall enumeration

	Number of		X_1		X	X_1		
	$x_i \geq \mu$	$x_i < \mu$	tp	tf	fp	fn	pr	re
X_1	9	18		18				
X_2	8	19	8		19	18	30.77%	29.63%
X_3	7	20	7		20	18	28.00%	25.93%

The balance between datasets, X_1 and X_2 or X_1 and X_3, determines the quality of the generic data model of the database based on the coverage of matrices tp, fp, tn, fn in the accuracy metric (ac) [32],

$$ac = \frac{tp + tn}{tp + tn + fp + fn} \tag{6}$$

With the precision and recall of X_2 and X_3 respectively as Table 2 and Eq. (6) states their respective accuracies are 41.27% and 39.68%.

X	1	2	3	4	5	6	7	8	9	10	11	12	13	14	15	16	17	18	19	20	21	22	23	24	25	26	27
1	3982	68	1854	60	121	160	14	9	0	57	243	31	579	527	37	370	21	6	285	6	11	63	52	18	1	191	148
2	0,015	656	61	40	1	1	30	1	0	7	43	3	62	79	4	2	6	1	72	7	30	31	1	1	40	348	0
3	0,158	0,006	9620	286	3502	501	82	0	1	22	351	143	2000	486	62	3669	78	21	1017	9	58	41	536	40	1	114	54
4	0,011	0,017	0,026	1725	168	19	65	185	0	11	648	23	271	61	14	4	46	4	51	0	1	23	11	3	39	272	0
5	0,014	0,000	0,325	0,027	4647	561	76	0	0	13	2	178	1832	237	4	2188	215	6	131	0	0	0	68	58	0	21	0
6	0,031	0,000	0,048	0,006	0,102	1389	34	0	0	17	0	105	248	170	14	53	200	12	86	6	0	0	108	93	0	9	0
7	0,003	0,026	0,008	0,030	0,015	0,018	513	36	0	0	21	34	180	23	3	3	56	85	36	2	3	0	1	58	2	160	0
8	0,002	0,001	0,000	0,105	0,000	0,000	0,051	230	0	0	4	2	34	2	0	0	0	14	1	0	0	0	0	0	0	15	0
9	0,000	0,000	0,000	0,000	0,000	0,000	0,000	0,000	17	0	0	0	0	0	1	0	0	0	6	0	0	0	0	0	0	0	0
10	0,013	0,007	0,002	0,005	0,003	0,010	0,000	0,000	0,000	404	3	40	78	260	0	4	16	1	4	0	0	5	3	5	0	32	0
11	0,050	0,025	0,034	0,320	0,000	0,073	0,013	0,003	0,000	0,002	1096	7	23	74	12	0	2	2	19	0	0	0	5	2	4	204	1
12	0,007	0,003	0,014	0,011	0,036	0,000	0,037	0,003	0,000	0,049	0,005	452	72	295	0	45	29	8	10	0	0	0	0	6	0	53	0
13	0,080	0,014	0,174	0,051	0,274	0,000	0,043	0,008	0,000	0,019	0,005	0,017	3875	302	17	938	409	59	499	4	2	3	253	145	0	179	13
14	0,098	0,031	0,044	0,017	0,037	0,000	0,009	0,001	0,000	0,125	0,025	0,141	0,055	1941	5	121	52	4	105	0	2	2	39	7	0	340	9
15	0,009	0,109	0,006	0,008	0,001	0,009	0,005	0,000	0,006	0,000	0,010	0,000	0,004	0,002	150	22	1	0	114	0	1	4	35	2	1	18	8
16	0,048	0,000	0,362	0,001	0,330	0,010	0,001	0,000	0,000	0,001	0,000	0,010	0,132	0,020	0,005	4176	11	1	375	0	12	0	106	3	0	18	17
17	0,005	0,005	0,008	0,020	0,042	0,108	0,050	0,000	0,000	0,015	0,001	0,027	0,099	0,020	0,001	0,002	658	34	26	0	1	0	11	157	0	1	0
18	0,001	0,001	0,002	0,002	0,001	0,008	0,149	0,039	0,000	0,002	0,002	0,014	0,015	0,002	0,000	0,000	0,044	144	2	0	2	0	2	41	0	19	0
19	0,042	0,020	0,087	0,011	0,017	0,020	0,010	0,000	0,002	0,001	0,005	0,003	0,078	0,022	0,037	0,055	0,008	0,001	3025	3	16	218	509	14	16	193	19
20	0,001	0,005	0,001	0,000	0,000	0,003	0,002	0,000	0,000	0,000	0,000	0,000	0,001	0,000	0,000	0,000	0,000	0,000	0,001	668	0	0	0	7	0	4	0
21	0,003	0,042	0,006	0,001	0,000	0,000	0,005	0,000	0,000	0,000	0,000	0,000	0,001	0,001	0,004	0,003	0,001	0,009	0,005	0,000	81	2	14	0	5	2	0
22	0,015	0,033	0,004	0,011	0,000	0,000	0,000	0,000	0,000	0,007	0,000	0,000	0,001	0,001	0,009	0,000	0,000	0,000	0,070	0,000	0,005	305	10	0	3	97	0
23	0,010	0,001	0,052	0,004	0,012	0,042	0,001	0,000	0,000	0,002	0,002	0,000	0,051	0,012	0,025	0,020	0,007	0,001	0,133	0,000	0,010	0,006	1308	0	0	13	3
24	0,004	0,001	0,004	0,001	0,012	0,056	0,070	0,000	0,000	0,006	0,001	0,007	0,035	0,003	0,004	0,001	0,103	0,085	0,004	0,007	0,000	0,000	0,000	379	0	3	0
25	0,000	0,054	0,000	0,021	0,000	0,000	0,003	0,000	0,000	0,000	0,003	0,000	0,000	0,000	0,004	0,000	0,000	0,000	0,005	0,000	0,024	0,007	0,000	0,000	129	67	0
26	0,032	0,144	0,010	0,076	0,003	0,003	0,065	0,006	0,000	0,013	0,068	0,021	0,031	0,092	0,008	0,003	0,000	0,009	0,039	0,001	0,001	0,042	0,004	0,001	0,031	2108	0
27	0,036	0,000	0,005	0,000	0,000	0,000	0,000	0,000	0,000	0,000	0,001	0,000	0,003	0,004	0,000	0,004	0,000	0,000	0,006	0,000	0,000	0,000	0,002	0,000	0,000	0,000	256

Fig. 3. Interaction between subject areas in [0, 1], Biotechnology documents in 1933–2023.

X	1	2	3	4	5	6	7	8	9	10	11	12	13	14	15	16	17	18	19	20	21	22	23	24	25	26	27
1	3982	68	1854	60	121	160	14	9	0	57	243	31	579	527	37	370	21	6	285	6	11	63	52	18	1	191	148
2	E	656	61	40	1	1	30	1	0	7	43	3	62	79	4	2	6	1	72	7	30	31	1	1	40	348	0
3	EH	ML	9620	286	3502	501	82	0	1	22	351	143	2000	486	62	3669	78	21	1017	9	58	41	536	40	1	114	54
4	ML	E	MH	1725	168	19	65	185	0	11	648	23	271	61	14	4	46	4	51	0	1	23	11	3	39	272	0
5	E	AL	AH	MH	4647	561	76	0	0	13	2	178	1832	237	4	2188	215	6	131	0	0	0	68	58	0	21	0
6	MH	AL	H	ML	EH	1389	34	0	0	17	0	105	248	170	14	53	200	12	86	6	0	0	108	93	0	9	0
7	L	MH	ML	MH	E	E	513	36	0	0	21	34	180	23	3	3	56	85	36	2	3	0	1	58	2	160	0
8	VL	AL	AL	EH	AL	AL	H	230	0	0	4	2	34	2	0	0	0	14	1	0	0	0	0	0	0	15	0
9	AL	AL	AL	AL	AL	AL	AL	AL	17	0	0	0	0	0	1	0	0	0	6	0	0	0	0	0	0	0	0
10	E	ML	VL	L	VL	ML	AL	AL	AL	404	3	40	78	260	0	4	16	1	4	0	0	5	3	5	0	32	0
11	H	MH	MH	AH	AL	H	E	L	AL	VL	1096	7	23	74	12	0	2	2	19	0	0	0	5	2	4	204	1
12	ML	VL	E	ML	MH	AL	MH	L	AL	H	L	452	72	295	0	45	29	8	10	0	0	0	0	6	0	53	0
13	H	E	EH	H	AH	AL	MH	ML	AL	E	L	E	3875	302	17	938	409	59	499	4	2	3	253	145	0	179	13
14	EH	MH	MH	E	MH	AL	ML	AL	AL	EH	MH	EH	H	1941	5	121	52	4	105	0	2	2	39	7	0	340	9
15	ML	EH	ML	ML	AL	ML	L	AL	ML	AL	ML	AL	L	VL	150	22	1	0	114	0	1	4	35	2	1	18	8
16	H	AL	AH	AL	AH	ML	AL	AL	AL	AL	AL	ML	EH	E	L	4176	11	1	375	0	12	0	106	3	0	18	17
17	L	L	ML	E	MH	EH	H	AL	AL	E	AL	MH	EH	E	AL	VL	658	34	26	0	1	0	11	157	0	1	0
18	VL	AL	VL	VL	AL	ML	EH	MH	AL	VL	VL	E	E	VL	AL	AL	MH	144	2	0	2	0	2	41	0	19	0
19	MH	E	H	ML	E	E	ML	AL	VL	AL	L	L	H	E	MH	H	ML	AL	3025	3	16	218	509	14	16	193	19
20	AL	L	AL	AL	AL	L	VL	AL	AL	AL	AL	AL	AL	AL	AL	AL	AL	AL	AL	668	0	0	0	7	0	4	0
21	VL	MH	ML	AL	AL	AL	L	AL	AL	AL	AL	AL	AL	AL	L	VL	AL	ML	L	AL	81	2	14	0	5	2	0
22	E	MH	L	E	AL	AL	AL	AL	AL	ML	AL	AL	AL	AL	ML	AL	AL	AL	H	AL	L	305	10	0	3	97	0
23	ML	AL	H	L	E	MH	AL	AL	AL	VL	VL	AL	H	E	MH	E	ML	AL	EH	AL	ML	ML	1308	0	0	13	3
24	L	AL	L	VL	E	H	H	AL	AL	ML	AL	ML	MH	L	L	AL	EH	H	L	ML	AL	AL	AL	379	0	3	0
25	AL	H	AL	E	AL	AL	L	AL	AL	AL	L	AL	AL	AL	L	AL	AL	AL	L	AL	MH	ML	AL	AL	129	67	0
26	MH	EH	ML	H	L	VL	H	ML	AL	E	H	E	MH	EH	ML	L	AL	ML	MH	VL	AL	MH	L	AL	MH	2108	0
27	MH	AL	L	AL	AL	AL	AL	AL	AL	AL	AL	AL	L	L	AL	L	AL	AL	ML	AL	AL	AL	VL	AL	AL	AL	256

Fig. 4. "Interaction between subject areas in categories, Biotechnology documents in 1933–2023."

2.3 An Approach Towards Method

To map the influence of scientific developments and the needs of other scientific fields on biotechnology, the number of documents, as shown in Fig. 2, representing interactions between subject areas, is transformed to form a strength relationship using a symmetric similarity metric, the Jaccard coefficient,

$$sim(x_i, x_j) = \frac{|x_i \cap x_j|}{|x_i| + |x_j| - |x_i \cap x_j|} \tag{7}$$

and produces values of $x_{ij} = sim(x_i, x_j)$ in $[0, 1]$, as shown in Fig. 3, where the values are presented symmetrically between the lower left and upper right sides of the diagonal [33,34]. To interpret each measurement result within $[0, 1]$, each value needs to be grouped according to the resulting strength relationship category based on the distribution concept that the smallest value will be the most [35], while the largest value will be the least. In other words, if $b = max(sim(x_i, x_j))$, then the limit of each category is

$$\ell_i = \frac{b}{2^i}, i = 1, \ldots, 8. \tag{8}$$

and produces a dynamic scale as in Table 3. Based on the measurements in Fig. 3, the application of the dynamic scale is obtained as in Fig. 4 [36]. Thus, by searching documents about Biotechnology, a number of documents are generated for 1933–2023 from the database of reputable indexer Scopus, and it produces interactions between two subject areas, for example, x_1 = "Chemistry" and x_2 = "Computer Science" with $|x_1 \cap x_2| = 34$ documents, where $|x_1| = 1389$ documents and $|x_2| = 513$ documents, based on Eq. (7) the strength of relation is $sim(x_1, x_2) = 0.018$, and the category is E (Equal). The overall categories for X_1 is as Fig. 4.

Table 3. Dynamic scale

Linguistic terms (C)	Linguistic values
Absolutely high (AH)	$1 > x_i \geq \ell_1$
Extremely high (EH)	$\ell_1 > x_i \geq \ell_2$
High (H)	$\ell_2 > x_i \geq \ell_3$
Medium high (MH)	$\ell_3 > x_i \geq \ell_4$
Equal (E)	$\ell_4 > x_i \geq \ell_5$
Medium low (ML)	$\ell_5 > x_i \geq \ell_6$
Low (L)	$\ell_6 > x_i \geq \ell_7$
Very low (VL)	$\ell_7 > x_i \geq \ell_8$
Absolutely low (AL)	$\ell_8 > x_i \geq 0$

So taxonomically, it is possible to describe studies - with titles containing the word - biotechnology as developing in two directions, namely chemistry and computer science [38].

3 Research Analysis

In the manner described above, data management on scientific documents in Biotechnology has presented its development since its inception, but the sustainability of that development requires behavioral testing by comparing it with

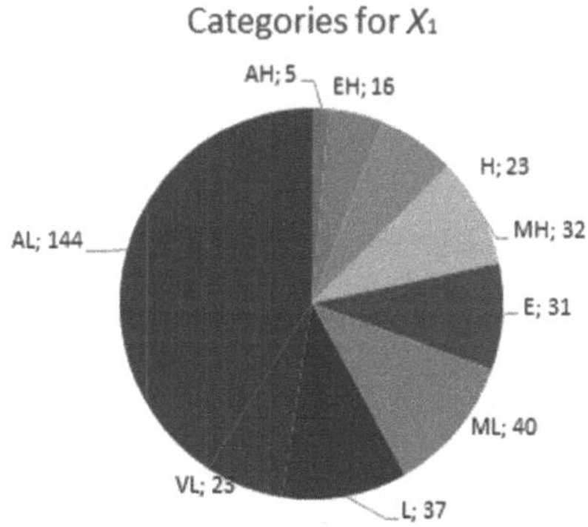

Fig. 5. Interaction between subject areas in categories.

the behavior of recent years [39]. Although at first glance it appears that the development of biotechnology follows an upward trend, Fig. 2, the measurement of the influence of that development on further development has been predicted as shown in Table 2 [40]. However, the spread of that influence across different scientific fields, Fig. 4, emphasizes the development of that science with the distribution of other sciences [41]. For a while, the distribution follows a pattern where all the small data gathers in a large number and then decreases little by little as the data size increases, Fig. 5.

With the data behavior revealed through the treatment, this behavior demonstrates that the treatment is an appropriate approach, and thus reveals a method for determining the role of IR in scientific development, which will then be tested in subsequent results [42,43]. The method is structured as follows:

1. Data behaves eligibly under appropriate measurements.
2. Data provides certainty regarding the definition of precision and recall as IR implementation metrics.
3. Information distribution mapping is tied to similarity metrics and grouped according to categories to map the development intent.
4. Validation of scientific development is carried out using the same IR role on comparative data that behaves as described in 1, 2, and 3.

4 Result

The flow of information about scientific development is recorded in a global scientific publication database. This flow flows through the interplay of scientific fields, as their development interacts, even though taxonomically, it originates from a field of thought [44], experiments in laboratory [45], discussion in research group [46], specifically a study program [47], for instance, such as Biotechnology. This information binds thought and determines the direction of scientific research and development, which sometimes deviates from the fundamental scientific interests of each scientist. Based on this, there are possible directions for the development of a science, and the attraction that determines that direction is tested by examining the strength of the relationship and determining its category (Fig. 8).

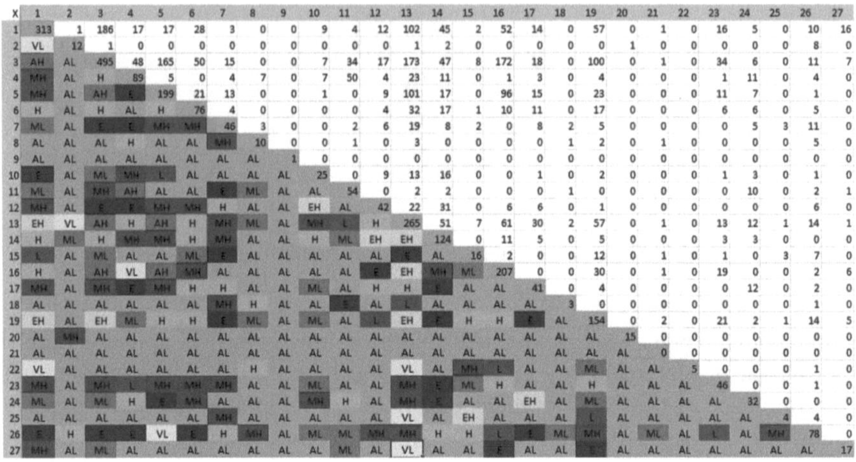

X	1	2	3	4	5	6	7	8	9	10	11	12	13	14	15	16	17	18	19	20	21	22	23	24	25	26	27
1	313	1	186	17	17	28	3	0	0	9	4	12	102	45	2	52	14	0	57	0	1	0	16	5	0	10	16
2	VL	12	1	0	0	0	0	0	0	0	0	0	1	2	0	0	0	0	0	1	0	0	0	0	0	8	0
3	AH	AL	495	48	165	50	15	0	0	7	34	17	173	47	8	172	18	0	100	0	1	0	34	6	0	11	7
4	MH	AL	H	89	5	0	4	7	0	7	50	4	23	11	0	1	3	0	4	0	0	0	1	11	0	4	0
5	MH	AL	AH	E	199	21	13	0	0	1	0	9	101	17	0	96	15	0	23	0	0	0	11	7	0	1	0
6	H	AL	H	AL	H	76	4	0	0	0	0	4	32	17	1	10	11	0	17	0	0	0	6	6	0	5	0
7	ML	AL	E	E	MH	MH	46	3	0	0	2	6	19	8	2	0	8	2	5	0	0	0	0	5	3	11	0
8	AL	AL	AL	H	AL	AL	MH	10	0	0	1	0	3	0	0	0	0	1	2	0	1	0	0	0	0	5	0
9	AL	AL	AL	AL	AL	AL	AL	AL	1	0	0	0	0	0	0	0	0	0	0	0	0	0	0	0	0	0	0
10	E	AL	ML	MH	L	AL	AL	AL	AL	25	0	9	13	16	0	0	1	0	2	0	0	0	1	3	0	1	0
11	ML	AL	MH	AH	AL	AL	E	ML	AL	AL	54	0	2	2	0	0	0	1	0	0	0	0	0	10	0	2	1
12	MH	AL	E	E	MH	MH	H	AL	AL	EH	AL	42	22	31	0	6	6	0	1	0	0	0	0	0	0	6	0
13	EH	VL	AH	H	AH	H	MH	ML	AL	MH	L	H	265	51	7	61	30	2	57	0	1	0	13	12	1	14	1
14	H	ML	H	MH	MH	H	MH	AL	AL	H	ML	EH	EH	124	0	11	5	0	5	0	0	0	3	3	0	0	0
15	L	AL	ML	AL	AL	ML	E	AL	AL	AL	AL	AL	E	AL	16	2	0	0	12	0	1	0	1	0	3	7	0
16	H	AL	AH	VL	AH	MH	AL	AL	AL	AL	AL	E	EH	MH	ML	207	0	0	30	0	1	0	19	0	0	2	6
17	MH	AL	MH	E	MH	H	H	AL	AL	ML	AL	H	H	E	AL	AL	41	0	4	0	0	0	0	12	0	2	0
18	AL	AL	AL	AL	AL	AL	MH	H	AL	AL	E	AL	L	AL	AL	AL	AL	3	0	0	0	0	0	0	0	1	0
19	EH	AL	EH	ML	H	H	E	ML	AL	ML	AL	L	EH	E	H	H	E	AL	154	0	2	0	21	2	1	14	5
20	AL	MH	AL	AL	AL	AL	AL	AL	AL	AL	AL	AL	AL	AL	AL	AL	AL	AL	AL	15	0	0	0	0	0	0	0
21	AL	AL	AL	AL	AL	AL	AL	AL	AL	AL	AL	AL	AL	AL	AL	AL	AL	AL	AL	AL	0	0	0	0	0	0	0
22	VL	AL	AL	AL	AL	AL	AL	H	AL	AL	AL	AL	VL	AL	MH	L	AL	AL	ML	AL	AL	5	0	0	0	1	0
23	MH	AL	MH	L	MH	MH	MH	AL	AL	ML	AL	AL	MH	E	ML	H	AL	AL	H	AL	AL	AL	46	0	0	1	0
24	ML	AL	ML	H	E	MH	AL	AL	AL	MH	H	AL	MH	E	AL	AL	EH	AL	ML	AL	AL	AL	AL	32	0	0	0
25	AL	AL	AL	AL	AL	AL	MH	AL	AL	AL	AL	AL	VL	AL	EH	AL	AL	AL	L	AL	AL	AL	AL	AL	4	4	0
26	E	H	E	E	VL	E	H	MH	AL	ML	ML	MH	MH	H	H	L	E	ML	MH	AL	ML	AL	L	AL	MH	78	0
27	MH	AL	ML	AL	AL	AL	AL	AL	AL	AL	ML	AL	VL	AL	AL	E	AL	AL	E	AL	AL	AL	AL	AL	AL	AL	17

Fig. 6. Interaction between subject areas in categories, Biotechnology documents in 2024.

4.1 Appeal Data Behavior

Regarding the scientific document data of Biotechnology from 1933 to 2023 (X_1), comparative data of 2024 (X_2) and 2025 (X_3) are submitted. The behavior of each is shown in Fig. 6 and Fig. 7. The behavior of the comparative data comes from all documents for 2024 and for 2025 for Biotechnology, as shown respectively on the upper right side, which are processed based on the Eqs. (7) and (8), and transformed following the dynamic scale category with results such as the lower left side of each figure.

A comparison between the distribution of categories for X_1, X_2, and X_3 reveals their respective behaviors, following a pattern where smaller data sizes are more numerous than larger data sizes are fewer. Thus, the overall trend for categories above E is upward, while the opposite is downward, Fig. 9. which means that the distribution of interactions is increasingly higher and more even between subject areas.

4.2 Prediction

By using the concept of Eqs. (4), (5), and (6), measurements are prepared as in the following Table 4.

X	1	2	3	4	5	6	7	8	9	10	11	12	13	14	15	16	17	18	19	20	21	22	23	24	25	26	27
1	324	6	175	23	14	14	2	0	20	20	4	4	68	61	3	35	3	0	29	1	1	0	11	1	0	10	8
2	E	21	6	5	3	3	1	0	0	0	3	0	5	3	0	0	0	0	1	2	0	1	0	0	4	15	2
3	AH	E	525	19	189	27	15	0	12	12	2	12	173	54	1	176	8	1	106	0	2	1	39	1	2	7	0
4	H	H	MH	51	4	4	6	6	7	7	16	2	23	4	0	1	0	0	6	0	0	0	2	0	0	12	0
5	MH	E	AH	E	246	27	16	0	3	3	1	13	133	19	0	91	16	1	49	0	0	0	5	9	0	1	0
6	AL	MH	H	MH	EH	66	6	0	0	0	1	5	11	8	1	1	11	0	6	0	0	0	7	4	0	1	0
7	L	E	MH	H	H	H	50	5	1	1	4	10	18	11	1	0	7	2	4	1	1	0	0	6	0	11	0
8	AL	AL	AL	EH	AL	AL	EH	11	0	0	0	0	5	0	0	0	0	1	2	0	0	0	0	0	0	7	0
9	AL	AL	AL	AL	AL	AL	AL	AL	1	0	0	0	0	0	0	0	0	0	0	0	0	0	0	0	0	0	0
10	H	AL	E	H	ML	AL	E	AL	AL	37	0	4	18	26	0	0	1	0	0	0	0	0	3	3	0	2	0
11	E	H	L	AH	L	ML	H	AL	AL	AL	27	2	4	6	0	1	0	0	2	0	0	0	2	0	2	8	1
12	E	AL	E	E	H	H	EH	AL	EH	H	MH	43	13	36	0	4	2	0	1	1	1	0	2	1	0	10	0
13	EH	E	AH	H	AH	MH	H	E	H	H	E	H	267	50	0	51	19	2	70	2	2	0	21	10	2	15	1
14	EH	E	H	E	H	MH	H	AL	AH	EH	MH	AH	EH	155	0	18	3	0	5	0	0	0	11	1	0	16	0
15	ML	AL	VL	AL	AL	E	E	AL	AL	AL	AL	AL	AL	AL	7	3	0	0	3	0	0	0	0	0	0	6	0
16	H	AL	AH	L	AH	L	AL	AL	AL	AL	L	E	AL	H	E	211	1	0	24	0	0	0	7	0	0	4	3
17	ML	AL	E	AL	H	EH	H	AL	MH	E	AL	MH	H	E	AL	L	38	0	1	0	0	0	0	11	0	0	0
18	AL	AL	VL	AL	L	AL	MH	H	AL	AL	AL	AL	ML	AL	AL	AL	AL	5	1	1	0	0	0	0	0	0	0
19	H	L	AH	MH	EH	MH	E	E	AL	AL	ML	L	AH	AL	E	H	ML	ML	170	1	1	6	29	0	0	8	0
20	AL	AL	VL	AL	AL	AL	AL	AL	AL	AL	AL	AL	L	AL	AL	AL	AL	H	AL	14	0	0	0	0	0	0	0
21	L	AL	L	AL	AL	AL	E	AL	AL	AL	AL	E	ML	AL	AL	AL	AL	AL	ML	AL	4	0	1	0	0	1	0
22	AL	MH	VL	AL	AL	AL	AL	AL	AL	AL	AL	AL	AL	AL	AL	AL	AL	AL	MH	AL	AL	7	0	0	0	2	0
23	MH	AL	H	E	E	H	AL	AL	H	MH	MH	E	H	H	AL	MH	AL	AL	EH	E	E	AL	55	0	0	1	1
24	L	AL	VL	AL	MH	H	EH	AL	EH	H	AL	E	MH	ML	AL	AL	EH	AL	AL	AL	AL	AL	AL	21	0	6	0
25	AL	EH	AL	MH	AL	AL	AL	AL	AL	AL	H	AL	ML	AL	AL	AL	AL	AL	AL	AL	AL	AL	AL	AL	6	0	0
26	MH	EH	E	EH	L	ML	EH	H	MH	E	H	H	H	H	H	E	AL	AL	MH	ML	E	MH	ML	AL	H	83	0
27	MH	AL	L	AL	AL	AL	AL	AL	AL	AL	MH	AL	L	AL	AL	E	AL	AL	AL	AL	AL	AL	E	AL	AL	AL	10

Fig. 7. Interaction between subject areas in categories, Biotechnology documents in 2025.

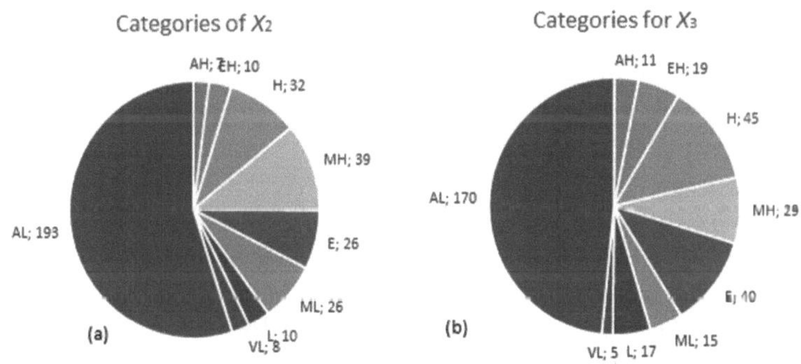

Fig. 8. Interaction between subject areas of X_2 and X_3 in categories.

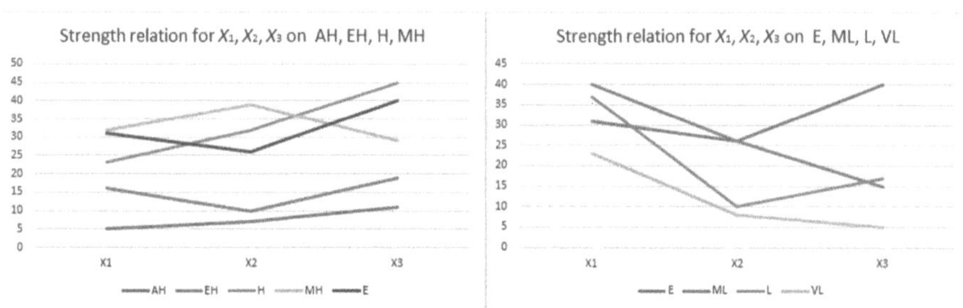

Fig. 9. Trend of categories X_1, X_2 and X_3.

Table 4. Prediction based on category

	Datasets			tp		fp		fn		tn		pr		re		ac		pt
C	X_1	X_2	X_3	X_2	X_3	X_2	X_3	X_2	X_3	X_2	X_3	X_2	X_3	X_2	X_3	X_2	X_3	
AH	5	7	11	5	5	0	0	2	6	0	0	1.000	1.000	0.714	0.455	0.714	0.455	0.847
EH	16	10	19	3	8	13	8	7	11	7	8	0.188	0.500	0.300	0.421	0.333	0.457	0.566
H	23	32	45	6	10	17	13	26	35	17	13	0.261	0.435	0.188	0.222	0.348	0.324	0.476
MH	32	39	29	11	9	21	23	28	20	21	20	0.344	0.281	0.282	0.310	0.395	0.403	0.564
E	31	26	40	8	8	23	23	18	32	18	23	0.258	0.258	0.308	0.200	0.388	0.360	0.529
ML	40	26	15	5	2	35	38	21	13	21	13	0.125	0.050	0.192	0.133	0.317	0.227	0.390
L	37	10	17	6	6	31	31	4	11	4	11	0.162	0.162	0.600	0.353	0.222	0.288	0.364
VL	23	8	5	0	1	23	22	8	4	8	4	0.000	0.043	0.000	0.200	0.205	0.161	0.261
AL	144	193	170	193	115	15	29	64	55	15	29	0.896	0.799	0.668	0.676	0.646	0.632	0.903
	351	351	351	173	164	178	187	178	187	111	121	0.493	0.467	0.493	0.467	0.444	0.432	0.619

1. $tp(X_2) = c(X_1) \cap c(X_2)$, and $tp(X_3) = c(X_1) \cap c(X_3)$, where c represents the category column, or by comparing the contents of cell in Fig. 4 with the contents of cell in Fig. 6 and the contents of cell in Fig. 4 with the contents of cell in Fig. 7, the resulting $tp(X_2) = \{5, 3, 6, 11, 8, 5, 6, 0, 193\}$ and $tp(X_3) = \{5, 8, 10, 9, 8, 2, 6, 1, 115\}$, respectively, correspond to their categories.
2. $fp(X_2) = c(X_1) - tp(X_2) = \{0, 13, 17, 21, 23, 35, 31, 23, 15\}$ and $fp(X_3) = c(X_1) - tp(X_3) = \{0, 8, 13, 23, 23, 38, 31, 22, 29\}$, where the values follow their respective categories.
3. $fn(X_2) = c(X_2) - tp(X_2) = \{2, 7, 26, 28, 18, 21, 4, 8, 64\}$ and $fn(X_3) = c(X_3) - tp(X_3) = \{6, 11, 35, 20, 32, 13, 11, 4, 55\}$, according to their respective categories.
4. $tn(X_2) = \{\min(fp(X_2), fn(X_2)\} = \{0, 7, 17, 21, 18, 21, 4, 8, 15\}$ and $tn(X_3) = \{\min(fp(X_3), fn(X_3)\} = \{0, 8, 13, 20, 23, 13, 11, 4, 29\}$, based on their respective categories.

and produced measurements of the provision of information pr, completeness of information re, and accuracy of information ac, which determine the prediction of science development with its distribution by category.

4.3 Discussion: Category Prediction

Information follows the reasoning of computational or measurement facts, while error follows computation that is comparable to accuracy, where it is related to translating keywords in queries that are comparable to the intended scientific development. In the method of building precision, categories, and accuracy, each category is a logical consequence of the category above it, because it follows the remaining percentages plus the percentage of error. A comprehensive understanding is not merely computational, but also a learning process, allowing for reduction, as represented by the potential measurement column (pt) as the resultant:

$$pt = \sqrt{ac(X_2)^2 + ac(X_3)^2 + ac(X_2)ac(X_3)\cos 90} \quad (9)$$

For this reason, all precision and recall in the dispersion category appear to be low, generally below 50%. Overall, the inference error leads to a low average computational result, $(\sum_{i=1}^{9} pr_i(X_2))/9 = 3.233/9 = 0.359$, even though all measurements are coupled, for example, each category alternates with values that are related to each other, as illustrated by the potential metric (Table 4), which provides an increase in predictive value.

By comparing the computational results against the trend, Table 2: $pr(X_2) = 30.77$; $pr(X_3) = 28.00$; $re(X_2) = 29.63$; $re(X_3) = 25.93$; $ac(X_2) = 41.27$; and $ac(X_3) = 39.68$, with the computational results based on category, Table 4: $pr(X_2) = 49.30$, $pr(X_3) = 46.70$; $re(X_2) = 49.30$; $re(X_3) = 46.70$; $ac(X_2) = 44.40$; and $ac(X_3) = 43.20$, it can be stated that the spread of a scientific field to another scientific field has a better impact on the development of science.

5 Conclusion

The role of IR has provided evidence of scientific development through document growth trends, albeit with low predictability. Improved predictability is seen when based on the distribution of categories that map the interactions of scientific development. The categories of the study object, Biotechnology, correspond to its dynamic characteristics. The development of metrics within IR follows the behavior of the data, where the treatment given enriches its meaning.

References

1. Abualrob, M.M.A., Daniel, E.G.S.: The Delphi technique in identifying learning objectives for the development of science, technology and society modules for Palestinian ninth grade science curriculum. Int. J. Sci. Educ. **35**(15), 2538–2558 (2013). https://doi.org/10.1080/09500693.2011.610381
2. Istambul, M. R.: The impact of i-performance in changing the work culture of lecturers to increase the productivity of three pilars (tri dharma) of higher education in Indonesia. Univ. J. Educ. Res. **7**(4), 15–21 (2019). https://doi.org/10.13189/ujer.2019.071403
3. Olivé, A., Cifuentes, I., Roca, J., Hashkes, P. J., Uziel, Y.: The fate of abstracts: without publication, science is deat [1] (multiple letters). J. Rheumatol. **31(5)**, 1007-1008 (2004). EID 2-s2.0-2342460315
4. Wong, C.-Y., Goh, K.-L.: The sustainability of functionality development of science and technology: papers and patents of emerging economies. J. Informet. **6**(1), 55–65 (2012). https://doi.org/10.1016/j.joi.2011.07.001
5. Nasution, M.K.M.: No research without publication: early mining. J. Phys: Conf. Ser. **987**(1) (2018). https://doi.org/10.1088/1742-6596/978/1/012063
6. Xu, W., Poole, A.: Academics without publications are just like imperial concubines without sons": the "new times" of Chinese higher education. J. Educ. Policy **39**(6), 861–878 (2024). https://doi.org/10.1080/02680939.2023.2288339

7. Hamid, N.A.W.A., Ismail, Z., Siyi, C.: An overview of artificial intelligence and its subsets. Artif. Intell. Comput. Fluid Dyn., 29–51 (2026). https://doi.org/10.1016/B978-0-443-29118-0.00002-5
8. Anh, N.H., Dien, T.T.: A comparative deep learning approach for image classification and retrieval in scientific publications. CTU J. Innov. Sustain. Dev. **18**(1), 156–168 (2026). https://doi.org/10.22144/ctujoisd.2026.016
9. Bauer, M., Reeves, P. M., Roney, J., Flore, S. M.: Expanding roles for research administration and research development professionals: a team science coaching program. J. Res. Adm. **55**(1), 7–27 (2025). EID 2-s2.0-105004359310
10. Bramantya, A. R., Rakhmawati, R.: Correlation of students final paper to archival scientific development: case study of archival science study program, vocational college, UGM. Rec. Libr. J. **8**(1), 185–198 (2022). https://doi.org/10.20473/rlj.V8-I1.2022.185-198
11. Nasution, R.M.W., Nasution, M.K.M.: A computational model of biotechnology. In: Lecture Notes in Networks and System, LNNS, vol. 723, pp. 122–133 (2023). https://doi.org/10.1007/978-3-031-35317-8_12
12. Chen, Z.: Enhancing database management to knowledge base management: the role of information retrieval technology. Inf. Process. Manag. **30**(3), 419–435 (1994). https://doi.org/10.1016/0306-4573(94)90054-X
13. Zhou, L., Zhang, D.: NLPIR: a theoretical framework for applying natural language processing to information retrieval. J. Am. Soc. Inf. Sci. Technol. **54**(2) (2002). https://doi.org/10.1002/asi.10193
14. Dasilva, E.J.: The colours of biotechnology: science, development and humankind. Electron. J. Biotechnol. **7**(3) (2004). EID 2-s2.0-12344295762
15. Rodrigues, L.C.: New biotechnologies and information retrieval: an empirical study the state-of-the-art of the biotechnology CRISPR/Cas9 of the funding agencies CAPES, CNPq, FAPERGS e FAPESP. Revista De Estudos Empiricos Em Direito **7**(3) (2020). https://doi.org/10.19092/reed.v7i3.343
16. Nasution, M.K.M., Nasution, R.M.W., Syah, R.B.Y, Elveny, M.: Biotechnology among computer science and data science: a review of scientific development. In: Lecture Notes in Networks and Systems, LNNS, vol. 597, pp. 903–911 (2023). https://doi.org/10.1007/978-3-031-21438-7_76
17. Prave, P., Sukatsch, D.A., Faust, U.: Biotechnology: the industrial production of natural substances. Interdisc. Sci. Rev. **1**(1), 85–103 (1976). https://doi.org/10.1179/isr.1976.1.1.85
18. Nasution, M.K.M.: Future directions in semantic technology, data analytics, and sustainability. Data Sci. (2026). https://doi.org/10.1201/9781003628569-11
19. Aldwairi, M., Jarrah, M., Mahasneh, N., Al-khateeb, B.: Graph-based data management system for efficient information storage, retrieval and processing. Inf. Process. Manag. **60**(2), 103165 (2023). https://doi.org/10.1016/j.ipm.2022.103165
20. Nasution, M.K.M.: Information retrieval optimization based on tree of social network. IOP Conf. Ser. Mater. Sci. Eng. **505**(1), 012099 (2019). https://doi.org/10.1088/1757-899X/505/1/012099
21. Singh, M.K.: Authorship pattern and collaboration coefficient of India in biotechnology research during 2001–2016: based on scopus database. Libr. Phil. Pract. **2017**(1) (2017). EID 2-s2.0-85021684243
22. Brzozowski, K., Matuszyk, E., Pieczara, A., Firlej, J., Nowakowska, A.M., Baranska, M.: Stimulated Raman scattering microscopy in chemistry and life science - development, innovation, perspectives. Biotechnol. Adv. **60** (2022). https://doi.org/10.1016/j.biotechadv.2022.108003

23. Alshraideh, H., Abdelgawad, M.: On the credibility of QS and THE ranking by subject area: misalighnment of subject mapping to academic disciplines. Scientometrics **130**(4), 2237–2272 (2025). https://doi.org/10.1007/s11192-025-05266-6
24. Sultan, Fitri, S.: The university students online reding behavior: gender, subject area, and academic achievement in the digital era. Asian EFL J. **27**(41), 276–290 (2020). EID 2-s2.0-85096055010
25. Nasution, M.K.M.: Data behavior: towards robust estimation. In: Lecture Notes in Networks and Systems, LNNS, vol. 1902, pp. 1–11 (2026). https://doi.org/10.1007/978-3-032-22091-2_1
26. Zvoleyko, E.V., Kalashnikova, S.A., Klimenko, T.K.: Assessment methods of personal learning outcomes in subject areas according to special basic educational program of primary education for students with mental retardation. Biosci. Biotechnol. Res. Asia **12**(3), 2695–2703 (2015). https://doi.org/10.13005/bbra/1951
27. Azura Md Nadzar, N.M., Bakri, A., Ibrahim, R.: Potential implementation of subject areas in Malaysia's research assessment co-word analysis study. In: IEEE Conference on e-Learning, e-Management and e-Services, IC3e2017, pp. 91–96 (2017). https://doi.org/10.1109/IC3e.2017.8409244
28. Baruah, J.B.: Inorganic molecular complexes: potential for growth of a new subject area in self-assembly. Top. Curr. Chem. **378**(2) (2020). https://doi.org/10.1007/s41061-020-0294-8
29. Ullah, A., Ameen, K.: Growth, subject areas, and application of research methods in user studies: a content analysis of articles produced by Paskitani authors. J. Librariansh. Inf. Sci. **55**(4), 1068–1077 (2023). https://doi.org/10.1177/09610006221124626
30. Pinheiro, M., Reigber, A.: Improving TamDEM-X DEMs accuracy using large-baseline data from the science phase. In: Proceedings of the European Conference on Synthetic Aperture Radar, EUSAR (2016). EID 2-s2.0-85001099610
31. Nasution, M. K. M., Herawati, E., Elveny, M.: Semantic information retrieval models. In: Data Science with Semantic Technologies: Deployment and Exploration, pp. 49–72 (2023). https://doi.org/10.1201/9781003310792-4
32. Upadhyay, R., Viviani, M.: Enhancing health information retrieval with RAG by prioritizing topical relevance and factual accuracy. Disc. Comput. **28**(1) (2025). https://doi.org/10.1007/s10791-025-09505-5
33. Chernov, A.V., Butakova, M.A., Deineko, O.V., Miroshnikov, A.M.: An approach to granulation and ontology similarity measurement for intelligent decision support in semi-structured subject areas. In: Proceedings of 2019 22nd International Conference on Soft Computing and Measurements, SCM 2019, pp. 146–148 (2019). https://doi.org/10.1109/SCM.2019.8903907
34. Nasution, M.K.M., Sitompul, O.S., Elveny, M., Syah, R., Rahmat, R.F.: A similarity for new meanings. In: 2020 International Conference on Data Science, Artificial Intelligence, and Business Analytics, DATABIA 2020 - Proceedings, pp. 37–44 (2020). https://doi.org/10.1109/DATABIA50434.2020.9190316
35. Nasution, M.K.M., Sitompul, O.S.: Enhancing extraction method for aggregating strength relation between social actors. Adv. Intell. Syst. Comput. **573**, 312–321 (2017). https://doi.org/10.1007/978-3-319-57261-1_31
36. Nasution, M. K. M., Syah, R., Elveny, M.: Enumerating countries' capabilities in artificial intelligence on categories. In: 2025 International Conference on Computing and Applied Informatics, ICCAI 2025 (2025). https://doi.org/10.1109/ICCAI65301.2025.11279294

37. Lin, B.-W., Chen, C.-J., Wu, H.-L.: Predicting citations to biotechnology patents based on the information from the patent documents. Int. J. Technol. Manag. **40**(1–3), 87–100 (2007). https://doi.org/10.1504/ijtm.2007.013528
38. Stephanopoulos, G.: Emerging directions in computer applications to biotechnology: upgrading the information content of biological data. Annu. Rev. Control. **23**, 61–69 (1999). https://doi.org/10.1016/S1367-5788(99)00008-5
39. Abdallah, C., Raoudha, K.: Knowledge management and risk control in the university environment: biotechnology center of Sfax-Tunisia. Stud. Syst. Decis. Control **489**, 797–805 (2024). https://doi.org/10.1007/978-3-031-36895-0_67
40. Li, J.: Analysis of research hotspots and trends in biotechnology and education based on CiteSpace: taking WOS data as an example. In: Proceedings of 2025 6th International Conference on Education, Knowledge and Information Management, ICEKIM 2025, pp. 234–239 (2025). https://doi.org/10.1145/3756580.3756618
41. Kruss, G.: Channels of interaction in health biotechnology networks in South Africa: who benefits and how? Int. J. Technol. Learn. Innov. Dev. **5**(1–2), 204–220 (2012). https://doi.org/10.1504/IJTLID.2012.044883
42. Sagynbayeva, A., Pyo, A., Yoon, S.-H., Yang, S.-B.: Evaluating user performance with RAG-based generative AI: a scenario-based experiment on AI-assisted information retrieval. Comput. Hum. Behav. **180** (2026). https://doi.org/10.1016/j.chb.2026.108952
43. Norhisham, N.N., Muhammad, N., Ismail, M.I.F., Fabregas, A.C.: The development of information retrieval model for Hadith data from knowledge graph. In: Proceedings - 2026 18th International Conference on Electronics, Computer, and Computation, ICECCO 2026 (2026). https://doi.org/10.1109/ICECCO67619.2026.11488814
44. Biotechnology. Nature **131**(3313) (1933). https://doi.org/10.1038/131597a0
45. Nuhriawangsa, A.M.P., Bachruddin, Z., Sajidan, S., Wibowo, A.: Production and characterization of crude intracelluler phytase from recombinant bacteria pEAS1AMP. J. Indon. Tropical Anim. Agric. **34**(4), 265–271 (2009). https://doi.org/10.14710/jitaa.34.4.265-271
46. Bruins, A.P., Pras, N.: Isolation of 3,4-dihydroxyphenylacetic acid produced from p-hydroxyphenylacetic acid by immobilized plant cells of Mucuna pruriens and its identification by liquid chromatography/mass spectrometry. Anal. Chim. Acta **163**(C), 91–100 (1984). https://doi.org/10.1016/S0003-2670(00)81497-X
47. Wulandari, D., Budiharjo, A., Putri, A.W.K., Setiarto, R. H. B.: Diversity of thermophilic bacteria isolated from extreme environments in Indonesia: a perspective in biotechnology applications. J. Tropical Biodiver. Biotechnol. **10**(3), 1–37 (2025). https://doi.org/10.22146/jtbb.19548

Knowledge Management

An Enhanced System Usability Scale for Evaluating Organizational Systems with Database and Knowledge Management Components

Siti Solehah Othman[1,2], A. Noraziah[1(✉)], Roslina Mohd Sidek[1], Julius Beneoluchi Odili[3,4], and Tutut Herawan[5,6]

[1] Faculty of Computing, Universiti Malaysia Pahang Al-Sultan Abdullah, Kuantan, Pahang, Malaysia
noraziah@umpsa.edu.my

[2] Digital Brain Sdn Bhd, PT530, Tingkat 1 Lembah Sireh, 15050 Kota Bharu, Kelantan, Malaysia

[3] Institute of Digital Humanities and Technologies, Anchor University Lagos, Lagos, Nigeria

[4] Department of Computer Science and Information Systems, Covenant University, Ota, Nigeria

[5] Department of Information Systems, Faculty of Computer Science and Information Technology, Universiti Malaya, Kuala Lumpur, Malaysia
tutut@um.edu.my

[6] AMCS Research Center, Yogyakarta, Indonesia

Abstract. When designing organizational systems, usability evaluation is crucial to show that the system is efficient to complete tasks and that the user is satisfied and can process information effectively. Though the traditional usability assessment techniques, such as the widely used System Usability Scale (SUS), have been applied to mobile applications, e-commerce systems and business platforms, they may not be sufficient to measure usability requirements in a database-supported and knowledge-intensive organizational setting. This study suggests an Extended System Usability Scale (Enhanced SUS) framework that adds some organizational usability attributes to the traditional SUS dimensions: accessibility, responsiveness, information accessibility, information retrieval, trustfulness, personalization and collaboration support. Thirty participants participated in a quantitative approach, which consisted of task-based usability testing and questionnaire-based assessment. Participants were then asked to perform tasks related to the organization that involved interacting with the database and retrieval of information and/or sharing of knowledge, prior to answering the SUS and Enhanced SUS questionnaires. Descriptive statistical analysis and SUS scoring techniques were used to analyse the collected data. The results show that the proposed Enhanced SUS framework could provide more comprehensive usability information than traditional SUS, as organizational usability attributes related to the management of information and database-supported environments have been added. The result of reliability analysis showed Cronbach's alpha value was 0.881, which was good internal consistency. However, further psychometric validation involving larger samples and factor analysis is required before the framework can be considered a fully validated usability measurement instrument.

N. Mohd Sharef et al. (Eds.): CAMP 2026, CCIS 3093, pp. 321–334, 2027.
https://doi.org/10.1007/978-981-92-4302-0_23

Keywords: Enhanced System Usability Scale · SUS · Usability Evaluation · Organizational Systems · Database Systems · Knowledge Management Introduction

1 Introduction

Digital systems are increasingly used to facilitate information processing, operational management, communication, and decision-making in organizations (Davenport & Prusak, 1998; Alavi & Leidner, 2001). Database-based organizational systems have been embraced and used to effectively store, retrieve and share information between departments (Dalkir, 2017; Rowley, 2007). As technology advances, making systems easy to use has become increasingly important because usability problems can lead to decreased productivity, errors, and negative user attitudes (Lazar et al., 2006).

Usability describes how well users can achieve intended objectives using a system with efficiency, effectiveness, and satisfaction under specific conditions of use (ISO, 2018). Earlier research has shown that usability is an important factor in user acceptance, operational efficiency, and system performance (Lund, 2005; Lazar et al., 2006). However, organizations continue to assess and refine the usability of their information systems (Nielsen, 2012).

Among numerous usability evaluation methods, the System Usability Scale (SUS) remains widely adopted because of its simplicity, reliability, and practical implementation across different application domains (Brooke, 1996). The application of SUS has been shown to be successful in several fields such as healthcare systems, mobile banking, e-commerce platforms and in the field of education technologies (Adilla et al., 2022; Ahmad et al., 2022; Alisya et al., 2023; Hamid et al., 2020; Hasudungan et al., 2024; Sani et al., 2024).

SUS is well suited for general usability evaluation but has its drawbacks when used for organizational systems that involve the use of databases and information management processes (Brooke, 1996; ISO, 2018). Traditional SUS does not adequately measure usability aspects related to information access, knowledge efficiency, responsiveness, collaboration, and trust in organizational settings (Dalkir, 2017; Alavi & Leidner, 2001). Modern organizational systems often integrate aspects of knowledge management functions like information sharing, retrieval, storage, and collaborative consumption (Nonaka & Takeuchi, 1995; Dalkir, 2017). These processes rely on the use of database systems that provide structured access to information (Davenport & Prusak, 1998). However, relying on a traditional SUS approach may not give a complete picture of the user experience in these situations (Lund, 2005; Rowley, 2007).

To accommodate these limitations, this study proposes an Enhanced System Usability Scale (Enhanced SUS) model based on the original SUS, with additional attributes of usability that are relevant to organizational systems that include database components and knowledge management systems (Dalkir, 2017; Alavi & Leidner, 2001). The proposed framework is not meant to be an alternative to SUS, but to add to the usability assessment dimensions that SUS offers in order to create a more complete capability for usability assessment (Brooke, 1996; Nielsen, 2012).

2 Literature Review

2.1 Description of Usability

Usability is the degree to which a specified user can use a product effectively, efficiently and satisfactorily to accomplish a specific task for a particular purpose in a specified context of use (ISO, 2018). Other usability aspects that have been found in earlier research are learnability, memorability, accessibility and trustfulness (Nielsen, 2012; Lund, 2005). If the usability is not good, it will be followed by decreased productivity, more errors, and less user acceptance (Lazar et al., 2006).

2.2 Description of System Usability Scale (SUS)

One of the most widely used and reliable usability assessment instruments is the Brooke's System Usability Scale (SUS) (Brooke, 1996). This is a 10 item Likert scale questionnaire which uses a standardised scoring scale (Brooke, 1996) to produce a usability score in the range 0 to 100. SUS has been implemented with good results in online shopping, mobile banking, digital wallets, education, supply chain systems, and political systems (Adilla et al., 2022; Ahmad et al., 2022; Alisya et al., 2023; Hamid et al., 2020; Hasudungan et al., 2024; Sani et al., 2024). Organizational usability needs involving information management, database interaction and knowledge-oriented activities (Dalkir, 2017), however, are not fully represented in traditional SUS.

2.3 Usability in Organizational Systems

Organizational systems support knowledge sharing, collaboration, decision-making, information storage, and information retrieval through database-supported environments (Alavi & Leidner, 2001; Davenport & Prusak, 1998; Nonaka & Takeuchi, 1995). For organizational users, systems must provide efficient access to information while remaining simple, reliable, and easy to use (Lund, 2005; Lazar et al., 2006). Poor usability may lead to reduced productivity, information retrieval delays, operational errors, lower adoption rates, and user dissatisfaction (Lazar et al., 2006; ISO, 2018).

Users frequently interact with databases to search, retrieve, update, and share organizational information (Davenport & Prusak, 1998). However, traditional usability evaluation methods often do not explicitly assess dimensions such as information accessibility, information retrieval efficiency, collaboration support, responsiveness, trustfulness, and personalization (Nielsen, 2012; Lund, 2005). These dimensions are particularly important in organizational and knowledge-oriented environments and can significantly influence overall usability perceptions (Dalkir, 2017). Therefore, extending the traditional SUS framework may provide a broader assessment of usability in database-supported organizational systems (Brooke, 1996).

2.4 Related Works and Research Gap

SUS and other usability techniques have been used in various domains (Brooke, 1996). Previous usability evaluation methods are summarized in Table 1.

Table 1. Previous Studies on SUS and Related Usability Methods

Study	Method	Main Attributes	Year
E-commerce usability studies	System Usability Scale (SUS)	Efficiency, Effectiveness, Satisfaction	2024
Mobile banking applications	System Usability Scale (SUS), Usability Testing (UT)	Learnability, Efficiency, Error Recovery	2022
Digital wallet systems	System Usability Scale (SUS), User Experience Questionnaire (UEQ)	Satisfaction, Dependability	2023
Organizational websites	User Acceptance Testing (UAT)	Clarity, Responsiveness	2020
Supply chain systems	System Usability Scale (SUS)	Satisfaction, Effectiveness	2022
Pension applications	System Usability Scale (SUS), Heuristic Evaluation	Navigation, Consistency	2024
Educational systems	System Usability Scale (SUS)	Learnability, Ease of Use	2013
Knowledge Management Systems	SUS, KM Evaluation	Information Accessibility, Collaboration Support	2025
Fintech and Digital Payment Systems	System Usability Scale (SUS)	Effectiveness, Efficiency, Satisfaction	2026

The System Usability Scale (SUS) and related usability evaluation methods have been widely applied across several areas, such as e-commerce platforms, mobile banking applications, digital wallet systems, organizational websites, supply chain systems, pension applications, educational systems, knowledge management systems, and fintech applications, as summarized in Table 1. Previous studies have shown that combining SUS with complementary methods such as Usability Testing (UT), User Experience Questionnaire (UEQ), Heuristic Evaluation (HE), and User Acceptance Testing (UAT) offers a more thorough comprehension of usability and user experience issues (Alisya et al., 2023; Abulfaraj & Steele, 2020; Adilla et al., 2022).

Although it is a popular approach, traditional SUS has been mainly concerned with effectiveness, efficiency, satisfaction and learnability (Brooke, 1996). Often, however, organizational systems include processes, repositories of information, and knowledge management activities that also demand other dimensions of usability, including information accessibility, efficiency of information retrieval, collaboration support, trustfulness, dependability, personalization, and responsiveness (Dalkir, 2017; Alavi & Leidner, 2001; Nielsen, 2012). The traditional SUS framework does not explicitly cover such aspects.

Therefore, there remains a gap in usability evaluation approaches for organizational systems that incorporate database and knowledge management components. This study proposes an Enhanced System Usability Scale (Enhanced SUS) framework that includes additional organizational usability attributes, which extends the traditional SUS dimensions and is a more extensive and comprehensive usability assessment for information-intensive organizational contexts.

3 Methodology

This study introduces an Enhanced System Usability Scale (Enhanced SUS) framework to assess usability in an organizational system, which incorporates the traditional SUS dimensions and other usability attributes related to information accessibility, responsiveness, trustfulness and database supported organizational activities (Brooke, 1996; Dalkir, 2017). The methodology uses a quantitative approach that includes the traditional SUS and an additional Enhanced SUS questionnaire to obtain more complete usability evaluation results (Brooke, 1996). The research process is divided into five stages: preparation of the research framework, participant selection, task execution, questionnaire distribution, and statistical data analysis (Brooke, 1996; Adilla et al., 2022). The approach proposed is to evaluate usability beyond the traditional SUS scale evaluation (Brooke, 1996).

3.1 Proposed Enhanced SUS Framework

The proposed framework builds on the classic SUS dimensions and adds usability attributes often found in organizational systems (Brooke, 1996). The traditional SUS commonly measures effectiveness, efficiency, satisfaction, and learnability (as discussed by Brooke, 1996). However, contemporary organizational structures demand a wide usability assessment that includes accessibility, responsiveness, efficiency in information retrieval, personalization and trustfulness (Nielsen, 2012; Lund, 2005).

Thus, the proposed Enhanced SUS framework has two components (Brooke, 1996). The first component is Traditional SUS Evaluation, which measures standard usability dimensions using the 10 SUS statements developed by Brooke (1996). Three main dimensions are effectiveness, efficiency, satisfaction, and learnability (Brooke, 1996). The second component is Enhanced Usability Attributes, which includes additional attributes adapted from previous usability studies including accessibility, responsiveness, trustfulness, personalization, navigation, clarity, memorability, dependability, information accessibility, collaboration support, recover from error, ease of use and interactivity quality (Nielsen, 2012; Lund, 2005; Abulfaraj & Steele, 2020).

The other dimensions are given to account for additional usability aspects not covered by the traditional measures of SUS (Brooke, 1996). The combination of these elements results in a more holistic assessment framework that has been applicable to an organizational context (Dalkir, 2017; Alavi & Leidner, 2001).

The additional attributes were selected based on three theoretical foundations: usability theory, organizational information systems, and knowledge management literature.

Attributes such as Information Accessibility, Information Retrieval Efficiency, and Collaboration Support were included because organizational systems rely heavily on information storage, retrieval, sharing, and database-supported activities. Alternative constructs were excluded if they overlapped conceptually with existing attributes or were not directly relevant to organizational information processing tasks.

The integration of traditional SUS dimensions and organizational usability attributes forms the proposed Enhanced SUS Framework. The framework is intended to provide a broader assessment of organizational systems by incorporating both general usability factors and organizational information management requirements.

Figure 1 illustrates the proposed Enhanced SUS Framework, which integrates traditional SUS dimensions with additional organizational usability attributes to provide a broader evaluation of organizational systems involving database and knowledge management components.

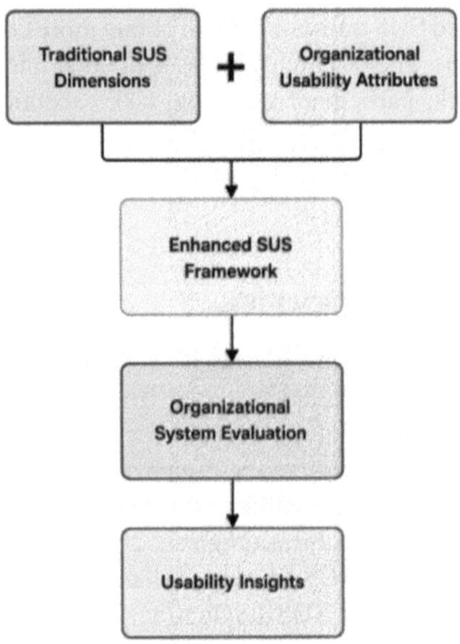

Fig. 1 Illustrates the proposed Enhanced SUS Framework

3.2 Research Design

A quantitative usability testing approach was employed in this study to evaluate organizational system usability using both traditional SUS and Enhanced SUS instruments. Participants are required to complete tasks within a defined organizational system, followed by questionnaires (Brooke, 1996).

The steps in the research process are participant briefing and consent, the completion of the predefined steps, distributing a SUS questionnaire, distributing an enhanced SUS questionnaire, capturing the data of the research steps, and conducting statistical analysis and interpretation (Brooke, 1996; Adilla et al., 2022).

3.3 Participants

The number of respondents is 30 people who are obtained by purposive sampling as in previous SUS usability studies (Adilla et al., 2022; Alisya et al., 2023). Participants have some experience of interacting with digital organizational systems. The selection criteria are knowledge of organizational information systems, basic experience with database-supported applications, and the ability to perform standard usability tasks. All participants agree to take part voluntarily prior to testing.

3.4 Experimental Setup

Testing is performed on a desktop or laptop device with stable internet connectivity. Google Forms, manual observation and statistical analysis using Microsoft Excel are used for data collection. Usability testing is carried out individually to reduce external influence.

3.5 Predefined Task Scenarios

Participants perform six predefined tasks representing common organizational activities (Table 2).

Table 2. Predefined Tasks

Task No	Task Description	Related Attributes
1	Log into the organizational system using provided credentials	Effectiveness, Learnability
2	Search for specific information within the system	Navigation, Accessibility
3	Retrieve and review stored information	Share information or collaborate using available functions
4	Update information records	Efficiency, Responsiveness
5	Correct an intentional error during submission	Recover from Error, User Control
6	Share information or collaborate using available functions	Collaboration Support, Trustfulness

The selected tasks simulate common organizational interactions involving information retrieval and database-supported processes.

3.6 Measurement Instruments

To evaluate usability, participants responded to ten Likert Scale statements. The scoring process applies different adjustments to positively and negatively worded items before calculating. The summed values are converted into the final SUS score through a multiplication factor of 2.5 (Brooke, 1996). Table 3 interprets SUS scores, where scores above 80 indicate excellent usability, 68 to 80 indicate good usability, 50 to 67 indicate acceptable usability, and usability is poor if the score is less than 50 (Brooke, 1996).

Table 3. Interpretation SUS Scores

SUS Score	Interpretation
>80	Excellent
68–80	Good
50–67	Acceptable
<50	Poor

The additional usability attributes were selected through a review of usability theory, organizational information systems literature, and knowledge management research. The traditional System Usability Scale focuses on general dimensions of usability, including effectiveness, efficiency, satisfaction and learnability. Unlike other software, organizational systems are often used for accessing and manipulating vast amounts of information stored in a repository, for executing knowledge-based processes, for accessing databases, and for working together in a collaborative workspace.

The attributes included were based on the premise that organizational users need to be able to retrieve, access and find information quickly and effectively for decision making and routine operations. Collaboration Support was added as many organizational systems can be used to share information and communicate with other users, departments, or integrated services. It was decided to include the characteristics of Trustfulness and Dependability, as organizational systems may be responsible for handling sensitive organizational data and thus demand reliable and trustworthy system behaviour. Personalization, Responsiveness, Accessibility, Navigation and Recover from Error were added to address usability elements that directly impact productivity, task completion and user experience in information-intensive contexts.

Constructs of alternative usability were considered but not added if they had a similar conceptual meaning as one of the existing attributes or if they were not directly related to the database-supported organizational activities. The final set of attributes was chosen to give a general measure of the usability of organizations, and to avoid conceptual duplication between the constructs (Table 4).

Table 4. Enhanced SUS Questionnaire

Attribute	Statement
Effectiveness	Measures successful task completion in organizational workflows
Efficiency	Measures speed and effort required to perform tasks

(continued)

Table 4. (*continued*)

Attribute	Statement
Satisfaction	Measures overall user perception and acceptance
Learnability	Measures ease of learning system functions
Accessibility	Measures access across devices and user capabilities
Responsiveness	Measures system reaction time during operations
Trustfulness	Measures confidence in system security and information integrity
Dependability	Measures system consistency and reliability
Information Accessibility	Measures ease of locating organizational information
Collaboration Support	Measures support for information sharing and interaction
Ease of Use	Measures perceived simplicity of interaction
Navigation	Measures movement through system functions and pages
Memorability	Measure's ability to recall previously learned functions
Recover from Error	Measure's ability to identify and recover from mistakes
Personalization	Measures adaptability to user preferences
Interactivity Quality	Measures quality of user-system interaction
Clarity	Measures understandability of content and interface elements
Information Retrieval Efficiency	Measures efficiency of retrieving stored information

3.7 Data Collcction and Analysis

The study goal and consent was explained to participants prior to undertaking the specified activities. Participants completed both the System Usability Scale (SUS) and Enhanced SUS surveys after finishing the tasks. The data collected from all the valid responses were analyzed by descriptive statistics such as mean scores, SUS scoring, and attribute level evaluation. The analysis aimed to identify usability strengths and weaknesses and determine whether the Enhanced SUS framework provides broader usability insights than the traditional SUS approach (Brooke, 1996).

3.8 Reliability and Validity Assessment

To assess the reliability of the proposed Enhanced SUS framework, internal consistency analysis was performed using Cronbach's Alpha. All questionnaire items were subject to reliability analysis to ascertain if the items measured the constructs of organizational usability reliably. Cronbach's Alpha was selected because it is widely used for assessing questionnaire reliability in usability and information systems research. A value above 0.80 indicate good reliability, while 0.70 implies average reliability.

4 Results and Discussion

This section presents the findings from the System Usability Scale (SUS) and the recently developed Enhanced SUS survey (Brooke, 1996). The data collection procedure involved the participation of thirty subjects in completing certain predefined organizational activities before filling out both scales (Brooke, 1996; Adilla et al., 2022). The analysis focuses on evaluating usability efficiency together with other aspects of usability that the new model considers.

4.1 System Usability Scale (SUS) Results

A total of thirty participants responded to the questions of the System Usability Scale (SUS) after performing the predefined tasks. Table 5 represents the scores of the SUS scale of all participants.

Table 5. System Usability Scale (SUS) summarizes scores obtained from all participants

Summary	Value
Mean	75.3
Min	68
Max	84

The average SUS score is 75.3, suggesting that the system being analyzed falls under the good usability class. It shows that the system is viewed by users as easy-to-use, efficient, and helpful in accomplishing the required tasks (Brooke, 1996). It implies that there are satisfactory levels of effectiveness, efficiency, and ease of learning (Brooke, 1996). However, SUS may not provide sufficient data about some aspects of organizational usability like accessibility of information and collaboration tools (Dalkir, 2017; Alavi & Leidner, 2001).

4.2 Enhanced SUS Results

Enhanced SUS survey described herein evaluates more attributes compared to the traditional SUS tools used for usability analysis (Brooke, 1996; Nielsen, 2012).

Attribute scores for the Enhanced SUS survey have been provided in Table 6.

Table 6. Enhanced SUS Attribute Scores

Attribute	Mean	SD
Effectiveness	3.92	1.06
Efficiency	3.99	1.09

(continued)

Table 6. *(continued)*

Attribute	Mean	SD
Satisfaction	3.98	0.99
Learnability	3.91	1.21
Accessibility	4.03	1.05
Responsiveness	3.91	1.09
Trustfulness	4.00	1.00
Dependability	3.92	1.08
Information Accessibility	4.13	0.97
Collaboration Support	3.93	0.97
Ease of Use	3.89	1.18
Navigation	3.91	1.08
Memorability	4.00	1.05
Recover from Error	3.96	1.03
Personalization	3.92	1.05
Interactivity Quality	3.94	1.06
Clarity	3.90	1.07
Information Retrieval Efficiency	4.00	1.05

With mean scores ranging from 3.89 to 4.13, the results show generally positive impressions across all usability aspects. The highest mean score was obtained for information accessibility (M = 4.13, SD = 0.97), suggesting that participants were generally able to locate and access information efficiently within the system. Information Retrieval Efficiency (M = 4.00, SD = 1.05), Trustfulness (M = 4.00, SD = 1.00), and Accessibility (M = 4.03, SD = 1.05) also received relatively high evaluations. The mean scores and standard deviations for Ease of Use (M = 3.89, SD = 1.18) and Learnability (M = 3.91, SD = 1.21) were comparatively low and high, respectively. The results indicated that there were wider differences among the participants on their perception of the ease of learning and using the system. Standard deviation values are also used to see how much the participants agree with one another; lower standard deviations will offer more insight into consistency of user responses while higher standard deviations will provide more insight into diversity of user perceptions.

4.3 Reliability Analysis of Enhanced SUS

To determine the internal consistency of the proposed Enhanced SUS instrument, a reliability analysis was carried out. The Cronbach's Alpha was used to analyse responses from 30 participants. The Cronbach's α coefficient resulting from the analysis was 0.881. This score is above the recommended score of 0.80, which signals good internal consistency reliability (Table 7).

Table 7. Reliability Analysis Results

Measure	Value
Number of Respondents	30
Number of Items	31
Cronbach's Alpha	0.881
Interpretation	Good Reliability

The result indicates preliminary evidence of the proposed Enhanced SUS framework having consistent measures for organizational usability construct.

4.4 Discussion

We found that our proposed Enhanced SUS framework evaluates usability more broadly than traditional SUS does, as it includes additional organisational usability dimensions. Whereas a traditional SUS score of $\geq$75.3 generally signifies good overall usability (Brooke, 1996), the Enhanced SUS provides more insightful views into what constitutes organisational usability requirements that are overlooked by conventional SUS (Dalkir, 2017).

Our proposed Enhanced SUS framework extends SUS dimensions with organisational usability features from usability theory, database-based information systems, and KM literature. Instead of focusing on general usability perceptions alone, the Enhanced SUS encompasses information accessibility, information retrieval efficiency, collaboration support, trustfulness, and dependability, which are essential in organisational environments. Higher scores in information accessibility and retrieval efficiency reveal that users can easily locate and retrieve required information, thereby enhancing their performance and satisfaction (Davenport & Prusak, 1998; Lazar et al., 2006). Trustfulness refers to users' faith in outputs and reliability of information, which is paramount in an organisational context (Alavi & Leidner, 2001). Lower scores in personalisation and memorability point towards possible improvements in the adaptability and long-term usability of interfaces (Nielsen, 2012; Lund, 2005). In summary, our results reveal that our proposed Enhanced SUS framework captures more organisational usability dimensions that are ignored by traditional SUS.

5 Conclusion

In this paper an Enhanced System Usability Scale (Enhanced SUS) framework was proposed to extend the traditional SUS with additional organisational usability attributes. The results of the reliability analysis showed that the questionnaire items had strong internal consistency, with a Cronbach's Alpha value of 0.881. With an average SUS score of 75.3, the assessed system demonstrated good usability performance (Brooke, 1996). The Enhanced SUS framework provided additional insights beyond traditional SUS, assessing organisational usability attributes such as accessibility, responsiveness, trustfulness, information accessibility, collaboration support, dependability, personalisation and information retrieval efficiency (Nielsen, 2012; Dalkir, 2017). Results suggest that the framework can support a broader evaluation of usability in database-supported and knowledge-intensive organisational environments (Alavi & Leidner, 2001; Lazar et al., 2006).

Though the sample size used in our study is sufficient for exploratory usability evaluation, it is still below the sample size commonly recommended for psychometric instrument validation.

Acknowledgments. The authors would like to express their appreciation to Universiti Malaysia Pahang Al-Sultan Abdullah (UMPSA) for funding this study under the Distinguish Research Grant, RDU253002. Special thanks also go to the grant RDU230350. Additionally, the authors thank UMPSA for its laboratory facilities and library services.

Competing Interests The authors state that they have no conflict of interest associated with this publication.

Ethics Approval This paper followed the necessary ethical guidelines, such as voluntary participation, consent, and confidentiality.

References

Brooke, J.: SUS: a quick and dirty usability scale. In: Jordan, P.W., Thomas, B., Weerdmeester, B.A., McClelland, I.L. (eds.) Usability Evaluation in Industry, pp. 189–194. Taylor & Francis, London (1996)

ISO: ISO 9241-11:2018: Ergonomics of human-system interaction – Part 11: Usability: Definitions and Concepts. International Organization for Standardization (ISO), Geneva (2018)

Nielsen, J.: Usability 101: Introduction to Usability. Nielsen Norman Group (2012). Retrieved from: https://www.nngroup.com/articles/usability-101-introduction-to-usability/

Hamid, S., Bawany, N.Z., Zahoor, K.: Assessment of e-commerce websites through usability and accessibility perspectives. In: Proceedings of the International Conference on Advanced Computer Science and Information Systems (ICACSIS), pp. 199–204 (2020)

Adilla, R., Jazman, M., Ahsyar, T.K., Hamzah, M.L.: Evaluation usability of mobile banking applications using usability testing (UT) method and system usability scale (SUS). In: Proceedings of the IEEE International Conference on Sustainable Engineering and Creative Computing, pp. 78–83 (2022)

Alisya, T., Hamzah, M.L., Saputra, E., Ahsyar, T.K.: Evaluation of user experience for the ShopeePay digital wallet using the System Usability Scale (SUS) and User Experience Questionnaire (UEQ). In: Proceedings of the International Conference on Emerging Smart Technologies and Applications, pp. 1–6 (2023)

Hasudungan, J., Arifianto, F., Achsan, H.T.: Application of the system usability scale in evaluating Shopee PayLater services. Jurnal Ilmiah Penelitian dan Pembelajaran Informatika. **9**(1), 416–423 (2024)

Ahmad, A.E., Kusrini, K., Sudarmawan, S.: Usability assessment of office stationery procurement and management systems using the System Usability Scale. In: Proceedings of the International Conference on Information Technology, Information Systems and Electrical Engineering, pp. 498–502 (2022)

Sani, A., et al.: Usability evaluation of e-commerce applications through system usability scale questionnaires. In: Proceedings of the International Conference on Cyber and IT Service Management, pp. 1–7 (2024)

Lazar, J., Jones, A., Hackley, M., Shneiderman, B.: Severity and effects of computer user frustration: comparison between student users and workplace users. Interact. Comput. **18**(2), 187–207 (2006)

Lund, A.M.: More than Ease of Use: An Introduction to the Usability Construct. Measuring Usability LLC (2005)

Nonaka, I., Takeuchi, H.: The Knowledge-Creating Company: How Japanese Companies Create the Dynamics of Innovation. Oxford University Press, New York (1995)

Dalkir, K.: Knowledge Management in Theory and Practice, 3rd edn. MIT Press, Cambridge (2017)

Rowley, J.: The wisdom hierarchy: representations associated with the DIKW hierarchy. J. Inf. Sci. **33**(2), 163–180 (2007)

Davenport, T.H., Prusak, L.: Working Knowledge: How Organizations Manage What they Know. Harvard Business School Press, Boston (1998)

Alavi, M., Leidner, D.E.: Knowledge management and related systems: conceptual foundations and associated research issues. MIS Q. **25**(1), 107–136 (2001)

Abulfaraj, A., Steele, A.: Detailed usability heuristics: breaking down usability heuristics to improve understanding among novice evaluators. In: HCI International 2020. Springer, Cham (2020)

Yasir, A., Khan, M., Rahman, S.: Evaluating the impact of knowledge management systems on organizational performance. J. Comput. Sci. Artif. Intell. Comput. **4**(1), 45–58 (2025)

Dewi, L.Y.P., Abdillah, L.A.: OVO fintech application analysis using the system usability scale. arXiv preprint. (2026) https://arxiv.org/abs/2601.11600

Designing a Just-in-Time (JIT), Microlearning, and Gamification Intervention for Phishing Awareness in Micro, Small, and Medium Enterprises (MSMEs)

Muhammad Naufal Daffa Ulhaq S(✉) and Masnizah Mohd

Faculty of Information Science and Technology, Universiti Kebangsaan Malaysia, 43600 Bangi, Selangor, Malaysia
p151337@siswa.ukm.edu.my

Abstract. Phishing remains a persistent threat, while conventional awareness training is often too general and detached from the moment of user risk. This paper proposes a phishing awareness intervention for MSMEs that combines phishing simulation, just-in-time learning, microlearning, and gamification. **Targeting MSME owners and employees** in Central Sulawesi, the study employs a quantitative single-group pretest-posttest design to evaluate changes in phishing-related knowledge and safe behavior. The intervention is structured as immediate post-exposure feedback, concise learning content, and quiz-based reinforcement. Instead of presenting final findings, the paper describes the logic of the intervention, its design components, and the planned evaluation approach. The proposed design contributes a realistic and measurable training model for MSME environments with limited time and resources.

Keywords: phishing awareness · MSMEs · phishing simulation · just-in-time learning · microlearning · gamification

1 Introduction

Phishing remains one of the most persistent cyber threats because it exploits social engineering through messages that appear normal in everyday digital communication. In organizational settings, malicious emails may resemble service notices, operational instructions, or transaction-related requests, making them difficult to distinguish at a glance. For that reason, phishing vulnerability cannot be reduced to a purely technical problem. It is also shaped by how users interpret message cues, assess credibility, and make decisions under limited attention and demanding work routines [1] . This human-centered perspective is reinforced by evidence showing that risky responses are closely related to how users process signals in messages and react under busy work conditions [2, 3].

Recent indicators also show that phishing exposure remains substantial at both global and national levels. Recent studies indicate continuing growth in reported phishing

N. Mohd Sharef et al. (Eds.): CAMP 2026, CCIS 3093, pp. 335–347, 2027.
https://doi.org/10.1007/978-981-92-4302-0_24

attacks internationally and highlight the rise of cyber incidents in Indonesia alongside the expansion of digital platforms, e-commerce, and electronic payments [4, 5]. In other words, the digitalization that supports business growth can also intensify exposure to phishing when protective awareness and safe routines do not develop at the same pace.

The issue becomes more pressing in MSMEs. Small firms typically operate with lean structures, limited role specialization, and strong dependence on owners or core staff in operational decision-making [6, 7]. Under such conditions, phishing risk does not stop at whether owners and staff can define phishing correctly. The more important question is whether that understanding can be translated into consistent preventive behavior in real work situations [7, 8]. This challenge is compounded by uneven security literacy, unsafe habits, limited training budgets, and high operational priorities, all of which make continuous awareness efforts difficult to sustain [9, 10].

At the training level, a central problem lies in the mismatch between the contextual, recurring nature of phishing and awareness approaches often delivered as one-time activities, with support not always available when users face moments of risk [1]. This creates a need for an intervention that is contextual enough to address risky decision moments, concise enough to remain realistic in MSME settings, and engaging enough to reinforce safer responses over time. In response, this study designs an intervention combining phishing simulation, just-in-time learning, microlearning, and gamification, focusing on MSME owners and employees in Central Sulawesi and limiting its evaluation to two measurable outcomes phishing-related knowledge and safe behavior through a quantitative pretest-posttest design.

This paper contributes a deployable study protocol for improving phishing awareness in MSMEs by integrating simulation-triggered just-in-time learning, microlearning, and gamified reinforcement. Specifically, we (1) provide an operational end-to-end workflow (phishing simulation → Just-In-Time (JIT) landing page → microlearning in multiple formats → quiz-based reinforcement) that can be implemented under MSME time and resource constraints; (2) specify clear operational definitions and branching rules (e.g., defining "affected" participants as those submitting data and providing follow-up learning for non-affected participants) to support reproducibility; and (3) outline a measurable evaluation plan using a one-group pretest-posttest design with planned statistical analyses for phishing knowledge and phishing-related safe behavior, accompanied by descriptive simulation indicators as supporting context

2 Literature Review

2.1 Phishing as a Social Engineering and Human-Factors Threat

Phishing is commonly understood as a social engineering-based fraud that uses electronic channels such as email, SMS, social media, or telephone to induce users to disclose sensitive information, including account credentials and financial data [11, 13]. In practice, the message is made to appear legitimate by imitating trusted sources and official communication formats, including the use of spoofed identities, compromised accounts, and copied visual elements such as logos, signatures, and disclaimers [14, 15]. Phishing content is also often tied to routine work matters, such as payroll, password resets, VPN updates, project access, or invoices, so that the request appears relevant to everyday workflow in more targeted attacks, the narrative may even be personalized to match the victim's role or ongoing tasks [14, 16].

The success of phishing cannot be explained only by technical deception. The literature positions phishing as a human-factors issue because these attacks exploit how users interpret cues, judge credibility, and make decisions during routine digital interactions [1, 17]. Even when technical filters are in place, harmful messages may still reach the inbox, leaving the user as an important point of defense in the final stage of decision-making [18]. This also explains why knowing about phishing does not automatically translate into safe behavior, since user responses remain influenced by habits, work conditions, and the way interventions are designed and evaluated [17, 19].

Phishing vulnerability also becomes more pronounced under cognitive constraints: when users work under high workload, multitasking, and time pressure, they inspect security cues less systematically and rely on quick heuristics such as whether a message looks official or urgent, while subtle warning signs are overlooked. The literature therefore links phishing failures not to individual weakness alone, but to the interaction between user, task, and work context [20].

2.2 Work Context and MSME Constraints

Phishing vulnerability in the workplace is shaped not only by what users know, but also by the context in which decisions are made. Information overload, time pressure, stress, and digital distraction can reduce vigilance and increase the likelihood of risky responses, even among users who have previously received cybersecurity training. When attention is divided across tasks, message evaluation becomes shallower, which increases the chance of mistaking phishing messages for legitimate communication [21, 22]. The literature also notes that multitasking and reduced focus can weaken users' ability to notice warning signs at specific moments of work activity [23].

These constraints are especially relevant in MSMEs. Small firms are typically characterized by lean structures, limited financial and human resources, and a strong dependence on owners or core staff in day-to-day operations and decision-making [6, 7]. In addition, employees and managers often perform multiple functions and rely more on learning through practice than on lengthy formal training. Under such conditions, cybersecurity is frequently subordinated to immediate operational priorities, which makes awareness interventions difficult to sustain unless they are concise, practical, and directly applicable to daily work [24, 25].

MSMEs also face interrelated barriers in cybersecurity control and training. Limited budgets, insufficient IT infrastructure, and a shortage of cybersecurity expertise make it difficult to adopt more comprehensive security programs [10, 26]. These barriers are intensified when cybersecurity is perceived as an additional cost rather than a business priority, especially in environments shaped by time pressure and multitasking [26, 27]. As a result, training in small firms is often ad hoc and difficult to maintain regularly, which is why the literature tends to recommend modular, realistic, and easily adoptable approaches that can be integrated into routine workflow rather than delivered as large stand-alone programs [28, 29].

2.3 Toward a Contextual and Flexible Training Approach

The phishing-awareness literature shows a shift from general awareness campaigns toward more contextual and behavior-oriented interventions. Training is increasingly viewed not simply as information delivery, but as an effort to influence user decisions at critical moments when suspicious messages are encountered. In that sense, training effectiveness is not determined solely by whether training exists, but also by how it is implemented, repeated, and evaluated in realistic user contexts [1].

Simulation-based training remains valuable because it places users in a realistic decision context and can reduce failures over repeated exposure. However, simulation alone does not eliminate vulnerability. The literature notes that some users remain vulnerable even after repeated rounds, that more realistic scenarios may increase click rates because the challenge becomes harder, and that feedback placement can influence training effectiveness [22, 30]. Simulation-based approaches also raise ethical and practical concerns when scenarios are perceived as deceptive or punitive, and the evidence base still includes studies with limited designs and short evaluation periods [31, 33].

Just-in-time learning becomes relevant because it places corrective input close to the moment of risky action. In this model, users who fail a phishing simulation immediately receive an educational page or brief training material, so the intervention is delivered when the event is still salient and meaningful [34]. The logic of this approach lies in temporal contiguity: when feedback is provided immediately after a mistake, it becomes easier for users to connect the action with its correction. At the same time, the literature notes that embedded just-in-time delivery may have limited reach because it primarily targets users who directly interact with the simulated attack [34, 35].

Microlearning addresses a different constraint, namely limited time and attention. It is described as a highly concise learning format focused on a specific objective and typically completed in seconds to minutes [36, 37]. Its bite-sized and often mobile-friendly format makes it suitable for dynamic work settings because it can be accessed flexibly and inserted between busy schedules [38, 39]. At the same time, the literature cautions that brevity alone does not guarantee effectiveness. When too many objectives are packed into a short unit, microlearning may oversimplify important ideas or even generate cognitive overload, making design quality and relevance essential [37, 40].

Gamification adds a reinforcement layer that can make training more engaging, especially when the objective extends beyond factual knowledge toward behavioral change [41, 42]. In cybersecurity training, gamification is commonly implemented through points, badges, leaderboards, feedback, and graded challenges, all of which help make progress visible and encourage repeated participation [42, 44]. Among these elements, feedback is particularly important because it immediately indicates whether a response is right or wrong, explains why an action is risky, and guides subsequent improvement [45]. In this sense, gamification is not treated merely as entertainment, but as a structured mechanism for reinforcement and sustained engagement [46].

2.4 Synthesis and Research Direction

Taken together, the literature indicates that phishing is shaped by social engineering, cognitive constraints, and work context, while MSMEs face resource limitations that make long and resource-intensive training difficult to sustain. The same literature also shows that training outcomes vary according to how interventions are designed, timed, reinforced, and evaluated [1]. This combination of findings supports the need for an intervention model that is contextual enough to address moments of risk, concise enough to remain feasible in MSMEs, and engaging enough to reinforce participation over time.

On that basis, integrating phishing simulation, just-in-time learning, microlearning, and gamification becomes theoretically justified. Within such a design, phishing simulation functions as the trigger, just-in-time learning delivers immediate corrective input, microlearning provides short and focused content, and gamification strengthens reinforcement through feedback and challenge. A narrower evaluation focus on phishing-related knowledge and safe behavior is also more defensible than broader claims, because these outcomes are directly aligned with the intervention logic and the intended research design.

It is important to note that the individual components of this design, and even partial combinations of them, are not novel in themselves. Phishing simulation, just-in-time feedback, microlearning, and gamification are each well established, and integrated solutions that bundle them have become increasingly common in commercial security-awareness platforms. However, such integrated solutions are predominantly developed for, and evaluated in, well-resourced organizational and enterprise environments, and they tend to report operational metrics such as click rates rather than theoretically grounded measures of learning. The contribution of this study therefore lies not in the combination of techniques per se, but in three under-addressed aspects. First, it adapts this integrated approach to micro and small enterprises in a developing-country regional setting (Central Sulawesi, Indonesia), a population that remains largely absent from the existing phishing-awareness literature despite its high exposure and limited security capacity. Second, it operationalizes the intervention using lightweight, low-cost, and widely accessible tools, so that the workflow remains feasible under the budget, infrastructure, and time constraints that characterize MSMEs, in contrast to the subscription-based enterprise platforms that are typically beyond their reach. Third, instead of relying on engagement or click-rate indicators alone, it evaluates outcomes through an indicator-based instrument grounded in the knowledge-attitude-behavior perspective, measuring both phishing-related knowledge and self-reported safe behavior before and after the intervention. Taken together, these aspects position the study as a context-sensitive and measurable adaptation of an integrated awareness model for an underserved, resource-constrained population, rather than as a claim of combining these techniques for the first time.

3 Methodology

This study employed a quantitative approach using a single-group pretest-posttest design to examine changes in phishing awareness after an intervention integrating phishing simulation, just-in-time learning, microlearning, and gamification. The design was selected to capture changes in participants' knowledge and behavioral indicators before and after the intervention in a context resembling real work situations, without using a control group. The overall methodological flow was organized into four phasesplanning, technical preparation, implementation, and analysis as illustrated in Fig. 1.

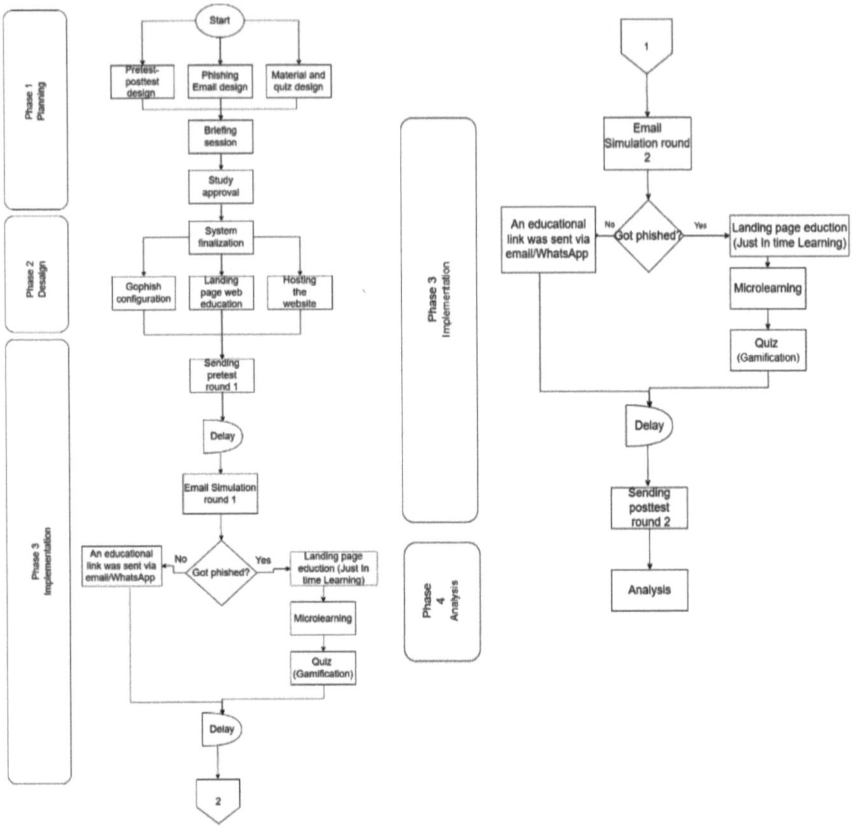

Fig. 1. Overall design of the proposed phishing awareness intervention and its evaluation workflow, including planning, technical preparation, implementation, and analysis phases

In the planning phase, the study prepared the pretest-posttest design, phishing email scenarios, and the microlearning and quiz materials used in the intervention. This was followed by technical preparation, which focused on configuring GoPhish, developing an educational landing page, and preparing the supporting website and hosting environment so that all intervention components were ready and integrated before field implementation began

The implementation phase began with the administration of a pretest in the first week to measure participants' initial level of understanding. After a one-day break, the first phishing email simulation was delivered. Based on the simulation outcome (see Fig. 1), participants were classified into two conditions: those affected by phishing and those not affected by phishing. In this study, "affected by phishing" was operationally defined as entering data on the simulation page. Participants who were affected were redirected to a just-in-time educational landing page and then continued to microlearning materials and gamification-based quizzes. Participants who were not affected still received the same educational links through email and WhatsApp as reinforcement material. The same general procedure was repeated in a second simulation round after an interval of 3 to 6 days, informed by the spacing effect, which holds that short inter-session intervals improve retention over massed repetition, with an optimal interval proportional to the retention period [47].

The study focused on individuals involved in MSME operational activities in Central Sulawesi, particularly those who used digital communication media such as email or WhatsApp for work purposes. The target participants included MSME owners and staff or employees, since both groups could encounter risky messages and make decisions at critical moments related to phishing. Sampling was conducted through purposive sampling, with participants selected from MSME owners and staff in Central Sulawesi who used digital communication media in their work and were willing to complete the full research process, including pretest, simulation, intervention, and posttest. The main evaluation focused on two outcomes: phishing knowledge and safe behavior related to phishing. Phishing knowledge was defined as the score representing participants' understanding of phishing based on the instrument items, whereas safe behavior referred to the score describing participants' tendency to act safely when encountering potential phishing situations.

The analysis phase was designed to assess changes in phishing knowledge and safe behavior by comparing paired pretest and posttest scores. The first step involved examining data completeness, identifying incomplete responses, duplicate entries, and unusual response patterns, and applying inclusion criteria so that only participants with usable paired data were analyzed. After that, the two variables were scored according to the instrument guidelines, with higher scores indicating better knowledge and safer behavior. Instrument reliability was then examined as supporting evidence for measurement quality, followed by descriptive statistics to illustrate score patterns before and after the intervention. For inferential testing, a paired-samples t-test was planned when the data met parametric assumptions, especially regarding the distribution of score differences otherwise, the Wilcoxon signed-rank test was used. In addition to score comparisons, behavioral data from phishing simulation outcomes and participant involvement in the learning sequence were analyzed descriptively as supporting indicators.

3.1 Measurement Instrument

Phishing awareness was operationalized through two measurable constructs aligned with the study outcomes: phishing-related knowledge and phishing-related safe behavior. A structured questionnaire was developed as the measurement instrument, consisting of two parts administered identically at pretest and posttest.

Part A (Knowledge) comprised 16 multiple-choice items, each with four options and a single correct answer, scored dichotomously (1 = correct, 0 = incorrect) for a total score ranging from 0 to 16. The items were organized around four knowledge indicators, with four items per indicator: (K1) recognizing the characteristics and types of digital fraud; (K2) understanding the risks and impacts of phishing; (K3) knowing preventive measures; and (K4) recognizing manipulation tactics. Item scenarios were framed around situations relevant to MSME operations, such as messages impersonating marketplaces, e-wallet services, or business contacts. For example, one K1 item presents a phone call from a supposed bank officer urgently requesting a one-time password and asks the respondent to identify the type of attack (correct answer: vishing); a K4 item presents a reward message with a countdown deadline and asks the respondent to identify the dominant manipulation tactic (correct answer: combining reward with time pressure). Part B (Safe Behavior) comprised 16 statements rated on a five-point Likert scale (1 = never to 5 = always), yielding a total score ranging from 16 to 80. The statements were organized around four behavioral indicators, with four items per indicator: (B1) verifying sender identity; (B2) exercising caution with links and attachments; (B3) protecting credentials; and (B4) reporting suspicious messages. All statements were phrased in a favorable (positive) direction, so higher scores consistently indicate safer behavior and no reverse scoring was required. For both parts, higher total scores reflect greater phishing knowledge and safer behavioral tendencies, respectively.

Each indicator was grounded in the phishing and security-awareness literature. The knowledge indicators draw on work identifying phishing cues, risks, preventive measures, and manipulation tactics [1, 12, 14, 19, 20], while the safe-behavior indicators verifying sender identity, cautious handling of links and attachments, protecting credentials, and reporting suspicious messages reflect established protective practices [1, 12, 17, 19, 32].

The instrument's content validity was established through expert review prior to deployment: items were derived from the defined indicators and the intervention's learning content, then evaluated by the supervising expert for relevance, clarity, and representativeness, with revisions applied accordingly. To further ensure measurement quality, internal-consistency reliability will be examined using Cronbach's alpha, with a threshold of 0.70 considered acceptable [48], reliability values will be reported alongside the study results as supporting evidence of measurement quality.

4 JIT Education Platform Design

The JIT education platform was designed as the immediate instructional response shown after participants interacted with the phishing simulation. Rather than functioning as a separate or delayed training session, the platform delivers corrective guidance at the teachable moment, when the simulated phishing event is still salient and directly connected to the participant's action. In this way, the platform supports the intervention logic by linking risky interaction, immediate feedback, short learning exposure, and reinforcement in one continuous sequence.

The platform consists of an educational landing page, concise microlearning content, and a quiz-based reinforcement component. The landing page serves as the first point of correction by informing participants that the previous message was part of a simulation and by redirecting them to short learning materials. The microlearning component is intended to provide focused content in a brief and accessible format, so that the intervention remains practical for MSME owners and staff who often work under limited time and multiple operational demands. The quiz component then reinforces the learning process by prompting participants to review key ideas and respond to short evaluative questions.

This design was intended to make phishing awareness training more contextual, concise, and actionable for MSME settings. By embedding learning directly after simulated exposure, the platform operationalizes the integration of phishing simulation, just-in-time learning, microlearning, and gamification within a single training flow. Figure 2 presents the interface of the JIT education platform used in the study.

Within this study, the gamification layer was implemented in a deliberately lightweight form suited to MSME constraints, centered on the reinforcement mechanism most strongly associated with behavioral change rather than on competitive features. After completing the microlearning content, participants proceed to a quiz that functions as retrieval practice, prompting active recall of key warning signs and safe responses rather than passive review. Upon submission, the quiz returns immediate item-level feedback: each item is marked correct or incorrect and accompanied by a short explanation of why a given response is safe or risky. This feedback design is intentional, as the literature identifies immediate, explanatory feedback as the gamification element most directly tied to learning and behavioral reinforcement [45]. Completion of the quiz sequence is further acknowledged through a digital certificate, which serves as a goal-based completion reward intended to support motivation and follow-through across the intervention series. Other common gamification elements, such as points, badges, and leaderboards [43], were intentionally excluded at this stage to preserve the brevity and low-friction nature required in MSME settings their integration is identified as a direction for future development

Fig. 2 Examples of microlearning pages in the JIT education platform, including video-based, text-based, and audio-based content

5 Conclusion

This paper presents the design of a phishing-awareness intervention for MSMEs that integrates phishing simulation, just-in-time learning, microlearning, and gamification. The study is motivated by the need for a training model that is not only informative, but also contextual, concise, and practical for MSME settings, where resource limitations and operational pressures often make lengthy security programs difficult to sustain. In line with the study objectives, the intervention is directed toward MSME owners and employees in Central Sulawesi and is evaluated through a quantitative pretest–posttest design with two core outcomes, namely phishing-related knowledge and safe behavior.

The literature basis and methodological design together support the argument that phishing awareness should be addressed closer to the moment of risk, supported by short, focused learning content, and reinforced through evaluative and motivational mechanisms. Within this framework, phishing simulation functions as the trigger, just-in-time learning provides immediate corrective input, microlearning delivers concise material, and gamification strengthens engagement through feedback and challenge. Rather than making broad claims beyond the limits of the design, the study deliberately narrows its evaluation to knowledge and safe behavior so that the analysis remains realistic and measurable.

Accordingly, the contribution of this study lies in proposing a more operational and context-sensitive phishing-awareness model for MSMEs, enriching the discussion on how intervention timing, concise material delivery, and reinforcement mechanisms can be combined into a single training flow, and providing a practical reference for measurable awareness-training implementation in small-business environments.

Although this paper does not report effectiveness results, it outlines a feasible workflow for MSMEs that delivers learning at a teachable moment and minimises training burden through microlearning and reinforcement. However, because the evaluation uses a one-group pretest-posttest design, causal claims will remain cautious, and observed changes may be influenced by testing effects or external exposure to cybersecurity information. In addition, behaviour observed in a simulated environment may differ from real attack contexts. Future work should therefore report effect sizes alongside pre-post comparisons and may strengthen the design using comparison groups, longer follow-up to assess retention, and systematic variation of microlearning and reinforcement elements. Future iterations may also expand the gamification layer beyond immediate feedback and completion rewards to include progressive elements such as points, badges, or leaderboards, allowing investigation of whether richer game mechanics further strengthen engagement and behavioral reinforcement without compromising the concise format that MSME contexts require.

References

1. Marshall, N., Sturman, D., Auton, J.C.: Computers & security exploring the evidence for email phishing training: a scoping review. Comput. Secur. **139**, 103695 (2024). https://doi.org/10.1016/j.cose.2023.103695

2. Gallo, L., Gentile, D., Ruggiero, S., Botta, A., Ventre, G.: Computers & security the human factor in phishing: collecting and analyzing user behavior when reading emails. Comput. Secur. **139**, 103671 (2024). https://doi.org/10.1016/j.cose.2023.103671
3. Jalali, M.S., Bruckes, M., Westmattelmann, D., Schewe, G.: Why employees (still) click on phishing links: investigation in hospitals. J. Med. Internet Res. **22**, 1–16 (2020). https://doi.org/10.2196/16775
4. Puspitasari, D., Tinggi, S., Ilmu, M., Mercusuar, K.: Digital risks in emerging economies: cyber threat escalation in Indonesia (2020–2023). Data J. Inform. Syst. Manag. **2**, 152–163 (2024)
5. Sukumar, A., Mahdiraji, H.A., Jafari-Sadeghi, V.: Cyber risk assessment in small and medium-sized enterprises: a multilevel decision-making approach for small e-tailors. Risk Anal. **43**, 2082–2098 (2023). https://doi.org/10.1111/risa.14092
6. Papathanasiou, A., Liontos, G., Katsouras, A., Liagkou, V.: Cybersecurity guide for SMEs: protecting small and medium-sized enterprises in the digital era. J. Inform. Sec. **16**, 1–43 (2025). https://doi.org/10.4236/jis.2025.161001
7. Wilson, M. et al.: It won ' t happen to me: surveying SME attitudes to it won't happen to me: surveying SME attitudes to cyber-security. J. Comput. Inf. Syst., 1–13 (2022). https://doi.org/10.1080/08874417.2022.2067791
8. Ajhari, A.A. et al.: Security awareness framework untuk Usaha Mikro. Kecil dan Menengah di Indonesia., 1–7 (2022)
9. Du, J., Kalafut, A., Schymik, G.: The health belief model and phishing: determinants of preventative security behaviors (2024)
10. Rupeika-Apoga, R., Petrovska, K.: Barriers to sustainable digital transformation in micro-, small-, and medium-sized enterprises. Sustainability. **14** (2022). https://doi.org/10.3390/su142013558
11. Catal, C., Giray, G., Tekinerdogan, B., Kumar, S., Shukla, S.: Applications of Deep Learning for Phishing Detection: A Systematic Literature Review. Springer, London (2022). https://doi.org/10.1007/s10115-022-01672-x
12. Naqvi, B., Perova, K., Farooq, A., Makhdoom, I., Oyedeji, S., Porras, J.: Computers & security mitigation strategies against the phishing attacks: a systematic literature review. Comput. Secur. **132**, 103387 (2023). https://doi.org/10.1016/j.cose.2023.103387
13. van Geest, R.J., Cascavilla, G., Hulstijn, J., Zannone, N.: The applicability of a hybrid framework for automated phishing detection. Comput. Secur. **139**, 103736 (2024). https://doi.org/10.1016/j.cose.2024.103736
14. Abroshan, H., Devos, J., Poels, G., Laermans, E.: Phishing happens beyond technology: the effects of human behaviors and demographics on each step of a phishing process. IEEE Access. **9**, 44928–44949 (2021). https://doi.org/10.1109/ACCESS.2021.3066383
15. Rathee, D., Mann, S.: Detection of E-mail phishing attacks – using machine learning and deep learning. Int. J. Comput. Appl. **183**, 1–7 (2022). https://doi.org/10.5120/ijca2022921868
16. Meda, S., Srinivas, V.S., Rao, K.C.B., Ramesh, R., Yamarthi, N.R.: A dual-phase deep learning framework for advanced phishing detection using the novel OptSHQCNN approach. PeerJ Comput. Sci. **11**, e3014 (2025). https://doi.org/10.7717/peerj-cs.3014
17. Prümmer, J., van Steen, T., van den Berg, B.: A systematic review of current cybersecurity training methods. Comput. Secur. **136**, 103585 (2024). https://doi.org/10.1016/j.cose.2023.103585
18. Bender, S., Horn, S., Loewenstein, G., Roberts, O.: Phishing feedback: just-in-time intervention improves online security. Behav. Public Policy. **1–13** (2024). https://doi.org/10.1017/bpp.2024.19
19. Alkhazi, B., Alshaikh, M., Alkhezi, S., Labbaci, H.: Assessment of the impact of information security awareness training methods on knowledge, attitude, and behavior. IEEE Access. **10**, 132132–132143 (2022). https://doi.org/10.1109/ACCESS.2022.3230286

20. Lu, X., Jiang, J., Head, M., Yang, J.: Phishing detection in multitasking contexts: the impact of working memory load, goal activation, and message framing cue on detection performance. Eur. J. Inf. Syst., 1–31 (2025). https://doi.org/10.1080/0960085X.2025.2548543
21. Rasool, T., Warraich, N.F., Sajid, M.: Examining the impact of technology overload at the workplace: a systematic review. SAGE Open. **12** (2022). https://doi.org/10.1177/21582440221114320
22. Rizzoni, F., Magalini, S., Casaroli, A., Mari, P., Dixon, M., Coventry, L.: Phishing simulation exercise in a large hospital: a case study. Digit. Heal. **8** (2022). https://doi.org/10.1177/20552076221081716
23. Mahdavi, N., Tapak, L., Darvishi, E., Doosti-Irani, A., Shafiee Motlagh, M.: Unraveling the interplay between mental workload, occupational fatigue, physiological responses and cognitive performance in office workers. Sci. Rep. **14**, 1–13 (2024). https://doi.org/10.1038/s41598-024-68889-4
24. Alayón, C.L., Säfsten, K., Johansson, G.: Barriers and enablers for the adoption of sustainable manufacturing by manufacturing SMEs. Sustainability. **14**, 1–34 (2022). https://doi.org/10.3390/su14042364
25. Ugbebor, F., Aina, O., Abass, M., Kushanu, D.: Employee cybersecurity awareness training programs customized for sme contexts to reduce human-error related security incidents. J. Knowl. Learn. Sci. Technol. **3**, 382–409 (2024). https://doi.org/10.60087/jklst.vol3.n3.p382-409
26. Arroyabe, M.F., Arranz, C.F.A., de Arroyabe, I.F., de Arroyabe, J.C.F.: The effect of IT security issues on the implementation of industry 4.0 in SMEs: barriers and challenges. Technol. Forecast. Soc. Change. **199**, 123051 (2024). https://doi.org/10.1016/j.techfore.2023.123051
27. Anatan, L.: Micro, Small, and Medium Enterprises' Readiness for Digital Transformation in Indonesia (2023)
28. Abdelmagid, A.M., Diaz, R.: Zero trust architecture as a risk countermeasure in small–medium enterprises and advanced technology systems. Risk Anal. **45**, 2390–2414 (2025). https://doi.org/10.1111/risa.70026
29. Fagbule, O.: Cyber Security Training in Small to Medium-sized Enterprises (SMEs): Exploring Organisation Culture and Employee Training Needs, 1–279 (2022)
30. Li, W., Manickam, S., Chong, Y.W., He, Y., Li, H.Y., Li, B.: ByteBait USB: a robust simulation toolkit for badUSB phishing campaign. J. King Saud Univ. Comput. Inf. Sci. **37** (2025). https://doi.org/10.1007/s44443-025-00067-6
31. Kävrestad, J., Hagberg, A., Nohlberg, M., Rambusch, J., Roos, R., Furnell, S.: Evaluation of contextual and game-based training for phishing detection. Futur. Internet. **14**, 1–16 (2022). https://doi.org/10.3390/fi14040104
32. Canham, M., Posey, C., Constantino, M.: Phish Derby: shoring the human shield through gamified phishing attacks. Front. Educ. **6**, 1–10 (2022). https://doi.org/10.3389/feduc.2021.807277
33. Alsaqer, A., Almajed, H., Alarfaj, K., Frikha, M.: Phishing simulation as a proactive defense: a customizable platform for training and behavioral analysis. Int. J. Adv. Comput. Sci. Appl. **16**(1064–1075) (2025). https://doi.org/10.14569/IJACSA.2025.01606104
34. Das Guptta, S., Shahriar, K.T., Alqahtani, H., Alsalman, D., Sarker, I.H.: Modeling hybrid feature-based phishing websites detection using machine learning techniques. Ann. Data Sci. **11**, 217–242 (2024). https://doi.org/10.1007/s40745-022-00379-8
35. Schöni, L., Carles, V., Strohmeier, M., Zimmermann, V.: Based on Users' Phishing Knowledge and Detection Skills You Know What? - Evaluation of a Personalised Phishing Training Based on Users' Phishing Knowledge and Detection Skills Categorisation (2025). https://doi.org/10.1145/3688459.3688460

36. Sankaranarayanan, R., Leung, J., Abramenka-Lachheb, V., Seo, G., Lachheb, A.: Microlearning in diverse contexts: a bibliometric analysis. TechTrends. **67**, 260–276 (2023). https://doi.org/10.1007/s11528-022-00794-x
37. Monib, W.K., Qazi, A., Apong, R.A.: Microlearning beyond boundaries: a systematic review and a novel framework for improving learning outcomes. Heliyon. **11**, e41413 (2025). https://doi.org/10.1016/j.heliyon.2024.e41413
38. Fidan, M.: The effects of microlearning-supported flipped classroom on pre-service teachers' learning performance, motivation and engagement. Educ. Inf. Technol. **28**, 12687–12714 (2023). https://doi.org/10.1007/s10639-023-11639-2
39. Lee, Y.M.: Mobile microlearning: a systematic literature review and its implications. Interact. Learn. Environ. **31**, 4636–4651 (2023). https://doi.org/10.1080/10494820.2021.1977964
40. Rof, A., Bikfalvi, A., Marques, P.: Exploring learner satisfaction and the effectiveness of microlearning in higher education. Internet High. Educ. **62**, 100952 (2024). https://doi.org/10.1016/j.iheduc.2024.100952
41. Van Steen, T., Deeleman, J.R.A.: Successful gamification of cybersecurity. Training. **24**, 593–598 (2021). https://doi.org/10.1089/cyber.2020.0526
42. Alothman, B.Y.: Cyber gamification: implementing gamified adaptive learning environments for effective cyber security teams education: enhancing team skills and engagement through interactive learning environments. ACM Int. Conf. Proceeding Ser., 33–40 (2024). https://doi.org/10.1145/3669947.3669953
43. Malone, M., Wang, Y., James, K., Anderegg, M., Werner, J., Monrose, F.: To gamify or not?: on leaderboard effects, student engagement and learning outcomes in a cybersecurity intervention. In: Proceedings of the 52nd ACM Technical Symposium on Computer Science Education, pp. 1135–1141 (2021). https://doi.org/10.1145/3408877.3432544
44. Shahzadi, A., Ishaq, K., Nawaz, N.A., Rosdi, F., Khan, F.A.: Unveiling personalized and gamification-based cybersecurity risks within financial institutions. PeerJ Comput. Sci. **11**, 1–36 (2025). https://doi.org/10.7717/PEERJ-CS.2598
45. Zolfaghari, Z., Karimian, Z., Zarifsanaicy, N., Farahmandi, A.Y.: A scoping review of gamified applications in English language teaching: a comparative discussion with medical education. BMC Med. Educ. **25** (2025). https://doi.org/10.1186/s12909-025-06822-7
46. Albaladejo-González, M., Nespoli, P., Gómez Mármol, F., Ruipérez-Valiente, J.A.. A multimodal and adaptive gamified system to improve cybersecurity competence training. Clust. Comput. **28** (2025). https://doi.org/10.1007/s10586-025-05264-6
47. Cepeda, N.J., Vul, E., Rohrer, D., Wixted, J.T., Pashler, H.: Spacing effects in learning: a temporal ridgeline of optimal retention. Psychol Sci. **19**, 1095–1102 (2008)
48. Nunnally, J.C.: Psychometric Theory. McGraw-Hill, New York (1978)

ERDucate: An AI-Powered ERD Assessment and Feedback Tool

Edrieanna Soraya Zuraidi Effendi, Mas Rina Mustaffa(✉), Lili Nurliyana Abdullah, and Nurul Amelina Nasharuddin

Multimedia Department, Faculty of Computer Science and Information Technology, Universiti Putra Malaysia, Serdang, Selangor, Malaysia
MasRina@upm.edu.my

Abstract. Evaluating Entity-Relationship Diagrams (ERDs) can be time-consuming, particularly when lecturers must review large numbers of submissions and provide feedback on modelling errors. This challenge becomes more apparent in larger classes. ERDucate was therefore developed as a web-based tool to support ERD assessment. The framework integrates multimodal diagram analysis, deterministic grading, automated feedback generation, and Human-in-the-Loop verification. Gemini 2.5 Flash is used to extract ERD elements, including entities, attributes, relationships, cardinalities, and key constraints, from submitted diagrams. Marks are then calculated separately using lecturer-defined assessment rules, while feedback is generated from the verified grading outcome. Testing involved 22 student submissions based on a Car Rental System assessment. Comparison with lecturer grading produced a Mean Absolute Error (MAE) of 0.86 on an eleven-mark scale. Most grading differences were associated with diagram extraction issues rather than rubric application. The results show that multimodal AI can support ERD assessment effectively when combined with deterministic grading and selective lecturer review.

Keywords: Entity-Relationship Diagram · Automated Assessment · Large Language Models · Human-in-the-Loop · Deterministic Grading · Educational Technology

1 Introduction

Students studying database design are commonly required to construct Entity-Relationship Diagrams (ERDs) to demonstrate their understanding of conceptual modelling. These diagrams describe entities, attributes, relationships, and constraints before implementation takes place at the database level [1, 2]. Consequently, ERDs are frequently used in coursework, laboratory exercises, and examinations.

Reviewing ERDs manually can be laborious. A lecturer must examine several diagram elements simultaneously while determining whether modelling decisions satisfy the assessment requirements. The effort required becomes more noticeable when the number of submissions increases. Providing meaningful feedback further adds to the

N. Mohd Sharef et al. (Eds.): CAMP 2026, CCIS 3093, pp. 348–358, 2027.
https://doi.org/10.1007/978-981-92-4302-0_25

assessment workload, despite its importance for student learning and improvement [3, 4].

Recent advances in multimodal AI have made it possible to analyze information that combines text and visual content. Unlike earlier approaches that required separate modules for image processing, text recognition, and interpretation, newer multimodal models can process diagrams directly from uploaded images [5, 6]. Such developments make it possible to analyze diagram-based learning artefacts, including Entity-Relationship Diagrams, without relying on multiple independent processing stages.

Existing ERD software primarily supports diagram creation and editing. Although tools such as Lucidchart and ERBuilder simplify modelling activities, the evaluation process still depends heavily on lecturer review. Recent AI technologies offer additional support for assessment tasks, but grading outcomes may differ when decisions rely solely on generated responses.

The work described in this paper originated from the need to manage increasing numbers of ERD submissions without compromising assessment consistency. To address this issue, ERDucate was developed as a web-based assessment tool that combines multimodal diagram interpretation, deterministic grading, automated feedback generation, and Human-in-the-Loop verification. Gemini 2.5 Flash is used to identify ERD components, whereas marks are determined separately through predefined grading rules. Feedback is generated only after grading has been completed, ensuring consistency between comments and awarded scores.

Three contributions are made in this study. First, a multimodal ERD assessment workflow is introduced for extracting diagram components directly from student submissions. Second, a deterministic grading mechanism is developed to ensure that identical extracted representations produce identical assessment outcomes. Third, Human-in-the-Loop verification is incorporated to allow lecturers to review uncertain detections before grades are finalized.

2 Related Works

2.1 ERD Learning and Diagram Understanding

Entity-Relationship Diagrams (ERDs) are widely used in database education because they allow students to represent entities, attributes, relationships, and constraints during conceptual design [1, 2]. Despite their importance, students frequently encounter difficulties in identifying appropriate entities, assigning attributes correctly, and modelling relationships and cardinalities [1]. Educational tools such as DiagrammeR have been developed to support ERD construction and learning activities [2]. Despite these developments, relatively few systems focus on assessing completed ERDs automatically.

Separate lines of research have investigated how information can be extracted and understood from diagrams. Earlier approaches typically relied on separate stages for image processing, text recognition, and structural reconstruction. More recently, language models have been explored for understanding knowledge representations and conceptual schemas. Insuasti et al. [7] demonstrated that BERT and Llama-based models can be used to interpret pre-conceptual schema representations. While such studies show promising capabilities for diagram understanding, they do not directly address grading and feedback generation in educational settings.

2.2 Multimodal LLMs and Automated Educational Assessment

Recent progress in multimodal Large Language Models (LLMs) has enabled systems to analyze both visual and textual information within a single framework. In database education, Riazi and Rooshenas [8] explored the use of LLM-generated feedback to support conceptual design learning, while Ramachandran et al. [9] investigated automated ERD and schema assessment using LLMs. The development of vision-language assistants such as YO'LLaVA further demonstrates the growing ability of multimodal systems to combine image understanding with contextual reasoning [10].

Automated feedback has long been recognized as a useful educational tool. Reviews by Deeva et al. [3] and Shi and Aryadoust [6] reported benefits related to scalability, personalization, and timeliness. Similar observations were reported by Fleckenstein et al. [11], who found positive effects of automated feedback on student performance. Interest in generative AI within education has increased rapidly in recent years. ChatGPT and similar technologies have been explored as tools for supporting learning and providing feedback to students [4, 12]. However, questions concerning grading consistency, reliability, and explainability remain unresolved when these technologies are applied to assessment tasks [13].

2.3 Human-in-the-Loop Assessment and Research Gap

Ambiguous diagram interpretations can present difficulties for fully automated assessment. HITL approaches allow lecturers to review uncertain cases before grading is completed, helping to maintain fairness, trust, and accountability in AI-assisted educational assessment [14, 15].

Despite recent advances in ERD learning tools and multimodal AI, several practical challenges remain. Most ERD platforms concentrate on diagram creation and editing, leaving assessment and feedback activities largely to lecturers. Existing diagram-understanding studies primarily focus on identifying diagram components [16], whereas educational assessment requires additional capabilities for grading and feedback generation. Furthermore, many recent LLM-based solutions rely heavily on model-generated decisions, which may introduce variations in grading outcomes. Limited attention has been given to integrating Human-in-the-Loop verification into ERD assessment workflows.

ERDucate was developed to address these limitations by combining multimodal diagram interpretation, deterministic grading, automated feedback generation, and Human-in-the-Loop verification within a single assessment framework.

3 Methodology

ERDucate was developed to reduce the amount of manual effort required when grading Entity-Relationship Diagram (ERD) assignments. During database courses, lecturers often need to inspect multiple diagram components individually, including entities, attributes, relationships, cardinalities, and key constraints. This process becomes increasingly time-consuming when large numbers of submissions must be evaluated within a short period.

To address this issue, a workflow was designed that combines multimodal diagram interpretation, rule-based grading, automated feedback generation, and lecturer verification. Figure 1 illustrates the overall assessment process implemented in ERDucate.

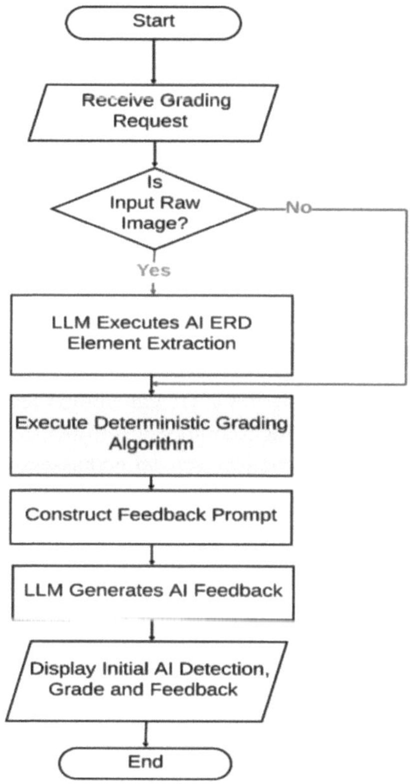

Fig. 1 Overall workflow of the ERDucate assessment framework

3.1 System Architecture

ERDucate was implemented as a web-based assessment platform for evaluating student ERD submissions. The system comprises several interconnected components responsible for ERD extraction, grading, feedback generation, and lecturer verification. ERDucate was implemented as a web-based platform for assessing student ERD submissions. After an ERD image is uploaded, Gemini 2.5 Flash identifies entities, attributes, relationships, cardinalities, and key constraints from the diagram. The extracted information is then used by the grading, feedback, and verification modules. Table 1 summarizes the technologies employed in the system.

Table 1 Main components of ERDucate

Component	Technology/Function
User Interface	React.js, Vite
Server-side Processing	Node.js, Express
Database	Firebase Firestore
Image Storage	Cloudinary
ERD Extraction	Gemini 2.5 Flash
Grading Module	Deterministic rule-based scoring
Feedback Generation	Gemini 2.5 Flash-Lite
Verification Module	Human-in-the-Loop review

3.2 Deterministic Grading Mechanism

In ERDucate, diagram interpretation and score calculation are handled independently. Gemini 2.5 Flash is used to identify ERD components from the submitted diagram, while marks are determined according to the lecturer-defined assessment rubric.

The extracted entities, attributes, relationships, cardinalities, and key constraints are stored in a structured JSON representation. This representation is subsequently compared with the lecturer's reference solution during grading.

Small differences in naming are tolerated during the matching process. In addition to exact matching, the system employs morphological normalization, bag-of-words comparison, and Levenshtein distance to avoid penalizing labels that differ only slightly in wording (Table 2).

Table 2 Example rubric criteria employed in the deterministic grading process

Criterion	Description	Marks
Entity Identification	Correct entity detection	2
Attribute Identification	Correct attribute detection	2
Relationship Modelling	Correct relationship representation	3
Cardinality Constraints	Correct cardinality assignment	2
Key Constraints	Correct primary/foreign key representation	2

For example, labels such as *Student* and *Students* are treated as equivalent. However, labels referring to different concepts remain distinguishable and receive different grading outcomes.

Since marks are calculated using predefined rules, the same extracted ERD representation always produces the same result. This makes the grading process easier to verify and reduces inconsistencies that may arise when marks are generated directly by a language model.

3.3 Feedback Generation Mechanism

Once grading has been completed, ERDucate prepares written feedback for the student. The feedback is generated using Gemini 2.5 Flash-Lite and is based on the assessment outcome rather than the uploaded diagram itself.

Information supplied to the model includes awarded marks, missing entities, incorrectly assigned attributes, relationship issues, cardinality mistakes, and other observations recorded during grading. Consequently, the comments generated remain consistent with the final assessment result.

During testing, the feedback generated was generally able to explain why marks were deducted and identify the diagram components involved. Missing entities, incorrect relationships, and invalid cardinalities were among the issues most frequently reported. Additional components that did not match the expected solution could also be highlighted for student review.

Generating feedback after grading helped prevent situations where comments contradicted the awarded score. Since the feedback module receives information from the completed assessment process, the generated explanations remain tied to the verified grading outcome.

3.4 Human-in-the-Loop Verification

Although multimodal models performed well in most cases, several diagrams encountered during development and testing proved more difficult to interpret automatically. Examples included overlapping labels, crowded layouts, relationship connectors crossing one another, and attributes positioned close to multiple entities.

Uncertain extraction results are reviewed by the lecturer before grading begins, as illustrated in Fig. 2.

During verification, incorrectly identified labels may be corrected, missing components may be added, and invalid detections may be removed. The updated representation is subsequently used during assessment.

Manual intervention is required only for cases where the extracted information remains ambiguous. Consequently, lecturer effort is concentrated on a small subset of submissions rather than the entire class, allowing most assessments to proceed automatically.

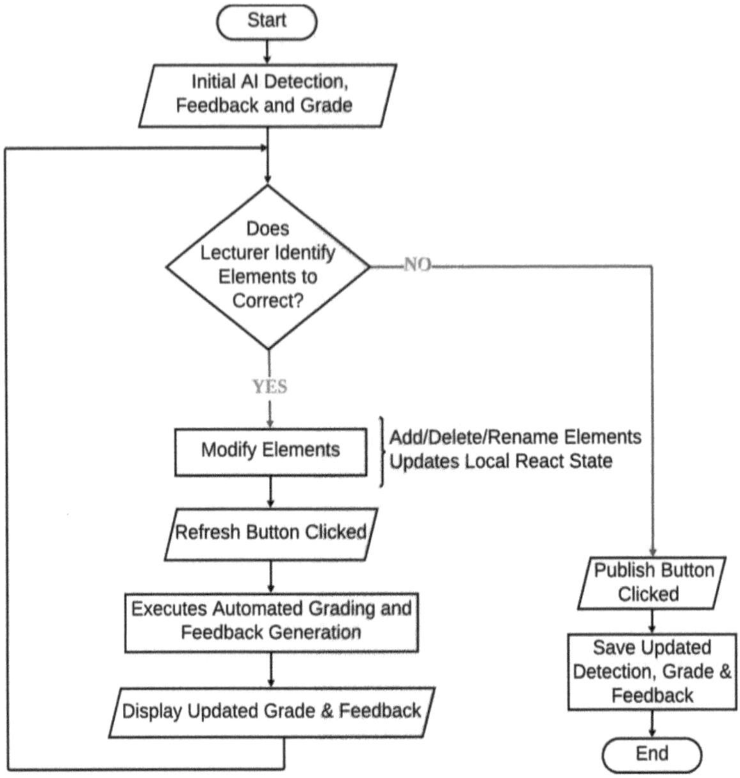

Fig. 2 Verification process used when extracted ERD components fall below the confidence threshold

4 Experimental Setup

Twenty-two ERD submissions obtained from a database design assessment were used during testing. All students answered the same Car Rental System question. The experimental configuration adopted in this study is summarized in Table 3.

Each submission was processed using the ERDucate workflow. Gemini 2.5 Flash extracted entities, attributes, relationships, cardinalities, and key constraints from the uploaded diagram before grading was performed using the lecturer-defined rubric. Lecturer verification remained available whenever uncertain detections were encountered.

The lecturer independently graded the same submissions using identical assessment criteria. Automated and manual scores were subsequently compared using Mean Absolute Error (MAE). The extracted ERD representations and feedback generated were also reviewed to identify common error patterns and situations requiring lecturer intervention.

Table 3 Experimental configuration

Item	Description
Number of submissions	22
Assessment scenario	Car Rental System
Diagram type	Entity-Relationship Diagram (ERD)
Automated evaluator	ERDucate
Human evaluator	Lecturer
Assessment rubric	11 marks
Evaluation metric	Mean Absolute Error (MAE)
Verification mode	Human-in-the-Loop

5 Results and Discussion

5.1 Grading Accuracy Performance

The grading performance of ERDucate was evaluated by comparing automated assessment results against human-verified grading outcomes. Figure 3 illustrates the relationship between ERDucate-generated scores and human-verified assessment scores, while Table 4 summarizes the overall grading accuracy achieved by the implemented work.

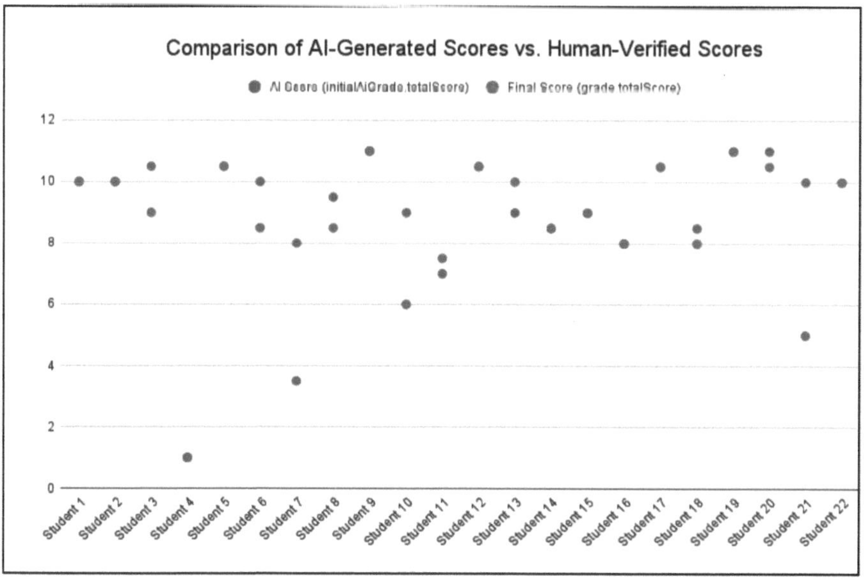

Fig. 3 Distribution of ERDucate-generated scores and manually assigned grades for the 22 evaluated submissions

Table 4 Grading accuracy performance of ERDucate against human-verified assessment

Metric	Description	Value
Mean Absolute Error (MAE)	Average Deviation from Human Grade	0.86
Exact Matches	Count of Submissions with 0.0 Deviation	54.5%
Near Matches	Count of Submissions with Grade Deviation ≤ 1.0	77.3%

The average difference between automated and manual grading was 0.86 marks on an eleven-mark scale. For many submissions, ERDucate and the lecturer assigned identical scores. Most of the remaining submissions differed by only a small margin.

The largest deviations were traced to extraction issues rather than rubric application. For instance, an attribute occasionally became associated with a neighboring entity when diagram elements were positioned closely together. In several cases, discrepancies could be traced to extraction errors rather than the grading rules. Once the extracted ERD representation had been corrected, the assigned score aligned with the lecturer's assessment. Another observation emerged during testing. Whenever the same ERD representation was assessed more than once, the resulting score remained unchanged. The use of fixed grading criteria ensured that identical inputs produced identical results.

5.2 Qualitative Analysis

The submitted ERDs contained a variety of modelling errors, including missing entities, incorrect attributes, relationship mistakes, and cardinality issues.

Extraction accuracy was generally higher for diagrams with clear layouts and sufficient spacing between elements. More challenging cases involved overlapping labels, intersecting connectors, and crowded diagram structures.

In these situations, lecturer verification helped resolve uncertain extraction results before assessment took place. The observations suggest that diagram readability plays a more important role in grading accuracy than the scoring mechanism itself.

6 Limitations and Future Work

Only 22 ERD submissions were available for evaluation in this study, all of which originated from the same database assessment task. While the dataset contained a range of modelling mistakes, additional testing involving different questions, student cohorts, and diagram styles would provide a broader picture of system performance.

The evaluation centered on grading results and their agreement with manual assessment. Factors such as usability, assessment speed, student perception, and the practical impact on lecturer workload were not explored during this phase of the work.

At the time of writing, ERDucate supports only ERD assessment and depends on external multimodal AI services. Further work could explore the use of the framework beyond ERD assessment, including other diagram formats and alternative deployment configurations.

7 Conclusion

ERDucate was developed to explore how multimodal AI can support the assessment of Entity-Relationship Diagrams in database design education. The framework combines automated diagram interpretation with lecturer verification and predefined grading criteria.

Results obtained from the evaluated submissions were generally close to those produced through manual assessment. Most discrepancies could be traced to diagram extraction difficulties rather than differences in rubric application.

Overall, the study demonstrates that AI can play a supporting role in ERD assessment. Lecturer involvement remains available for cases requiring additional verification, while routine assessment activities can be handled automatically.

References

1. Rashkovits, R., Lavy, I.: Mapping common errors in entity relationship diagram design of novice designers. Int. J. Database Manag. Syst. **13**(1), 1–19 (2021)
2. Jaimez-González, C., Martínez-Samora, J.: DiagrammER: a web application to support the teaching-learning process of database courses through the creation of ER diagrams. Int. J. Emerg. Technol. Learn. **15**(19), 4–21 (2020)
3. Deeva, G., Bogdanova, D., Serral, E., Snoeck, M., De Weerdt, J.: A review of automated feedback systems for learners: classification framework, challenges and opportunities. Comput. Educ. **162**, 104094 (2021)
4. Mustaffa, M.R., Mansor, E.I.: Facilitating students' learning through M-LMS for tablet device. Int. J. Comput. Theory Eng. **7**(3), 236–240 (2015)
5. Kasneci, E., Seßler, K., Küchemann, S., Bannert, M., Dementieva, D., Fischer, F., et al.: ChatGPT for good? On opportunities and challenges of large language models for education. Learn. Individ. Differ. **103**, 102274 (2023)
6. Shi, H., Aryadoust, V.: A systematic review of AI-based automated written feedback research. ReCALL. **36**(2), 187–209 (2024)
7. Insuasti, J., Roa, F., Zapata-Jaramillo, C.M.: Computers' interpretations of knowledge representation using pre-conceptual schemas: an approach based on the BERT and Llama 2-Chat models. Big Data Cogn. Comput. **7**(4), 182 (2023)
8. Riazi, S., Rooshenas, P.: LLM-driven feedback for enhancing conceptual design learning in database systems courses. In: Proceedings of the 56th ACM Technical Symposium on Computer Science Education, vol. 1, pp. 1001–1007 (2025)
9. Ramachandran, R., Vijayan, P., Anilkumar, A., Gangadharan, V.: AI assisted system for automated evaluation of entity-relationship diagram and schema diagram using large language models. Big Data Cogn. Comput. **10**(1), 2 (2025)
10. Nguyen, T., Liu, H., Li, Y., Cai, M., Ojha, U., Lee, Y.J.: YO'LLaVA: your personalized language and vision assistant. In: Advances in Neural Information Processing Systems, vol. 37, pp. 40913–40951 (2024)
11. Fleckenstein, J., Liebenow, L.W., Meyer, J.: Automated feedback and writing: a multi-level meta-analysis of effects on students' performance. Front. Artif. Intell. **6**, 1162454 (2023)
12. Tlili, A. et al.: What if the devil is my guardian angel: ChatGPT as a case study of using chatbots in education. Smart Learn. Environ. **10**(1), 15 (2023)

13. Dwivedi, Y.K., Kshetri, N., Hughes, L., Slade, E.L., Jeyaraj, A., Kar, A.K., et al.: So what if ChatGPT wrote it? Multidisciplinary perspectives on opportunities, challenges and implications of generative conversational AI for research, practice and policy. Int. J. Inf. Manag. **71**, 102642 (2023)
14. Fajardo-Ramos, D.C., Chiappe, A., Mella-Norambuena, J.: Human-in-the-loop assessment with AI: implications for teacher education in Ibero-American universities. Front. Educ. **10**, 1710992 (2025)
15. Holmes, W., Miao, F.: Guidance for Generative AI in Education and Research. UNESCO Publishing, Paris (2023)
16. Mustaffa, M.R., Zheng, A.G.: Region-based convolutional neural network and tesseract optical character recognition for recognising entity relationship diagram. In: 2023 10th International Conference on Electrical Engineering, Computer Science and Informatics (EECSI), pp. 237–241. IEEE (2023)

Smartnotes: AI-Powered Summarization and Flashcard Generation for Learning Database Principles

Aleeya Sofea Sh. Salim, Mas Rina Mustaffa(✉), Lili Nurliyana Abdullah, and Nurul Amelina Nasharuddin

Multimedia Department, Faculty of Computer Science and Information Technology, Universiti Putra Malaysia, Serdang, Selangor, Malaysia
masrina@upm.edu.my

Abstract. Database Principles is a core component of many computing programmes, covering topics such as entity-relationship modelling, normalization, relational schemas, and Structured Query Language (SQL). Revising these topics often requires students to work through extensive lecture materials and identify important concepts from large amounts of information. To support this process, SmartNotes was implemented as a mobile application that generates summaries, flashcards, and multiple-choice quizzes from uploaded lecture notes. The application utilizes Natural Language Processing (NLP) and Large Language Model (LLM) to transform PDF-based learning materials into revision resources for topic review, active recall, and self-assessment. An evaluation involving 22 university students was conducted to examine the usability of the application and its usefulness during revision. Participants reported that the summaries generated helped them identify key concepts more efficiently. The flashcard and quiz functions were also viewed favorably, particularly for supporting self-testing and repeated revision activities. The evaluation results provide preliminary evidence that AI-generated learning materials can reduce part of the manual effort involved in preparing study resources and may encourage more active engagement with lecture content. However, the generated content should still be treated as learning support rather than a replacement for lecturer-provided materials, particularly because AI-generated outputs may require verification. Several areas remain open for further development on improving content reliability, supporting more document formats, and evaluating the effect of SmartNotes on actual learning performance.

Keywords: Artificial Intelligence · Database Principles · Flashcard Generation · Large Language Models · Mobile Learning · Natural Language Processing · Text Summarization

1 Introduction

The use of artificial intelligence in higher education has expanded considerably in recent years. What began primarily as learning analytics and recommendation systems has gradually evolved into tools capable of generating learning content, providing feedback,

N. Mohd Sharef et al. (Eds.): CAMP 2026, CCIS 3093, pp. 359–369, 2027.
https://doi.org/10.1007/978-981-92-4302-0_26

and supporting student learning activities. Recent developments in large language models (LLMs) have further accelerated this trend, allowing educational applications to generate explanations, summaries, questions, and other forms of learning material directly from textual content [1–4].

Although these technologies have created new opportunities for learning support, challenges remain in subjects that involve abstract concepts and technical terminology. Database Principles is an example of such subject. Students are expected to understand topics such as entity-relationship modelling, relational schemas, normalization, data integrity, and Structured Query Language (SQL). Many students find themselves repeatedly returning to lecture notes, examples, and diagrams when revising these topics, especially when trying to connect related concepts and rules.

Many students incorporate flashcards, quizzes, and self-testing activities into their revision routines. Such practices are aligned with retrieval-practice theory, which proposes that actively recalling information can improve long-term retention more effectively than passive reading alone [5]. Despite these advantages, preparing revision resources often requires additional effort. Students may spend time identifying important concepts, condensing lecture notes, and creating their own practice materials before meaningful revision can take place. As a result, a considerable portion of study time may be devoted to preparation rather than learning itself.

Recent advances in large language models (LLMs) have created new possibilities for automating some of these tasks. By analysing lecture materials, AI systems can generate summaries, flashcards, and assessment questions with minimal user input. Previous studies have reported promising applications of LLMs in educational content generation and personalised learning support, while also highlighting concerns related to reliability, transparency, and accuracy [3, 4]. These concerns are particularly relevant in technical subjects, where inaccuracies in generated explanations or assessment items may affect students' understanding of important concepts.

SmartNotes was developed as a mobile learning application for Database Principles that converts uploaded lecture notes into AI-generated summaries, flashcards, and multiple-choice quizzes. In addition to content generation, the application supports self-assessment and learning progress monitoring within the same platform. Rather than replacing lecturer-provided learning materials, SmartNotes is intended to complement existing revision practices by providing additional learning resources for independent study. The application was subsequently evaluated to obtain feedback on its usability and perceived usefulness, while also identifying practical considerations associated with the use of AI-generated content in educational settings.

The remainder of this paper is organized as follows. Section 2 reviews related work on AI-assisted learning, text summarization, question generation, flashcard-based learning, and existing learning applications. Section 3 describes the methodology and system design. Section 4 presents the SmartNotes application, user needs analysis, discussion, and system limitations. Finally, Sect. 5 concludes the paper and outlines future work.

2 Related Works

The increasing availability of large language models has encouraged the development of educational applications that can generate and organize learning materials automatically. Recent studies have explored the use of AI in areas such as content creation, assessment support, tutoring, and personalized learning [6–8]. These technologies provide new opportunities to reduce the effort required to prepare learning resources and to support more flexible learning experiences. At the same time, researchers continue to highlight concerns related to content accuracy, transparency, and appropriate educational use [4, 7].

Automatic text summarization is frequently used to reduce the amount of information that learners need to review during revision. Lecture notes and educational documents often contain substantial textual content, making it difficult for students to locate key ideas efficiently. Text summarization aims to reduce the amount of information that learners need to process by producing shorter representations of longer documents while preserving key content. El-Kassas et al. [9] reviewed a range of summarization approaches and reported substantial improvements brought by deep learning techniques in capturing semantic relationships within textual data. Zhang et al. [10] further noted that transformer-based abstractive models can generate summaries with stronger contextual coherence than conventional extractive methods. As a result, automated summarization has become increasingly applicable in educational contexts where students must review substantial amounts of textual content.

Automatic question generation represents another area of educational AI research. By generating assessment items directly from learning materials, question generation systems can support self-assessment without requiring educators or students to prepare questions manually. Kurdi et al. [11] highlighted significant advances in educational question generation driven by developments in natural language processing. More recent studies have examined the use of large language models for this purpose. Scaria et al. [12] reported that LLM-generated questions could address multiple cognitive levels, while Wang et al. [13] demonstrated that prompt design plays an important role in determining the quality and relevance of generated questions. These findings indicate the potential of AI-generated quizzes as a supplementary revision mechanism.

The relationship between AI and learning support has also been explored from the perspective of knowledge retention and personalised learning. Retrieval-practice theory emphasises the importance of recalling information during learning activities rather than relying solely on repeated reading [5]. Shu et al. [14] introduced KARL, a framework that combines retrieval mechanisms with large language models to generate adaptive flashcards based on learner progress. Their findings suggest that learning resources tailored to individual learners can contribute positively to knowledge retention.

The growing use of smartphones and tablets has expanded the ways in which learning resources can be accessed. Revision activities are no longer confined to conventional classroom settings, as students can engage with learning materials through mobile devices in a variety of environments. Previous studies have identified accessibility, usability, and learner engagement as important factors influencing the adoption of mobile learning systems [15]. Earlier work by Mustaffa and Mansor [16] similarly demonstrated the potential of mobile learning platforms in supporting student learning. These developments make mobile devices a suitable medium for delivering AI-assisted revision resources.

Several learning applications, including Quizlet, Anki, and Brainscape, support revision through flashcards, quizzes, and progress monitoring. Although these platforms are widely used for active recall activities, the preparation of learning materials generally remains the responsibility of the user. The automatic conversion of lecture notes into summaries, flashcards, and quizzes is less commonly supported. A comparison between these applications and SmartNotes is presented in Table 1. Unlike the existing platforms, SmartNotes combines AI-generated summaries, flashcards, quiz generation, and performance tracking within a mobile application designed specifically for Database Principles revision.

Table 1 Comparison of SmartNotes with existing mobile learning applications

Feature	Quizlet	Anki	Brainscape	SmartNotes
Automatic Flashcard Generation from PDF	✗	✗	✗	✓
AI-based Summarization	✗	✗	✗	✓
AI-generated Quiz Questions	✗	✗	✗	✓
Performance Tracking	✓	✓	✓	✓
Personalized Learning Materials	Limited	Limited	Limited	✓
Database Principles Focus	✗	✗	✗	✓
Mobile Learning Support	✓	✓	✓	✓

The reviewed studies demonstrate that AI can support a wide range of educational activities, including summarization, question generation, adaptive learning, and mobile learning. Nevertheless, these capabilities are rarely available on a single application. Students often need to switch between different tools or prepare learning materials manually before revision can begin. This observation motivated the development of SmartNotes, which brings together summary generation, flashcards, quizzes, and learning progress monitoring in one mobile learning platform designed for Database Principles revision.

3 Methodology

SmartNotes was developed to assist students who found revision preparation both time-consuming and repetitive. Rather than requiring students to manually create summaries, flashcards, or practice questions from lecture notes, the application was designed to generate these resources automatically. Development of the system began with a study of student requirements, followed by application implementation and integration of AI services responsible for content generation.

3.1 Requirement Analysis

The development of SmartNotes was preceded by a questionnaire involving students enrolled in Database Principles. The survey was conducted to obtain a clearer understanding of students' revision practices, difficulties encountered during learning, and expectations regarding AI-assisted educational tools.

Survey findings showed that revision often involved considerable preparation. Students frequently reported spending time reviewing lecture notes, identifying key concepts, and producing their own study materials before meaningful revision could take place. Difficulties in retaining information from lengthy learning materials were also reported. Interest was also expressed in tools that could support summarization, self-assessment, and more interactive learning activities.

The survey findings informed the design of SmartNotes. Functions such as lecture note upload, summary generation, flashcards, quizzes, and learning progress tracking were selected based on the needs identified by the participants.

3.2 System Design and Implementation

SmartNotes was implemented as a mobile application supported by cloud-based services and AI-assisted content generation. The architecture includes a mobile interface for user interaction, backend services for processing, and a large language model used to generate learning resources from uploaded lecture notes.

Flutter was used to develop mobile applications, allowing deployment across multiple mobile platforms. User authentication and data storage were managed through Firebase. Communication between the application and backend services was handled through FastAPI, while Google Cloud Run hosted the server-side components.

Learning resources were generated through a large language model accessed via the Groq API. After a lecture note was uploaded, the extracted text was submitted for processing, and the generated outputs were returned to the application for presentation to the user.

3.3 AI-Assisted Learning Content Generation

The content generation workflow begins with the upload of a PDF lecture note through the mobile application. Text extracted from the document is forwarded to the large language model for analysis.

The uploaded content is then transformed into several forms of revision resources, including summaries, flashcards, and multiple-choice quiz questions. Each resource supports a different revision activity, ranging from topic review to self-assessment and active recall.

The generated summaries highlight key concepts and important information from the uploaded material. Flashcards are presented in a question-and-answer format, while quiz questions allow students to evaluate their understanding of the topic independently.

Once generated, the learning resources are stored within the application and remain available for future revision sessions. As a result, a single lecture note can be converted into multiple forms of revision support without requiring students to prepare the materials manually.

4 Results and Discussion

4.1 SmartNotes Application

SmartNotes functions as a mobile platform for Database Principles revision. Students upload lecture notes through the application, after which AI-generated learning resources become available for review and self-assessment.

The workflow is illustrated in Fig. 1. Once processing is completed, the uploaded content is transformed into summaries, flashcards, and quiz questions, each intended to support a different aspect of the revision process.

One of the main functions of SmartNotes is the automatic generation of revision resources from lecture materials. Instead of preparing summaries or practice questions manually, students can obtain learning materials directly from the uploaded document. This allows revision activities to begin immediately after the lecture notes have been processed.

The summary module provides a concise overview of the uploaded content, allowing important concepts to be reviewed more efficiently. For students revising multiple topics, this reduced the need to repeatedly search through extensive lecture notes for key information.

The flashcard module encouraged a more active form of revision by presenting concepts in a question-and-answer format. This allowed students to check their understanding while revising the material. Instead of repeatedly reading notes, students are encouraged to recall information actively during study sessions. The generated quizzes provided an additional opportunity for self-assessment. Students could identify concepts that required further attention based on their quiz performance.

Revision activities and quiz results were recorded by the system, allowing students to monitor their learning progress throughout the revision process. This information allows students to observe their revision progress over time and provides an additional source of feedback during independent study.

Selected interfaces of the application are shown in Fig. 2. The screenshots illustrate the registration and login modules, PDF upload functionality, summary and flashcard views, quiz interface, performance tracking dashboard, and user profile management screens.

4.2 Preliminary Survey Findings

Prior to developing SmartNotes, a survey involving 22 students enrolled in Database Principles was conducted to better understand their revision practices and learning challenges. Several trends emerged from the responses.

The survey responses revealed several challenges associated with revising Database Principles. More than half of the participants reported moderate to high levels of difficulty when studying the subject, while 63.7% indicated that remembering important concepts after revision remained problematic. These findings suggest that retaining database-related knowledge continues to be a challenge for many students.

Considerable effort was also associated with revision preparation. Approximately 77.3% of respondents reported spending substantial time producing notes and other

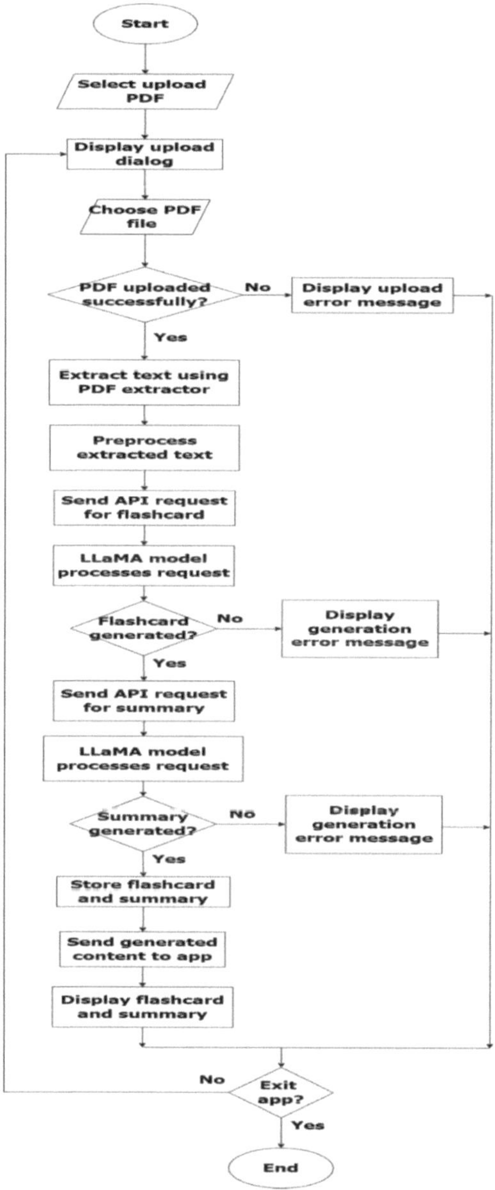

Fig. 1 PDF upload and AI-assisted learning content generation workflow

study materials before revision activities could begin. The same proportion expressed a preference for learning approaches that encourage greater interaction with the learning material rather than relying solely on conventional note reading.

The responses also reflected strong interest in AI-assisted learning support. More than four-fifths of the participants indicated interest in applications capable of generating

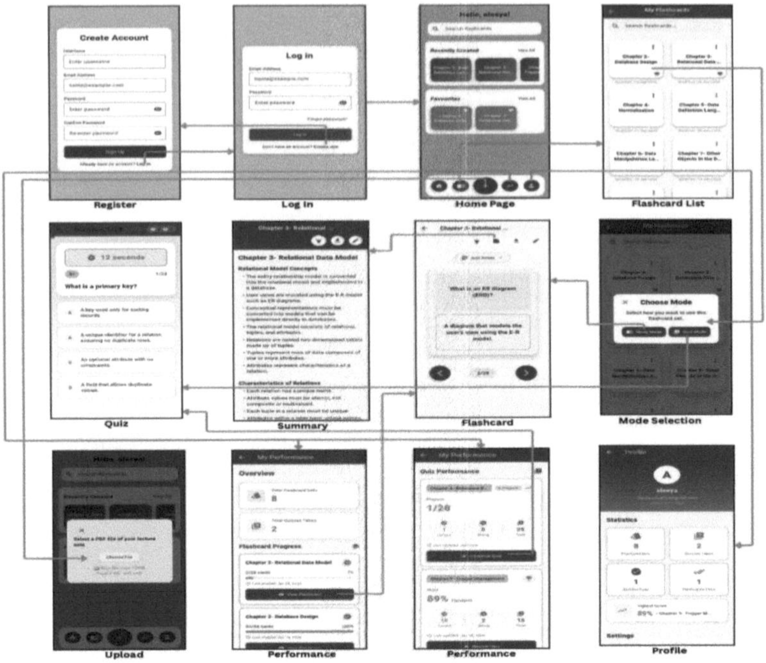

Fig. 2 Screenshots of the SmartNotes mobile learning application showing the registration, login, upload, summary, flashcard, quiz, performance tracking, and profile modules

summaries and flashcards automatically from lecture notes. Similarly, 72.7% believed that such functionality could help them understand Database Principles more effectively. Progress-monitoring capabilities were also viewed favorably, with 86.4% indicating that tracking learning performance would be useful during revision.

Overall, the findings suggest that students are receptive to tools that reduce the effort associated with preparing revision materials while supporting self-assessment and more engaging revision activities.

4.3 Discussion

Responses collected during the evaluation suggest that SmartNotes was perceived positively as a revision support tool. Participants indicated that the application could reduce part of the effort associated with revision preparation by generating learning resources directly from uploaded lecture notes. Instead of producing summaries, flashcards, and practice questions manually, students could obtain these materials automatically through the application.

The generated summaries provided a condensed view of the lecture content and were particularly useful when revising multiple topics within a limited period. Important concepts could be identified more quickly before referring to the original lecture notes for additional explanation and detail.

Flashcards and quizzes were used as self-assessment activities during revision. By requiring students to recall information from memory, these activities reflect the principles of retrieval-practice theory, which associates active recall with improved learning and retention [5].

Table 1 highlights several differences between SmartNotes and existing learning applications. Platforms such as Quizlet, Anki, and Brainscape provide useful support for revision; however, users are generally required to prepare their own learning materials. In contrast, SmartNotes generates summaries, flashcards, and quiz questions directly from uploaded lecture notes, reducing the amount of preparation needed before revision can begin. In SmartNotes, summaries, flashcards, and quiz questions are generated directly from uploaded lecture notes, reducing the amount of preparation required before revision activities can take place.

Access through mobile devices allows revision activities to take place in a variety of learning settings and at different times. Since smartphones are widely used by university students, learning materials can be accessed in different settings and at different times according to individual study preferences.

Overall, SmartNotes combines content generation, revision support, self-assessment, and progress monitoring within a single mobile environment. While the present evaluation focused primarily on user perceptions, the results indicate that integrating large language models into mobile learning applications represents a promising approach for supporting revision in technical subjects such as Database Principles.

4.4 Limitations

While SmartNotes showed encouraging results as a revision support tool, several factors should be considered when interpreting the results of this study.

The quality of the generated learning resources depends heavily on the quality of the uploaded lecture notes. Since summaries, flashcards, and quiz questions are produced directly from the supplied content, incomplete or poorly organized materials may affect the accuracy and usefulness of the generated outputs. Students should therefore treat the generated content as supplementary revision materials and refer to the original lecture notes whenever clarification is required.

The scope of the current study is another limitation. SmartNotes was developed and evaluated using learning materials from the Database Principles course. While the underlying approach is not restricted to a single subject area, its effectiveness for other disciplines has not yet been examined. Different subjects may require different styles of content generation, question formulation, or learning support.

The application also depends on cloud-based AI services to perform content generation. As a result, internet connectivity is required when processing lecture notes. Response time may vary depending on network conditions and service availability. Also, the quality of generated output remains dependent on the capabilities and behavior of the underlying language model.

The evaluation conducted in this study focused primarily on user perceptions of usability and usefulness. As a result, the findings cannot be used to determine whether SmartNotes leads to improvements in academic performance or knowledge retention.

Examining how students use the application over a longer period may provide further insight into its contribution to learning and revision practices.

These limitations should be considered when interpreting the results and planning future enhancements to the application.

5 Conclusion and Future Work

The development of SmartNotes was motivated by feedback obtained from students enrolled in Database Principles. Survey responses indicated that many students experienced difficulties retaining important concepts and spent considerable time preparing revision materials before revision could begin. These findings highlighted the need for a tool capable of supporting revision through AI-generated learning resources and interactive study activities.

SmartNotes represents one example of how large language models can be incorporated into a mobile learning environment to support revision activities. Nevertheless, the quality of the generated learning resources remains dependent on the quality of uploaded materials and the performance of the underlying language model. In addition, the current evaluation was limited to a single academic subject and involved a relatively small group of participants.

Future enhancements may focus on extending SmartNotes to support a wider range of academic subjects and learning materials. Several opportunities remain for extending the current implementation. These include adaptive learning features, personalized content generation, broader document support, and the use of newer language models. The current evaluation concentrated on user perceptions; consequently, further investigation may focus on how students utilize the generated learning resources during revision and how such usage relates to learning outcomes over time.

References

1. Crompton, H., Burke, D.: Artificial intelligence in higher education: the state of the field. Int. J. Educ. Technol. High. Educ. **20**(1), 1–22 (2023)
2. Chen, L., Chen, P., Lin, Z.: Artificial intelligence in education: a review. IEEE Access. **8**, 75264–75278 (2020)
3. Kasneci, E., Seßler, K., Küchemann, S., Bannert, M., Dementieva, D., Fischer, F., et al.: ChatGPT for good? On opportunities and challenges of large language models for education. Learn. Individ. Differ. **103**, 102274 (2023)
4. Yan, L., Sha, L., Zhao, L., Li, Y., Martinez-Maldonado, R., Chen, G., et al.: Practical and ethical challenges of large language models in education: a systematic scoping review. Br. J. Educ. Technol. **55**(1), 90–112 (2024)
5. Karpicke, J.D., Blunt, J.R.: Retrieval practice produces more learning than elaborative studying with concept mapping. Science. **331**(6018), 772–775 (2011)
6. Xia, Q., Weng, X., Ouyang, F., Lin, T.J., Chiu, T.K.: A scoping review on how generative artificial intelligence transforms assessment in higher education. Int. J. Educ. Technol. High. Educ. **21**(1), 40 (2024)
7. Guizani, S., Mazhar, T., Shahzad, T., Ahmad, W., Bibi, A., Hamam, H.: A systematic literature review to implement large language model in higher education: issues and solutions. Discov. Educ. **4**(1), 35 (2025)

8. Cibu, B.R., Crăciun, L., Molănescu, A.G., Cotfas, L.A.: Exploring the educational applications of large language models: a systematic review and topic analysis. Electronics. **14**(23), Article 4683 (2025)
9. El-Kassas, W.S., Salama, C.R., Rafea, A.A., Mohamed, H.K.: Automatic text summarization: a comprehensive survey. Expert Syst. Appl. **165**, 113679 (2021)
10. Zhang, M., Zhou, G., Yu, W., Huang, N., Liu, W.: A comprehensive survey of abstractive text summarization based on deep learning. Comput. Intell. Neurosci. **2022**, Article 7132226 (2022)
11. Kurdi, G., Leo, J., Parsia, B., Sattler, U., Al-Emari, S.: A systematic review of automatic question generation for educational purposes. Int. J. Artif. Intell. Educ. **30**(1), 121–204 (2020)
12. Scaria, N., Chenna, S.D., Subramani, D.: How good are modern LLMs in generating relevant and high-quality questions at different Bloom's skill levels for Indian high school social science curriculum? In: Proceedings of the 19th Workshop on Innovative Use of NLP for Building Educational Applications (BEA 2024), pp. 1–10 (2024)
13. Wang, L., Song, R., Guo, W., Yang, H.: Exploring prompt pattern for generative artificial intelligence in automatic question generation. Interact. Learn. Environ. **33**(3), 2559–2584 (2025)
14. Shu, M., Balepur, N., Feng, S., Boyd-Graber, J.L.: KARL: Knowledge-aware retrieval and representations aid retention and learning in students. In: Proceedings of the 2024 Conference on Empirical Methods in Natural Language Processing (EMNLP 2024), pp. 14161–14178 (2024)
15. Alrasheedi, M., Capretz, L.F., Raza, A.: A systematic review of the critical factors for success of mobile learning in higher education (university students' perspective). J. Educ. Comput. Res. **52**(2), 257–276 (2015)
16. Mustaffa, M.R., Mansor, E.I.: Facilitating students' learning through M-LMS for tablet device. Int. J. Comput. Theory Eng. **7**(3), 236–240 (2015)

User-Centred Design of a Virtual Reality Eco-Literacy Game Based on Learner Feedback

Noor Hafizah Hassan[1(✉)], Nurul Aiman Abdul Rahim[2], Javid Iqbal Thirupattur[3], and Angela Lee Siew Hoong[3]

[1] Department of Smart Computing and Cyber Resilience, Faculty of Engineering and Technology, Sunway University, No. 5, Jalan Universiti, Bandar Sunway 47500, Selangor, Malaysia
nhafizahh@sunway.edu.my

[2] The Design School, Faculty of Innovation and Technology, Taylor's University, Lakeside Campus, No. 1, Jalan Taylor's, 47500 Subang Jaya, Selangor Darul Ehsan, Malaysia
nurulaiman.abdulrahim@taylors.edu.my

[3] Department of Data Science and Artificial Intelligence, Faculty of Engineering and Technology, Sunway University, No. 5, Jalan Universiti, Bandar Sunway 47500, Selangor, Malaysia
{javidt,angelal}@sunway.edu.my

Abstract. Virtual reality (VR) has emerged as a promising educational intervention for addressing global environmental challenges. The technological advancement that incorporated immersion, interactivity, and emotional engagement viewed as feasible tools for sustainability education delivery. This study employed a mixed-method approach to gather empirical data from an existing VR application, followed by the development of a prototype using the Unity engine and the ADDIE model. The two phases study involved testing session with twenty-one participants with existing VR recycle game called Trash Time provided at SteamVR. This group of participant feedback was gathered in qualitative semi-structured interview questions. The data collected was incorporated for new development. Data gathered associated with effectiveness and engagement factors. The development of new sustainable game improvised many aspects like clear instructions, new immersive environments, recognizable objects or assets, multiple players, and real-time scoring. The findings offer valuable insights into users' experiences with the Virtual Reality Eco-Literacy Game, providing important implications for the design and development of immersive game-based learning environments that promote eco-literacy across local, national, and global communities. These insights can be further leveraged to formulate strategies for fostering ecological mindsets, enhancing sustainability awareness, and generating initiative ideas for policymakers and non-profit organizations. This study aligns with United Nation Sustainable Development Goals (SDGs): SDG 3 (Good Health and Well-being), SDG 4 (Quality Education), SDG 9 (Industry, Innovation and Infrastructure), SDG 11 (Sustainable Cities and Communities), SDG 12 (Responsible Consumption and Production), SDG 13 (Climate Action).

Keywords: eco-literacy · game-based education · sustainability education · sustainability development goal · virtual reality

N. Mohd Sharef et al. (Eds.): CAMP 2026, CCIS 3093, pp. 370–381, 2027.
https://doi.org/10.1007/978-981-92-4302-0_27

1 Introduction

Current global environmental challenges have worsened. Increasing waste generation, global warming, and air and water pollution have contributed to rising rates of premature deaths worldwide. Biodiversity loss and water scarcity were in worrying state [1]. United Nation Sustainable Development Goals (SDGs) has reported critical areas and actionable targets to mitigate key areas. Abundant strategies could be undertaken by many other stakeholders like researchers, government and non-governmental agencies (NGOs), to support the problem solutions. The exploration of current gaps still high and all stakeholders are urged to actively addressing the current challenges [2, 3].

With the rise of global environmental challenges, the development of eco-literate communities has become increasingly important. This is due to weak recycling behaviours among public communities [4, 5]. Despite the availability of information on why and how recycling should be exercised, many individuals remain indifferent to its significance, resulting in a behavioural gap. Empirical findings revealed that environmental behaviour developments in the current state are more short-term initiatives rather than long-term. Therefore, embedded new patterns of education are required to encourage people to adopt new mindsets and habits [6–8].

Virtual Reality (VR) found to be one of the options for interactive education. This technology promoted interactive and engaging game-based education. VR offers a promising approach that offered immersive experiences by simulating waste-sorting and nurture eco-literacy behaviours [9]. With several benefits that VR has demonstrated in supporting sustainability education, such as being engaging, interactive, and immersive, the number of studies and applications investigating and developing ecological games remains limited. Numerous studies have highlighted existing gaps or under-explored areas in VR ecological games-based education [10]. These include limited number of ecological games, even complimentary games for public, low response by users towards the existing games, even low retention time. Many existing games are limited in scope, focusing only on basic environmental concepts or recycling practices. This displayed non balance behaviour between games and real-world scenarios. Human actions in the real world were worse than the games education. This gap highlighted that eco-literacy communities development should be taken seriously and strategies on eco-literacy development should be continuously developed [10, 11].

Based on the above discussion this study aims to examine the current situation and propose a prototype that addresses the feedback given by users. The study started by investigating users feedback on the existing VR recycle games called Trashtime. The qualitative findings were further incorporated into the development of new recycle games in Unity. These findings very helpful for developers to design and add necessary items that could enhance the effectiveness and engagement as well as the development of eco-literate individuals or communities. The novelty and contribution of this study comprise of the following: (1) primary data on users' feedback of existing VR game, (2) features added in the proposed prototype immersive environment that close and familiar to users (campus environment), real-time score and feedback, and multiplayers feature.

2 Theoretical Background

2.1 Sustainability Education Issues and Challenges

Sustainability education or also known as green education referred as environmental knowledge and practice that individual acquired through education that associated with diverse ecological topics such as climate, recycle, earth and sciences. It emphasizes on the need to balance environmental and social well-being for future generations [12]. By integrating sustainability into educational curriculum, learners are equipped with the skills and the motivation to address real-world environmental challenges such as climate change, resource conservation, and many others. Ecological knowledge and practice certainly available in informal and formal education [12, 13]. As part of education curriculum this knowledge is considered a top-down approach. Merely to traditional approaches with the usage of textbooks, lectures, field-based or classroom-based activities, role-playing, storytelling, and any outdoor activities. The current method and delivery able to achieve the learning outcomes yet lack to sustain learners' ecological mindset which called eco-literacy. The recycling participation is shaped not only by infrastructure or economic incentives but also by social and behavioural factors [14].

The traditional method of green education found to have several areas lacking that need to be further investigated. The aims of sustainability education are to bring change to societal behaviour toward environmental awareness and practice. The existing traditional approaches have been criticised for their inability to inspire more significant ecological mindsets and attitudes [14, 15]. The current global environmental situation has worried various stakeholders. Day by day the situation has worsened. From the rise of global warming, high waste disposition and management, biodiversity loss, greenhouse gas emissions, pollutions and many more. This has negatively affected any nation with high cost of maintenance, that even harm the public health in general. It is hard and impossible for any nation to ignore the alarms and even let the public harm more the environment that increase the risks for future generations [16].

2.2 Eco-literacy

In regard to sustainability education the need of eco-literacy communities development is vital. Eco-literacy defined as individual that capable to learn and understand on environmental ecosystem, share and exchange the knowledge, constantly practice and influence others to sustain ecological behaviours. This individual able to inspire others and develop ecological responsible attitudes that able to make informed decisions on sustainability. Education efforts in promoting eco-literacy have appeared lacking and past studies found the possibilities of technological education might help [17]. For example, the utilization of online gamification especially for young learners. Most users in Malaysia found to have basic recycle knowledge but it only covered lower levels of eco-literacy [17, 18]. The need to nurture eco-literacy is highly needed. One of the main challenges in achieving eco-literacy is lack of users experiences that limit users' exploration on skills acquisition. Past findings also revealed how the interest and effectiveness of educational technologies have enhanced the delivery of ecological education [18]. Advance technologies such as online learning, blended learning, gamification, has increased the comprehension of basic eco-literacy. This positive opportunity has inspired games developers on the development ecological game-based education [18, 19].

2.3 Virtual Reality Game Based Education

Prior studies exposed the challenges of the adoption of emergent technologies due to several factors like costly, resistance to change and the need of technical training for developers even users [20, 21]. However, along the time the usage of emergent technologies like Artificial Intelligence (AI), Virtual Reality (VR) and Augmented Reality (AR) and many more. The Virtual Reality (VR) is a transformative technology that allows users to interact with and explore computer-generated environments. Unlike traditional interfaces, VR completely immerses users by transporting them into alternative virtual worlds through advanced hardware, such as head-mounted displays (HMDs). The usage of VR technology has risen to be essential tools for various industries, offering immersive and interactive experiences that redefine how we learn, train, and solve real-world problems [22].

VR offers numerous benefits in educational settings. Past studies result evidently highlighted the usage of VR has improved the users or learners' knowledge, self and response efficacy as well as social intentions.it also improvised complexity practical or theoretical content into real self-experiences that reduce the difficulties level. VR facilitated better demonstration has constantly enhanced along the time since it was introduced in education field [22]. Along the time VR technologies has evolved in wireless connectivity, hardware affordability and motion tracking making it increasingly accessible to broader global audiences or learners The advantages of VR on visualization of environmental, economic, political and social situation and implications of any material offered engagement in educational settings [22, 23].

Despite its immense potential challenges such as costs of maintenance, compatibility and connectivity issues in some areas, motion sickness for some users or new users. VR also received resistance to change by some users all over the world. The need for transformation of hardware and software has become a barrier to the adoption of this technology. This followed by the requirement of human training and development has complicated even further. Due to insufficient training and change resistance it has consequently led to time consuming besides being financial constraints.

3 Method

3.1 Research Design

This study employed two types of method to increase the accuracy and effectiveness of the findings and prototype developed. The two phases of data collection procedures presented in the following section.

3.1.1 Phase 1 Procedure

Phase 1 involved primary data collection through a qualitative approach, in which a group of participants interacted with and evaluated the existing Virtual Reality (VR) game, Trash Time, developed by Dalibor Bartoš. The input from this phase includes the data provided by the developer that consisted of Trash Time Steam Review. Additionally, feedback was collected from twenty-one Malaysian participants aged between 18 and 35 years who tested the game and responded to thirteen semi-structured questions designed to evaluate its engagement, efficiency, and effectiveness. The participants possessed basic recycling knowledge and represented a diverse range of VR experience levels, including individuals who had never used a VR headset, those who had used one once, and those with multiple prior VR experiences. This data collected very critical for second phase of the research on design and development of new game that addressing all lacking areas findings (Table 1).

Table 1: Effectiveness and engagement semi-structured interview questions

Questions	Measurement
The session/game was easy to learn and play – Yes or No, why? The session/game scenario was realistic – Yes or No, why? 1. The session/game was interesting. – Yes or No, why? 2. I am enjoyed the session/games – Yes or No, why? 3. The session/games helped me make links between the content with real world (me as human to recycle) 4. The session/games helped me to recognize and differentiate various items for recycle – Yes or No, why? 5. How immersed did you feel in the visual elements of the VR environment? 6. The visual aspects of the virtual environment involved me	Effectiveness [24]
7. The sense of moving around inside the virtual environment was compelling (feel the sense of moving) 8. I was involved in the virtual environment experiences 9. To what extent did you feel a sense of presence and interaction with virtual objects in the VR space? 10. How convincing was your sense of movement and navigation within the VR environment? 11. How long were you able to stay engaged in the VR experience before feeling fatigued?	Engagement [25, 26]

3.1.2 Phase 2 Procedure

Based on the input from Phase 1, The ADDIE Model (refer Fig. 1) was employed to prepare Development Framework (refer Fig. 2) was chosen to guide the development. This model is well-recognized for its iterative nature, ensuring that feedback and modification can be incorporated at different phases [27]. The development activities employed set of digital tools.

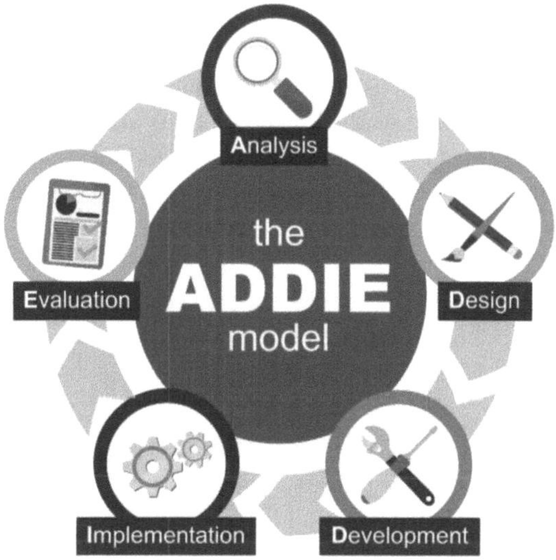

Fig. 1: ADDIE Model

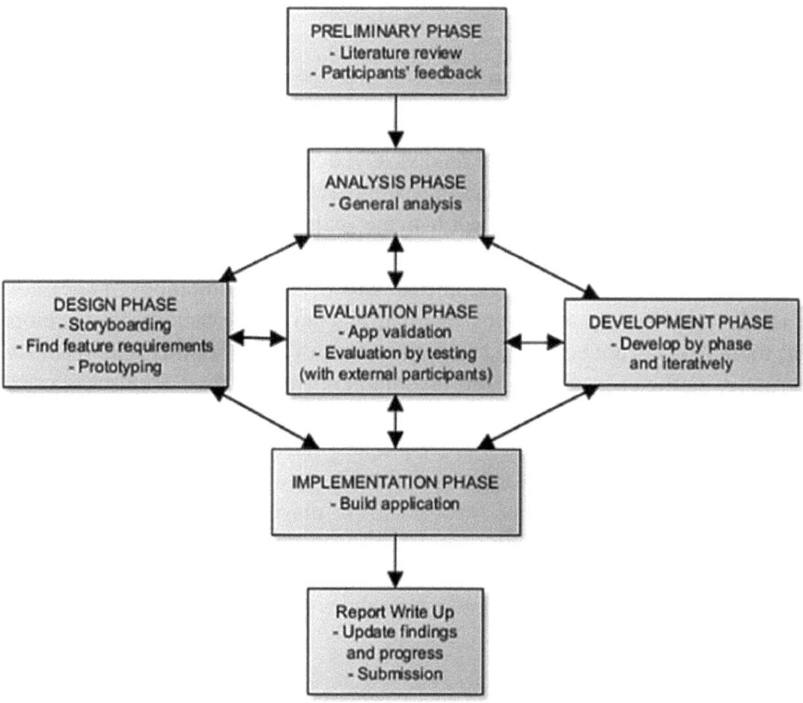

Fig. 2: VR eco-literacy games development framework

Figma was utilized for initial storyboarding and interface design, for map visualization, user interactions and narrative flow (refer Fig. 3). For prototyping and application development, the Unity engine is used. The reputable platform for immersive building. As for 3D model assets Blender was selected due to its visual fidelity and virtual environment optimization [28]. In the hardware part, Meta Quest 3 headset was used for testing and deployment which enabled researcher connectivity and configuration within a fully immersive setting.

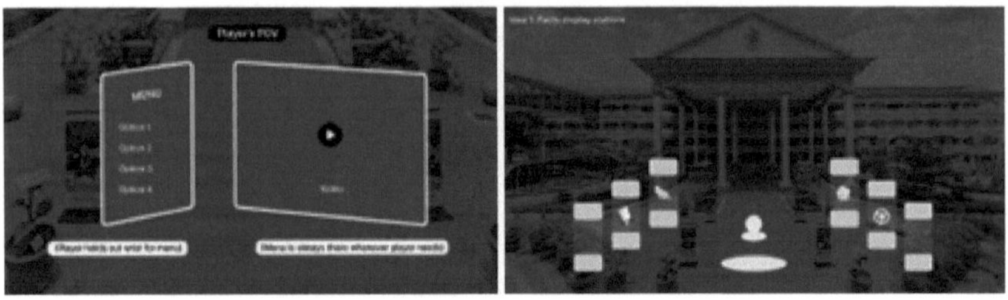

Fig. 3: Visualisation of prototype design in Figma

4 Results and Discussion

4.1 Findings of Qualitative Data

Phase 1 aimed to evaluate the usability, engagement, and effectiveness of the existing VR application Trash Time based on user feedback (n = 21) and developer review data [29]. Data gathered became a valuable input for developers to develop new game that addresses several lacking areas. However, according to the analysis of the developer reviews, there were several common problems that emerged (refer Table 2). The most frequent one is VR compatibility (n = 4), followed by graphics quality (n = 3), instructions (n = 2), controls (n = 2) and sound issues (n = 1). Thus, this state that the key obstacles to the development of educational games are connected with the issues of compatibility and visualization. Furthermore, inefficiency of compatibility and performance may demotivate users and spoil their experience. Future developers should carefully consider this factor when developing new games. This followed by graphic visual quality of the game. Players suggested for auto-detection setting provided in the game to improve this functionality. Controls and sound were another feedback from players that required more control and movement as well as better audio and finally the clear instructions were demanded by players when onboarding because is very vital for players to comprehend prior the session. This also will increase the interest in players retention (frequencies to play the game).

Table 2: Technical areas summary

Technical	Frequency	Sample of Comment
VR Compatibility	4	*Neat concept and there's some important lessons to take away from this game. Think about what you're doing and how overwhelming the trash gets with the task of sorting it. -played on vive cosmos | Experienced on the Oculus Rift with Touch Controllers I'm giving a neutral recommendation*
Graphics	3	*Great game. I like the nostalgia I got from it from playing old Roblox tycoons back in 4th grade. Nice concept. Has a few details that need to be rechecked*
Instructions	2	*You earn money by sorting the stuff right. Gets hectic and some stuff don't fit correctly in sorting bins. Has teleport movement mode. Did Not have instructions*
Controls	2	*Has upbeat background music on a loop, didn't see a mute or volume control. Has several map, with 3 days each. You earn money by sorting the stuff right. Gets hectic and some stuff don't fit correctly in sorting bins*
Sound	1	*Has upbeat background music on a loop, didn't see a mute or volume control. Has several map, with 3 days each*

The qualitative data of twenty-one participants exposed key areas that could be improvised. Table 3 summarized effectiveness and engagement review of Trash Time. The findings highlighted that Most participants agreed that the game was easy and enjoyable to play. However, some areas that the players like to have more like clear instructions in the beginning of the game. The sense and movement have slightly lower positive response because they prefer that more movement allowed. They also have variety duration of fatigue and motion sickness it is flexible if the game allowed players to stop at any stage so it would not prolong their sickness. The graphics or visuals even the items ID could be clearer and objects more realistic for players easy recognition to differentiate items for recycle or bin. Because the gameplay is repetitive, it may eventually lead to boredom. Therefore, any enhancement to address this issue could be implemented.

Table 3: Summary of effectiveness & engagement lacking areas

Area	Lacking Aspect	Example Issue
Tutorial	No instructions	Users confused how to start
Graphics	Needs improvement	Visuals not clear/realistic
Movement	Feels restricted	Some fear, not smooth
Fatigue	Users tired quickly	Short play sessions
Item ID	Hard to sort items	Some items unclear
Variety	Gets repetitive	Gameplay boring over time

4.2 Prototype of New Game Development

The prototype was developed based on the findings obtained in Phase 1. Prototype development has successfully completed intended designs outlined (refer Fig. 4). Some of key features that added to address the lacking areas include clear instructions added in the beginning of the game for players. This is to ensure that players clearly understand how to play and recognise the different types of recycle items (objects) in the game. Real time scoring motivates players to play more rounds which address the boredom found in the existing VR recycle games. The visual and audio components were also enhanced based on the feedback received. The multiplayer feature enables multiple users to participate simultaneously, thereby increasing engagement and social interaction. Indirectly increased awareness when players invited their friends and families to play together.

The User Acceptance Testing (UAT) was performed using 11 participants to assess usability, engagement, and user experience. As a result of the test, it was found that more than half of the participants (>50%) have positive impressions of the prototype, specifically in such aspects as ease of use, clarity of instructions, and engagement level. For instance, real-time scoring mechanism implemented into the prototype to increase player motivation and promote continued engagement with the game. Due to motion sickness the game allowed players to stop anytime they like. Items or objects, assets included provided with clearer labels and colour for ease of recognition. To reduce boredom and increase player retention, the new game has added variety of items for recycle in different context or level which consequently added awareness, educate, familiarization on recycle items, increase interest and exploration.

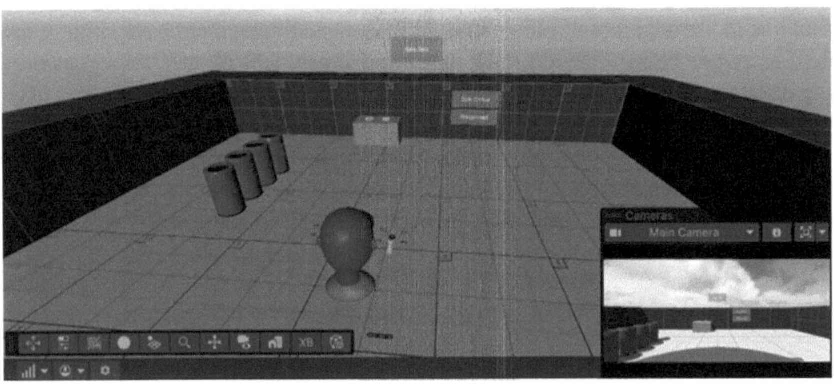

Fig. 4: Prototype in Unity Engine

In regard to sustainability education, virtual reality learning, and eco-literacy development, findings from the User Acceptance Testing (UAT) revealed that participants responded positively to the Virtual Reality Eco-Literacy Game prototype, particularly in terms of its usability, engagement, and educational value. Most participants expressed strong interest in utilizing VR as a tool for sustainability education and demonstrated a willingness to recommend VR-based learning experiences to others. Participants highlighted immersive 3D environments, reward-based features, interactive guidance, and realistic learning scenarios as key elements that enhanced their engagement and learning experience.

Furthermore, the findings suggest that game mechanics such as real-time scoring and reward systems can effectively motivate users, encourage active participation, and sustain engagement throughout the learning process. Participants also emphasized the importance of clear instructions, practical recycling activities, and immediate feedback in helping them understand proper waste separation practices and the environmental impacts of waste and pollution. Several recommendations were also identified, including the incorporation of multiplayer functions, more engaging narratives, enhanced 3D representations of recyclable objects, and improved interaction design to further strengthen the learning experience. Collectively, these findings demonstrate the potential of immersive VR game-based learning environments to support sustainability education, foster eco-literacy, and encourage environmentally responsible behaviours among diverse user communities

5 Conclusions

This study aimed to develop a Virtual Reality (VR) sustainability game that can be used by individuals of different ages and backgrounds worldwide. The existing game called Trash Time became subject of study in Phase 1 to ensure the new game developed will address some key lacking areas. This game intended not only to create awareness and interest among public but also to nurture eco-literacy among citizens. Eco-literate communities are not only knowledgeable about environmental issues but are also equipped with the skills and motivation required to engage in sustainable practices. Take it as lifelong learning and part of their lives. In the future, the game could be commercialized, protected through intellectual property registration, and integrated into educational curricula. This study parallel with United Nation Sustainable Development Goals (SDGs) that aims for global sustainable well-being. Several SDGs mapped with the study objectives and outcomes namely SDG 3, promoted healthy lifestyles with 3RC (reduce, reuse, recycle and compost). SDG 4, integrating ecological education that improved that quality of life. Followed by SDG 9, encourage industry involvement that support global, national and regional directions. SDG 11, moving towards smart cities and communities that sustaining balance ecological life. SDG 12, which consequently supports healthy and well-balance consumption and production. Finally, support the global and nation mitigating climate issues from various perspectives and initiatives.

Acknowledgments. The authors would like to express their sincere gratitude to the Faculty of Engineering and Technology (FET), Sunway University, for their invaluable support and contribution to this study, made possible through two research grants: GRTIN-ECR-DCIS-04-2024 and GRTIN-RAG-DDSAI-03-2025. Special appreciation is extended to Mr. Dalibor Bartoš, developer of the Trash Time VR application, for generously providing access to the platform and enabling us to explore its potential in advancing green education through immersive virtual reality.

Ethics Approval Approval was granted by Sunway University Research Ethics (2024/REC1013)

References

1. Demirci, S., Reid, A., Teksöz, G., Sahin, E.: Systems literacy: towards a conspectus and model for environmental and sustainability education. Environ. Educ. Res. **31**(4), 856–891 (2025). https://doi.org/10.1080/13504622.2024.1234567
2. Rushton, E.A.C., Walshe, N.: Curriculum making and climate change and sustainability education: a case study of school teachers' practices from England, UK. Environ. Educ. Res. **31**(5), 1083–1097 (2025). https://doi.org/10.1080/13504622.2025.2471990
3. Quirós-Alpera, S., Roets, A.O.S.B., Robina-Ramírez, R., Leal-Solís, A.: Bridging theory and practice: challenges and opportunities in dual training for sustainability education at Spanish universities. Societies. **15**(2), 32 (2025). https://doi.org/10.3390/soc15020032
4. Hassan, T.H. et al.: Hierarchical seniority vs. innovation in hospitality and tourism sustainability education: a social exchange theory perspective. Sustainability. **17**(18), 8100 (2025). https://doi.org/10.3390/su17188100
5. Ionescu-Feleagă, L. et al.: Business simulation games from the perspective of accounting and management professors: implications for sustainability education in universities. Int. J. Manag. Educ. **23**(2), 101147 (2025). https://doi.org/10.1016/j.ijme.2025.101147
6. Novaes, A.L.: Enhancing sustainability education in higher education through simulation-based learning: integrating sustainable development goals. Int. J. Sustain. High. Educ. **26**(4) (2025). https://doi.org/10.1108/IJSHE-08-2024-0571
7. Firinci Orman, T.: Significance of lived experiences in developing eco-literacy: pedagogical insights for climate education. Clim. Liter. Educ. **3**(1) (2025). https://doi.org/10.24926/cle.v3i1.6408
8. Roy, U.: Green skills and eco-literacy: reimagining curriculum for a sustainable world. In: Green skills for the 21st century learner: educational perspectives and practices, p. 51 (2025). https://doi.org/10.5281/zenodo.1234567
9. Aguilar-Cruz, P.J., Olaya-Marín, E.J.: Gamification and eco-literacy: a bibliometric analysis of research trends. Int. J. Serious Games. **12**(3) (2025). https://doi.org/10.17083/8mzab045
10. Gestiada, R.J., Tisoy, F.J., Lasala Jr., N.: The 360° view: contextualized virtual reality tours as innovative teaching tool in ecology for elementary school students. J. Bas. Educ. Res. **6**(1), 35–48 (2025). https://doi.org/10.37251/jber.v6i1.1213
11. Ahmadov, T. et al.: A two-phase systematic literature review on the use of serious games for sustainable environmental education. Interact. Learn. Environ. **33**(3), 1945–1966 (2025). https://doi.org/10.1080/10494820.2024.2414429
12. Onyeaka, H., Akinsemolu, A.A.: The role of green education in achieving the sustainable development goals: a review. Renew. Sust. Energ. Rev. **210**, 115239 (2025). https://doi.org/10.1016/j.rser.2024.115239
13. Velempini, K.: Assessing the role of environmental education practices towards the attainment of the 2030 sustainable development goals. Sustainability. **17**(5), 2043 (2025). https://doi.org/10.3390/su17052043
14. Lubis, B.N.A.: Multimodal-based approaches in teaching ecological literacy: enhancing environmental awareness through textual and visual integration. STAIRS Eng. Lang. Educ. J. **6**(1), 53–68 (2025). https://doi.org/10.21009/stairs.6.1
15. Satrio Binusa, S., Risa, J., Redi Sigit, F.: Developing a sustainable educational curriculum: integration of green technology and environmental awareness in schools. J. Paddisengeng Technol. **1**(3), 129–138 (2025). https://doi.org/10.17323/Paddisengeng.2023.216
16. World Resources Institute; Climate Analytics and Systems Change Lab: State of Climate Action 2025. World Resources Institute. Retrieved October 31, 2025, from https://www.wri.org/research/state-climate-action-2025. (2025)

17. Salimi, M., Karsono, Hidayah, R., Surya, A., Fauziah, M.: Concepts, aspects and methods for developing ecoliteracy: a systematic literature review. Multidisciplinar (Montevideo). **3**, 214 (2025). https://doi.org/10.62486/agmu2025214
18. Carmona-Galindo, V.D., Velado-Cano, M.A., Groat-Carmona, A.M.: The ecology of climate change: using virtual reality to share, experience, and cultivate local and global perspectives. Educ. Sci. **15**(3), 290 (2025). https://doi.org/10.3390/educsci15030290
19. Bao, H., Lo, T.T., He, Y., Li, Y.: Transforming waste plastic into play: an interactive gaming installation utilizing immersive technology. In: Proceedings of the 30th International Conference of the Association for Computer-Aided Architectural Design Research in Asia (CAADRIA 2025), Hong Kong, pp. 437–446 (2025). https://doi.org/10.5281/zenodo.113008
20. Pratama, R., Nugroho, A.: Raising awareness of 3R (reduce, reuse, recycle) using Revin the environment hero serious game for elementary schoolers. J. Educ. Technol. Innov. **8**(2), 145–158 (2025). https://doi.org/10.5281/zenodo.1234568
21. Venturi, S., Zulauf, K., Cuel, R., Wagner, R.: Trash to treasure: gamification and informed recycling behavior. Res. Conserv. Recycl. **215**, 108108 (2024). https://doi.org/10.1016/j.resconrec.2024.108108
22. Schillinger, D., Kroll, S.: Cooperation in virtual reality: exploring environmental decision-making through a real-effort threshold public goods game. J. Environ. Psychol. **90**, 102145 (2025). https://doi.org/10.1016/j.jenvp.2024.102145
23. Rezaei Shandiz, R. et al.: Utilizing gamification in smart waste management: a participatory approach integrating green schools, the Internet of Things (IoT), and social marketing. Quant. Res. **3**(1) (2025). https://doi.org/10.5281/zenodo.2570000
24. Álvarez-Nieto, C., Álvarez-García, C., Parra-Anguita, L., Sanz-Martos, S., López-Medina, I.M.: Effectiveness of scenario-based learning and augmented reality for nursing students' attitudes and awareness toward climate change and sustainability. BMC Nurs. **21**(1), 245 (2022). https://doi.org/10.1186/s12912-022-00991-7
25. Škola, F. et al.: Virtual reality with 360-video storytelling in cultural heritage: study of presence, engagement, and immersion. Sensors. **20**(20), 5851 (2020). https://doi.org/10.3390/s20205851
26. Rubini, B., Pursitasari, I.D., Suriansyah, M.I., Ramadhanti, G.N., Rachman, I.: Improving students' eco-literacy through the development of electronic interactive teaching materials on climate change. Jurnal Penelitian dan Pembelajaran IPA. **9**(2), 288–308 (2023). https://doi.org/10.29303/jppipa.v9i2.4000
27. Li, R.-Y., Sun, J.C.-Y.: Identifying key factors of dynamic ADDIE model for instructional virtual reality design: an exploratory study. Interact. Learn. Environ. **32**(10), 6984–6996 (2024)
28. Unity Technologies: VR Curricular Framework: A resource for educators and administrators to bring virtual reality design and development to the classroom. Retrieved November 15, 2025, from https://connect-prd-cdn.unity.com/educator-hub/VR_CurricularFramework_English_2025.pdf. (2025)
29. Hassan, N.H., Rahim, N.a.A., Lee, A.S.H.: Sustainability education and eco-literacy using virtual reality (VR): trash time application engagement data analysis. In: Lecture notes in networks and systems, pp. 245–255 (2026). https://doi.org/10.1007/978-3-032-08246-6_19

Author Index

N. Mohd Sharef et al. (Eds.): CAMP 2026, CCIS 3093, pp. 383–384, 2027.
https://doi.org/10.1007/978-981-92-4302-0

MIX
Papier aus verantwortungsvollen Quellen
Paper from responsible sources
FSC® C105338

If you have any concerns about our products,
you can contact us on
ProductSafety@springernature.com

In case Publisher is established outside the EU,
the EU authorized representative is:
Springer Nature Customer Service Center GmbH
Europaplatz 3, 69115 Heidelberg, Germany

Printed by Libri Plureos GmbH
in Hamburg, Germany